T0298694

Genetic Diversity of Cultivated Tropical Plants

Genetic Diversity of Cultivated Tropical Plants

Scientific editors
Perla Hamon
Marc Seguin
Xavier Perrier
Jean Christophe Glaszmann

 CRC Press
Taylor & Francis Group
Boca Raton London New York

CRC Press is an imprint of the
Taylor & Francis Group, an **informa** business
A SCIENCE PUBLISHERS BOOK

CIRAD

Centre de coopération internationale en recherche
agronomique pour le développment, France

Preface

The development of agriculture marked the beginnings of plant improvement. It started from the Paleolithic, about ten to twelve thousand years ago, when communities that had till then drawn their subsistence from hunting, fishing, and gathering began to form settlements. Between that era and the middle of the 19th century, almost all of the plants cultivated today were domesticated. These plants were transformed from wild species to those of a traditional ecotype or variety by progressively acquiring a set of characteristics more or less different from those of their ancestors. This is called the 'domestication syndrome'. Domestication took place either within closely limited areas—as in the case of maize and potato, in pre-Colombian times—or across several continents, as in the case of rice and sorghum. Cultivated species thus evolved under the pressures of natural and human selection linked to cumulative as well as divergent processes, of permanent introgression and diversification. Humans exploited this diversity by improving first the resources of their immediate environment and then, from the 16th century onwards, those resources made accessible by the rapid development of trade and transport. An infinite number of cultivation systems were thus developed, each of which was an original response to the particular needs and constraints of a community.

Till the 1930s, modern improvement of varieties based on selection procedures relied solely on the use of traditional varieties or ecotypes and the notion of genetic resources was limited to cultivated plants. Following the works of Vavilov, this notion was extended to related wild species and then to increasingly distant genera. Nevertheless, the importance of genetic resources became apparent only with the threat of their extinction. This phenomenon was particularly marked in Europe after World War II. Indeed, the natural environmental balance was seriously disturbed by the considerable extension of urban areas at the expense of farmland, the cultivation of new areas that might be unsuitable or fragile, with or without deforestation, and the degradation of vast natural regions that had been preserved until that time.

Conscious of this risk, the international scientific community was mobilized during the 1960s and 1970s to collect and conserve as many genetic resources as possible, giving priority to food species, which have a major economic value on the global scale. A large number of collections were thus assembled throughout the world. The FAO published a synthesis on this subject, the *Report on the state of the world's plant genetic resources for food and agriculture* (1997), on the occasion of the International Conference at Leipzig in June 1996. At present, most of these collections have become so large that they are hard to maintain and the taxonomic characterization of their accessions has become difficult. Their management is now an acute problem. It has become necessary to find a balance between conservation, evaluation, and use of genetic resources. Without evaluation, no rational use is possible. Without utility, the conservation is not justified from the point of view of institutions that are not specifically mandated to maintain them.

To respond to these concerns, Frankel and Brown introduced in the 1980s the concept of *core collection*, which they defined as a limited sample of accessions taken from a larger collection, called the base collection, and chosen to best represent the existing spectrum of diversity. Such a core collection can serve several purposes. It allows scientists to identify the material that ought to be conserved as a priority, but also allows reasonably good access to the genetic diversity available in the base collection, which facilitates the search for new sources of useful characteristics for selection.

In general, the base collections contain passport data, including the geographic and ecological origin of the accessions. This information is complemented as needed by morphotaxonomic, agronomic, and genetic data generated from biochemical markers (e.g., isozymes, polyphenols) or molecular markers (e.g., RFLP, RAPD).

During the compilation of a core collection, the useful characteristics are most important for the breeder, but they are difficult to evaluate directly and may conceal complex genetic controls. Moreover, this evaluation may be burdened by a difficulty in predicting the constraints that may weigh on the culture in the future. On the other hand, molecular genetic markers have no direct utility, but they express general structures of diversity, which may in turn serve as a basis for constituting a core collection. Markers sometimes reveal groups of accessions partly isolated from each other, which may have fixed distinct alleles for useful characteristics, by foundation effect, by genetic drift, or even under the action of various pressures of selection.

The relationships between the two levels of variability—that of molecular genetic markers, probably mostly neutral, and those of agronomic characters, generally more complex and subject to natural or human selection—are poorly understood. They are of central concern to the scientific community involved in the management of genetic resources, as appears in the conclusions of the colloquium on plant and animal genetic resources and the methodologies of their study and management, organized by INRA and BRG at Montpellier, in September 1993.

These relationships probably vary as a function of the population structure. They are generally strong if there is a serious gametic disequilibrium. To proceed from this reflection, however, we must tackle specific questions: Are the different types of molecular markers equivalent? Are strong structures at the molecular scale, which control the gametic disequilibrium generalized on the genome as a whole, systematically associated with strong structures for agronomic characters? Do the structures at these two levels therefore coincide?

Several methods have been proposed to construct the core collections. As a general rule, they are based on a stratification of the base collection and then on a random sample within each group thus defined, according to various modalities. When data are available for morphoagronomic evaluation, it is useful to consider the structure of the genetic diversity established with the help of genetic markers in order to maximize the agronomic diversity and minimize the loss of alleles that are rare overall but occasionally significant.

The constitution of a core collection must therefore be founded on an excellent description of the populations and on a sound understanding of their structure. Statistical tools are indispensable here. For each type of marker, it is important to use a measure of difference between the relevant taxonomic units (individuals or populations) from the perspective of their mathematical properties and their interpretation in genetic terms. In the same way, the choice of algorithms of representation of dissimilarities must be based on an equilibrium between maximum efficiency and minimum complexity, in order to deal with large tables covering more than 100 individuals. By comparing the structures observed with the help of various types of descriptors, we can envisage overall the organization of a plant's genetic diversity. This question can be tackled in various ways, among others, by looking for structures common to two or several trees. In these conditions, how can we construct consensus trees or common minimum trees? What is their reliability and their biological significance?

This work provides the elements of an answer to these questions, beginning with the study of eleven types of plants, chosen so as to cover a wide range of biological characteristics (perennial, annual, autogamous, allogamous, etc.): asian rice, banana, cacao, cassava, citrus, coconut, coffee, pearl millet, rubber tree, sorghum, and sugarcane. Three methodological chapters complement these studies. The first is devoted to the use of biological and molecular markers to analyse the diversity of collections, the second addresses data analysis, and the third describes a method for constituting core collections based on maximization of variability.

List of Abbreviations

AAT	aspartate aminotransferase
ADH	alcohol dehydrogenase
AFLP™	amplified fragment length polymorphism
AMP	aminopeptidase
BT	binary table
CA	correspondence analysis
CAT	catalase
cDNA	complementary DNA
cpDNA	chloroplastic DNA
DAF	DNA amplification fingerprinting
DNA	deoxyribonucleic acid
EST	esterase
FISH	fluorescent *in situ* hybridization
GISH	genomic *in situ* hybridization
GOT	glutamate oxaloacetate transaminase
HC	hierarchical clustering
ICD/IDH	isocitrate dehydrogenase
ISSR	inter-simple sequence repeat
LAP	leucine aminopeptidase
MCA	multiple correspondence analysis
MDH	malate dehydrogenase
MFA	multiple factorial analysis
mtDNA	mitochondrial DNA
PCA	principal components analysis
PCoA	principal coordinates analysis
PCR	polymerase chain reaction
PER	peroxydase

PGD	6-phosphogluconate dehydrogenase
PGI	phosphogluco-isomerase
PGM	phosphoglucomutase
PIC	polymorphism information content
QTL	quantitative trait loci
RAPD	randomly amplified polymorphic DNA
rDNA	ribosomal DNA
RFLP	restriction fragment length polymorphism
RNA	ribonucleic acid
SKDH	shikimate dehydrogenase
SSR	simple sequence repeat
STMS	sequence tagged microsatellite site
UPGMA	unweighted pair group method with average
VNRT	variable number of tandem repeat
WPGMA	weighted pair group method using average

List of Contributors

Albar, Laurence, IRD, BP 64501, 34394 Montpellier Cedex 5, France

Anthony, François, CATIE, Apartado 59, 7170 Turrialba, Costa Rica

Baudouin, Luc, CIRAD, département des cultures pérennes, TA 80/03, 34398 Montpellier Cedex 5, France

Berthaud, Julien, IRD, BP 64501, 34394 Montpellier Cedex 5, France

Bezançon, Gilles, IRD, BP 64501, 34394 Montpellier Cedex 5, France

Bonnot, François, CIRAD, département des cultures pérennes, TA 80/03, 34398 Montpellier Cedex 5, France

Bourdeix, Roland, CIRAD, département des cultures pérennes, représentation du CIRAD, résidence les Acacias, 01 BP 6483, Abidjan 01, Côte d'Ivoire

Carreel, Françoise, CIRAD, département des productions fruitières et horticoles, station de Neufchâteau, Sainte-Marie, 97130 Capesterre-Belle-Eau, Guadeloupe

Chantereau, Jacques, CIRAD, département des cultures annuelles, TA 70/01, 34398 Montpellier Cedex 5, France

Clément-Demange André, CIRAD, département des cultures pérennes, TA 80/PS1, 34398 Montpellier Cedex 5, France

Colombo Carlos, Instituto Agronômico de Campinas, Caixa postal 28, 13001-970 Campinas, SP, Brazil

Combes, Marie-Christine, IRD, BP 64501, 34394 Montpellier Cedex 5, France

Courtois, Brigitte, CIRAD, département d'amélioration des méthodes pour l'innovation scientifique, TA 40/03, 34398 Montpellier Cedex 5, France

Deu, Monique, CIRAD, département des cultures annuelles, TA 70/03, 34398 Montpellier Cedex 5, France

D'Hont, Angélique, CIRAD, département des cultures annuelles, TA 40/03, 34398 Montpellier Cedex 5, France

Dubois, Cécile, CIRAD, département des productions fruitières et horticoles, TA 50/PS4, 34398 Montpellier Cedex 5, France

Dussert, Stéphane, IRD, BP 64501, 34394 Montpellier Cedex 5, France

Flori, Albert, CIRAD, département des cultures pérennes, TA 80/03, 34398 Montpellier Cedex 5, France

Ghesquière, Alain, IRD, BP 64501, 34394 Montpellier Cedex 5, France

Glaszmann, Jean Christophe, CIRAD, département d'amélioration des méthodes pour l'innovation scientifique, TA 40/03, 34398 Montpellier Cedex 5, France

Grivet, Laurent, CIRAD, département des cultures annuelles, TA 40/03, 34398 Montpellier Cedex 5, France

Hamon, Perla, Centre universitaire, rue du Dr Georges Salan, 30000 Nîmes, France

Hamon, Serge, IRD, BP 64501, 34394 Montpellier Cedex 5, France

Horry, Jean-Pierre, BP 153, 97202 Fort de France, Martinique

Jacquemond, Camille, INRA, station de recherche agronomique de San Giuliano, 20230 San Nicolao, France

Jacquot, Michel, CIRAD, département des cultures annuelles, TA 70/03, 34398 Montpellier Cedex 5, France

Jannoo, Nazeema, Centro de biologia molecular e engenharia genetica, Universidade Estadual de Campinas, Caixa Postal 6109-13083-970 Campinas SP, Brasil

Jenny, Christophe, CIRAD, département des productions fruitières et horticoles, station de Neufchâteau, Sainte-Marie, 97130 Capesterre-Belle-Eau, Guadeloupe

Lanaud, Claire, CIRAD, département d'amélioration des méthodes pour l'innovation scientifique, TA 40/03, 34398 Montpellier Cedex 5, France

Lashermes, Philippe, IRD, BP 64501, 34394 Montpellier Cedex 5, France

Lebrun, Patricia, CIRAD, département des cultures pérennes, TA 80/03, 34398 Montpellier Cedex 5, France

Legnaté, Hyacinthe, CNRA, 01 BP 6483, Abidjan 01, Côte d'Ivoire

Luce, Claude, CIRAD, département des cultures annuelles, Station de Roujol, 97170 Petit Bourg, Guadeloupe

Luro, François, INRA, station de recherche agronomique de San Giuliano, 20230 San Nicolao, France

Montagnon, Christophe, CIRAD, département des cultures pérennes, représentation du CIRAD, résidence les Acacias, 01 BP 6483, Abidjan 01, Côte d'Ivoire

Motamayor, Juan-Carlos, USDA-ARS, Subtropical Horticultural Research Station 13601 Old Cutler Road, Miami, FL 33158, USA

N'cho, Yavo-Pierre, CNRA, station Marc Delorme, 07 BP 13, Abidjan 07, Côte d'Ivoire

Noirot, Michel, IRD, BP 64501, 34394 Montpellier Cedex 5, France

Noyer, Jean-Louis, CIRAD, département d'amélioration des méthodes pour l'innovation scientifique, TA 40/03, 34398 Montpellier Cedex 5, France

Ollitrault, Patrick, CIRAD, département des productions fruitières et horticoles, TA 50/PS4, 34398 Montpellier Cedex 5, France

Perrier, Xavier, CIRAD, département des productions fruitières et horticoles, TA 50/PS4, 34398 Montpellier Cedex 5, France

Raffaillac, Jean-Pierre, IRD, CIRAD, département d'amélioration des méthodes pour l'innovation scientifique, laboratoire Ecotrop, 34398 Montpellier Cedex 5, France

Second, Gérard, mission IRD, Whimper 442 y Corua, apartado 17-12-857, Quito, Ecuador

Seguin, Marc, CIRAD, département des cultures pérennes, TA 80/03, 34398 Montpellier Cedex 5, France

Sounigo, Olivier, CIRAD, département des cultures pérennes, TA 80/PS3, 34398 Montpellier Cedex 5, France

Tézenas du Montcel, Hugues, CIRAD, département des productions fruitières et horticoles, BP 153, 97202 Fort-de-France, Martinique

Tomekpe, Kodjo, CIRAD, département des productions fruitières et horticoles, BP 2995, Douala, Cameroun

Trouslot, Pierre, IRD, BP 64501, 34394 Montpellier Cedex 5, France

Contents

Preface   v

List of abbreviations   ix

List of contributors   xi

Biochemical and molecular markers   1
    Laurent Grivet and Jean-Louis Noyer

Methods of data analysis   31
    Xavier Perrier, Albert Flori and François Bonnot

A method for building core collections   65
    Michel Noirot, François Anthony, Stephane Dussert and
    Serge Hamon

Asian rice   77
    Jean-Christophe Glaszmann, Laurent Grivet, Brigitte Courtois,
    Jean-Louis Noyer, Claude Luce, Michel Jacquot, Laurence Albar,
    Alain Ghesquière and Gérard Second

Banana   99
    Christophe Jenny, Françoise Carreel, Kodjo Tomekpe,
    Xavier Perrier, Cécile Dubois, Jean-Pierre Horry and
    Hugues Tézenas du Montcel

Cacao   125
    Claire Lanaud, Juan-Carlos Motamayor and Olivier Sounigo

Cassava   157
    Gérard Second, Jean-Pierre Raffaillac and Carlos Colombo

Citrus   193
    Patrick Ollitrault, Camille Jacquemond, Cécile Dubois and
    François Luro

Coconut   219
    Patricia Lebrun, Yavo-Pierre N'cho, Roland Bourdeix and
    Luc Baudouin

Coffee (*Coffea canephora*)   239
    Stéphane Dussert, Philippe Lashermes, François Anthony,
    Christophe Montagnon, Pierre Trouslot, Marie-Christine
    Combes, Julien Berthaud, Michel Noirot and Serge Hamon

Pearl millet   259
    Gilles Bezançon

Rubber tree (*Hevea brasiliensis*)   277
    Marc Seguin, Albert Flori, Hyacinthe Legnaté and
    André Clément-Demange

Sorghum   307
    Monique Deu, Perla Hamon, François Bonnot and
    Jacques Chantereau

Sugarcane   337
    Jean-Christophe Glaszmann, Nazeema Jannoo, Laurent Grivet and
    Angélique D'Hont

Biochemical and Molecular Markers

Laurent Grivet and Jean-Louis Noyer

Biochemical and molecular markers have several applications in plant genetics. They allow us to observe closely the polymorphism of DNA sequences at a certain number of sites or loci spread across the genome. More precisely, biochemical markers reveal the polymorphism of sequences of certain proteins and thus, indirectly, the polymorphism of DNA sequences from which they are translated. Molecular markers directly reveal the polymorphism of DNA, the targeted sequences corresponding or not corresponding to the coding sequences.

Because of their properties, biochemical and molecular markers are a powerful tool to study the structure of genetic variability within a species and trace its evolutionary history. They are relatively unaffected by the environment or genetic basis. We can thus use them to compare individuals that were studied in different experiments or that are present in different collections. It is generally acknowledged that biochemical and molecular markers reveal a neutral polymorphism, i.e., one that is not subject to selection. They are relatively insensitive to homoplasy: there is little chance of observing two identical alleles that result from different mutational histories.

In this chapter we specify the characteristics of different biochemical and molecular markers to study the diversity of collections of plant genetic material. There are now more than ten techniques of genetic marking (see, for example, Weising et al., 1995; Karp et al., 1996, 1997, 1998; Santoni, 1996; de Vienne and Santoni, 1998). We describe here five widely used and promising techniques: isozymes, RFLP (restriction fragment length polymorphism), RAPD (random amplified polymorphic DNA, also called rapid), AFLP (amplified fragment length polymorphism), and microsatellites. The principle of each technique is briefly described in the appendix to this chapter. After summarizing the structure of the plant genome, we compare the five techniques in methodological terms (target sequences, nature and level of polymorphism detected, genetic similarity) and practical terms (cost, quickness of result, infrastructure needed).

ORGANIZATION AND VARIABILITY OF PLANT GENOMES

The Organization of the Genome

The genome of a plant is distributed in three cell compartments: the mitochondria, the chloroplasts, and the nucleus.

The mitochondrial genome is composed of a circular molecule of master DNA of 200 to 2500 kilobases (1 kilobase = 1 kb = 10^3 base pairs) depending on the species, carrying 100 to 120 genes. The chloroplast genome is also circular and its size is about 150 kb. It has about 100 genes. The chloroplast and mitochondrial genomes are, most often, inherited through the maternal side.

The nuclear genome is composed of a definite number of chromosomes comprising DNA linear molecules. The genes are spread on a non-coding DNA matrix, essentially constituted of repeat DNA. The size of the nuclear genome varies considerably from one species to another: it is about 400 megabases (1 megabase = 1 Mb = 10^6 base pairs) in rice and about 16,000 Mb in wheat. The largest known nuclear genome among the angiosperms, that of *Fritillaria assyriaca*, contains 600 times more DNA than the smallest, that of *Arabidopsis thaliana* (Bennett and Smith, 1991). These variations are due to differences in ploidy and especially to differences in the quantity of non-coding dispersed repeat DNA. It is accepted that the number of genes carried by a nuclear genome on the plant is between 20,000 and 50,000. Therefore, the information contained in the nucleus, even if diluted, thus remains clearly more important than that contained in the cytoplasmic organelles. The nuclear genome has a biparental heredity via meiosis and fertilization.

DNA Polymorphism

The polymorphism of DNA results from the accumulation of mutations, that is, of modifications of sequences under the action of endogenous or exogenous factors. The mutations may appear in the form of visibly large rearrangements on the cytogenetic scale (deletion, translocation, inversion) or in the form of occasional modifications of sequences.

Biochemical and molecular markers are essentially used to detect occasional variations in sequences. Two types of variations can generally be distinguished: mutations corresponding to tmsubstitution of one base for another and mutations by insertion or deletion of a short fragment of DNA.

DNA polymorphism generated by occasional mutations is very commonly observed. It affects the entire genome, at every level, depending on the type of sequence and genomic compartment. For example, it is high in the repeat nuclear sequences of the micro- or minisatellite type, but low in the chloroplast DNA. The sequencing, over 1933 base pairs, of the nuclear gene *Opaque-2* of maize in 21 lines representing the diversity of temperate germplasm showed that the 21 sequences are entirely different. In total, 14% of the bases present a polymorphism resulting from occasional mutations and 26 phenomena of insertion-deletion have been detected (Henry and Damerval, 1997).

A particular type of DNA polymorphism is exploited by microsatellite markers. A microsatellite is generally composed of a tandem repeat of a motif formed of a small number of base pairs, generally two or three. Microsatellites are numerous in the eukaryote genomes and are spread throughout the chromosomes. Each is flanked by sequences specific to it. For each microsatellite site, mutations produced at a high frequency cause a variation in the number of repetitions of the base motif, which results in high allelic diversity. These mutations can be due to a gliding of the polymerase arising from unequal replication or crossing-over (Jarne and Lagoda, 1996, for review).

Sometimes there are significant differences in the total quantity of nuclear DNA between related interfertile species or even between genetic groups within the same species. These differences are probably due to variations in the quantity of repeat DNA. They can be revealed by flow cytometry (Dolezel, 1991; Bennett and Leitch, 1995). They may sometimes carry elements useful in differentiating groups within a species or within a species complex.

INFORMATION REVEALED BY DIFFERENT GENETIC MARKERS

In this section, the methodological aspects that differentiate the five types of genetic markers considered are covered, then the differences are examined to see how they affect the estimation of genetic diversity and the similarities between individuals.

The Nature and Genetic Interpretation of Polymorphism

Each type of marker is associated with a methodology that determines the nature of the information obtained. Two major categories of markers can be distinguished: (1) markers that can be used to reveal a series of several alleles for each locus studied (multiple allele markers) and (2) markers that allow detection of the presence or absence of a single allele for each locus, simultaneously for a large number of loci (genetic fingerprinting).

MULTIPLE ALLELE MARKERS

Multiple allele markers comprise isozymes, RFLP, and microsatellites. The alleles detected are most often codominant: the two homologous alleles can be observed in heterozygous individuals.

Isozymes and RFLP

For the isozymes, a locus is defined by a catalytic function towards a specific substrate. For the RFLP, a locus corresponds to the region of the genome that hybridizes with a probe, the size of which is a few hundred to a few thousand base pairs. In both cases, the alleles are distinguished by their electrophoretic mobility. In the simplest case, an allele is materialized by a single band of specific molecular weight. Nevertheless, several factors may complicate this scenario.

❏ For the RFLP, an allele may correspond to a combination of several bands if there is a restriction site or sites for the restriction enzyme in the DNA sequence homologous to that of the probe.

❏ For the isozymes, the polymer enzymes may generate new bands in the heterozygotes.

❏ A substrate (isozymes) or a probe (RFLP) may simultaneously reveal the alleles of several paralogous loci, that is, of loci presenting the same DNA sequence or very similar sequences.

❏ For the RFLP, if several restriction enzymes are used in combination with a single probe, each allele corresponds in principle to a specific combination of bands detected with different enzymes. In practice, alleles are rarely reconstructed from analysis of multienzyme polymorphism because they are too complex. In this case, it is enough to note the presence or absence of each band or of each profile.

With isozymes and RFLP, therefore, it is often thus necessary to study the heredity of bands in controlled populations to interpret the profiles observed in terms of loci and alleles. Genetic interpretation is generally simpler in autogamous plants, for which each individual is most often homozygous for all its loci.

For the isozymes, the target sequences are by definition genes coding for the enzymes. The polymorphism is essentially due to occasional mutations that induce the replacement of one amino acid by another, modifying thus the overall electric charge of the protein or its molecular mass and, in consequence, its electrophoretic behaviour. However, several mutations are silent because of the degeneracy of the genetic code and a mutation inducing a change of amino acid does not systematically cause a modification of the electrophoretic mobility of the protein. This has been proved, for example, for the a-amylases of *Drosophila* (Inomata et al., 1995).

For the RFLP, the nature of the target sequences depends on the type of probe selected. It is possible to use probes that reveal the polymorphism of repeat sequences or single copy coding or non-coding sequences, of nuclear or cytoplasmic sequences. We are essentially interested in single copy nuclear sequences. The polymorphism observed is due to mutations affecting the homologous sequence of the probe and above all that of its adjacent regions over about 10 kb. It may be a question of occasional mutations in the restriction sites of the enzyme or even insertions or deletions between these sites. The relative importance of the two phenomena may vary from one species to another. The occasional mutations that do not affect the restriction site of the enzyme or the small insertions-deletions may go unnoticed. Whether it is for the isozymes or for RFLP, an allele corresponds potentially to a group of several sequences, to the extent that these two techniques do not have the power of resolution to detect all the mutations that affect the sequence of the target locus.

Microsatellites
For the microsatellites, a locus is defined by a microsatellite site accompanied by its flanking sequences. Defined primers in these sequences, from either side of the microsatellite, allow us to amplify the locus to the exclusion of any other in the genome. This specificity is guaranteed by the length of the primers (20 to 25 nucleotides). It renders infinitely low the probability of amplifying another sequence in the genome purely at random. The source of the polymorphism being looked for is variation in the number of repetitions of the base motif of the microsatellite site. In most cases, each allele will be materialized by a band of specific molecular weight, which may be translated into a number of repetitions of the base motif of the microsatellite if the gel is sufficiently resolutive. In some cases, however, part of the alleles arise from variations of sequence between one of the primers and the microsatellite site (Orti et al., 1997). The rates of mutation associated with the microsatellite sites are 100 to 1000 times higher than for the isozymes (Jarne and Lagoda, 1996). Whatever the nature of the polymorphism, the profiles observed can be directly interpreted in terms of locus and alleles.

GENETIC FINGERPRINTING

Two techniques of genetic fingerprinting are presented: RAPD and AFLP. The term 'genetic fingerprinting' is commonly used because these techniques allow us to reveal simultaneously the polymorphism of a large number of loci so well that each individual has ample chance of having a multilocus profile of its own, as with human fingerprints.

    With RAPD and AFLP techniques, the similarity of sequence between bands revealed by a single pair of primers does not allow us to establish relations of homology. It covers 20 base pairs at most, corresponding to the sum total of the sequence of the two primer sites, for RAPD. The similarity is less for AFLP, to the extent that the only common point between all the bands revealed by a pair of primers is the sequence of two flanking restriction sites and that of the selective bases, which represents at most 16 base pairs.

Typically, a RAPD amplification allows the detection of 5 to 20 bands and an AFLP amplification a few tens to 100 bands. For the genetic interpretation of profiles observed, each band is considered as an allele of a particular locus. The identity of the locus and of the allele depends entirely on the electrophoretic mobility of the band. For each locus, the only other allele possible is the absence of the band. The band is thus a dominant allele since the homozygous individuals that present the band cannot be distinguished from heterozygotes. For each locus, the absence of the band probably corresponds to a heterogeneous set of sequences.

Even if the bands amplified by the same pair of RAPD or AFLP primers have little chance of corresponding to homologous sequences, the two different bands may, in a small number of cases, materialize two alleles from a single locus. It generally cannot be perceived in studies of diversity, but this phenomenon can be observed when these techniques are used to make a genetic map.

In principle, AFLP and RAPD allow the targeting of all compartments of the genome and all types of sequences. The polymorphism revealed by RAPD corresponds somewhat to the modification of a primer site following the substitution of one base for another or to an insertion-deletion phenomenon. It may also be a matter of one insertion between two primer sites that are sufficiently far apart to make the amplification impossible (Williams et al., 1990). The polymorphism observed with AFLP corresponds probably to the modification of a restriction site of one of the two restriction enzymes following an occasional mutation (Vos et al., 1995).

Several other techniques of genetic fingerprinting give profiles that can be interpreted in genetic terms in the same way that RAPD and AFLP are. Among these are DAF (DNA amplification fingerprinting; Caetano-Añolles and Gresshoff, 1991), which is a variant of RAPD, ISSR (inter-simple sequence repeat analysis; Zietkiewicz et al., 1994), which consists of amplifying the sequences located between microsatellite sites, or even the use of microsatellite probes in RFLP (Rus-Koretekaas et al., 1994).

Comparison of the Five Techniques

In this section the polymorphism and genetic similarities measured with the different markers are compared on the basis of experimental data.

POLYMORPHISM OF MULTIPLE ALLELE TECHNIQUES

It is possible to compare the polymorphism obtained with different multiple allele techniques by using the usual parameters of population genetics such as the percentage of polymorphic loci, the mean number of alleles per locus, and the Nei diversity index, applied to the same sample of individuals.

Table 1. Comparison of rates of polymorphism, Nei diversity and the average number of alleles per locus between isozymes (Iso) and RFLP and between microsatellites (SSR) and RFLP, in various studies. For RFLP, a single restriction enzyme was used, unless otherwise mentioned

Species	Technique	No. of individuals studied	No. of loci studied*	Rate of polymorphism (%)	Nei diversity	Av. no. of alleles per locus	Reference
Brassica campestris	Iso	285	5	60	0.41	2.7	Mitchell and Quiros (1992)
	RFLP	277	4	100***	0.60	8.5	
Hordeum vulgare	Iso	268	7	100	0.44	4.9	Zhang et al. (1993)
	RFLP	240	13	100***	0.47	4.2	
Populus tremuloides	Iso	118	14	77	0.25	2.8	Liu and Furnier (1993)
	RFLP	91	41	71	0.25	2.7	
Populus grandidenta	Iso	96	14	29	0.08	1.4	Liu and Furnier (1993)
	RFLP	75	37	65	0.13	1.8	
Zea mays	Iso	21	22	68	-	2.1	Messmer et al. (1991)
	RFLP	21	144(79)*	94	-	3.34**	
Zea mays	Iso	31	27	-	-	2.2	Gerdes and Tracy (1994)
	RFLP	43	71	-	-	4.1	
Zea mays	Iso	445	20	75	0.23	2.4	Dubreuil and Charcosset (1998)
Hordeum vulgare	RFLP	285	35	100	0.61	6.3	Russell et al. (1997)
	RFLP	18	114(42)*	-	0.32	2.62**	
	SSR	18	13	100	0.57	5.38	
Glycine max	RFLP	19	114	100***	0.38	2.15	Morgante et al. (1994)
	SSR	19	4	100	0.52	4.25	
Glycine sp.	RFLP	12	110	25	0.41	—	Powell et al. (1996)
	SSR	12	36	100	0.60	4.28	
Oryza sativa	RFLP	14	18	33	0.10	2.3	Wu and Tanksley (1993)
	SSR	14	8	100	0.70	5.2	
Oryza sativa	RFLP	20	12	83	0.32	2.5	Olufowote et al. (1997)
	SSR	20	10	100	0.62	7.4	
Zea mays	RFLP	12	96	-	0.58	—	Taramino and Tingey (1996)
	SSR	12	34	100	0.76	6.56	

*Number of enzyme-probe combinations, value in parentheses is number of probes.
** Average number of bands per enzyme-probe combination.
****RFLP probes preselected for their polymorphism.

Examples of comparisons between isozymes and RFLP and between RFLP and microsatellites are given in Table 1. These results must be interpreted with caution to the extent that samples of individuals are generally small and to the extent that the numbers of loci observed are often not equal among the techniques. In most of the studies mentioned, the RFLP polymorphism is observed with a single enzyme. Even with this restriction, RFLP appears at least as polymorphic as the isozymes and sometimes clearly more so. This result agrees with what is known of the nature of polymorphism detected. Isozymes allow us to detect the polymorphism of gene sequences only across the fine filter of translation and degeneracy of the genetic code. The RFLP allows detection of polymorphism of coding and non-coding regions. Moreover, when the probes corresponding to the genes are used, it is essentially the polymorphism of the regions flanking the target genes that is revealed.

In Table 1, microsatellites appear systematically much more polymorphic than RFLP, which is consistent with their particularly high rates of mutation. These observations are confirmed by several other studies. Saghai-Maroof et al. (1994), for example, have found up to 28 different alleles for a microsatellite locus in barley.

POLYMORPHISM REVEALED BY ALL THE TECHNIQUES

The number of alleles per locus and the Nei diversity index lose part of their genetic significance when fingerprinting techniques are applied. The number of alleles per locus is by nature limited to two (presence and absence of the band). In consequence, the Nei diversity index reaches a maximum at 0.5. Moreover, calculation of the Nei index is based on the knowledge of allele frequencies. Because markers are dominant, these frequencies are directly accessible in only two particular cases: if one makes the hypothesis that there high probability of pairing individuals (the Hardy-Weinberg equilibrium is respected) or even if there is autogamy and one makes the hypothesis that all the individuals are homozygous for all the loci. If these hypotheses are not acceptable, the calculation of allelic frequencies is not possible at least to make an independent estimation of the fixation index F_{IS} (Lynch and Milligan, 1994; Kremer, 1998). If one of these hypotheses is overused, the estimation of the Nei index may be biased. This index is nevertheless sometimes used indicatively on markers of the genetic fingerprinting type. In a study by Liu and Furnier (1993) on two species of poplar, the Nei indexes calculated for the RAPD markers are about 0.3 for each species. These values are higher than those obtained for isozymes and RFLP on the same individuals (Table 1). Nevertheless, the hypothesis made on the Hardy-Weinberg equilibrium prevents a clear conclusion as to the better aptitude of RAPD in revealing a polymorphism.

In the particular case of autogamous species, the allele frequencies may be directly estimated from the frequency of genotypes and the Nei index may be calculated. On soybean, for example, Powell et al. (1996) found Nei indexes of 0.31 and 0.32 for RAPD and AFLP, respectively. These values are lower than those obtained for RFLP and microsatellites on the same sample of individuals (Table 1).

Genetic fingerprinting techniques also reveal simultaneously the polymorphism of several loci per experimental unit. An experimental unit corresponds to a strip on a gel. To take into account this characteristic, which has great practical importance for the experimenter, it may be useful to compare the polymorphism not at the level of the locus but at the level of the experimental unit. Several measurements of diversity may thus be considered, for example, the average number of polymorphic bands per experimental unit, the PIC (polymorphism information content; Weir, 1990), and, for autogamous species, the marker index (Powell et al., 1996). The PIC is written as the Nei index, replacing the allele frequencies by frequencies of multilocus profiles. The marker index is defined as the product of the Nei diversity index by the number of polymorph bands per experimental unit. It is interpreted as the average number of bands per experimental unit differentiating two randomly selected individuals. On the basis of these criteria, comparisons of techniques used on barley (Russell et al., 1997), soybean (Powell et al., 1996), and peas (Lu et al., 1996) show that AFLP is distinguished by an average number of polymorphic bands per experimental unit, a PIC, and a marker index clearly higher than those for other techniques. AFLP is useful not because it reveals more diversity in each locus, but because it allows us to observe simultaneously a large number of polymorphic loci.

GENETIC SIMILARITIES

When a group of individuals is observed for several loci with a given technique, it is possible to calculate a genetic similarity for all the pairs of individuals (Perrier et al., 1999). The set of similarities can be synthesized by a multivariant analysis or a branched representation. It is legitimate to ask whether the various markers give the same estimate of the genetic proximity of individuals.

There are so far few experimental results that allow us to address these questions. Some are grouped in Table 2 in the form of correlations between indexes of similarity calculated for different pairs of techniques. Examples of relations between distributions are given in Fig. 1. The absence of homogeneity between the tests, the levels of significance used, the ranges of grouping between individuals, and the power of parameters (number of individuals studied) complicate the synthesis of observations. Nevertheless, as a general rule, the correlations between the similarities calculated from several markers are significant but of variable intensity.

The low correlation values may be explained for the most part by the locus sampling bias. This is particularly true for isozymes in that the number

Table 2. Correlation between similarities calculated from different techniques of molecular markers

Reference	Species	No. of individuals used to calculate correlation	Test*	Comparison**	Correlation (r)	Threshold (P)
Lu et al. (1996)	peas	10	Tm	RFLP-RAPD	0.5	<0.05
				RFLP-AFLP	0.7	<0.05
				RAPD-AFLP	0.6	<0.05
Powell et al. (1996)	soybean	10	Tm	RFLP-RAPD	0.24	ns
				RFLP-AFLP	0.42	<0.01
				RFLP-SSR	0.18	ns
				RAPD-AFLP	0.45	<0.01
				RAPD-SSR	0.15	ns
				AFLP-SSR	0.14	ns
Beer et al. (1993)	Avena sterilis	177	Tm	RFLP-Iso	0.26	<0.005
Heun et al. (1994)	Avena sterilis	24	Tm	RFLP-Iso	0.36	<0.01
Peakall et al. (1995)	Buchloe dactyloides	48	Tm	RAPD-Iso	0.63	<0.001
Messmer et al. (1991)	maize	21	Cp	RFLP-Iso	0.23	<0.01
Gerdes and Tracy (1994)	maize	31	Cp	RFLP-Iso	0.26	<0.01
Prabhu et al. (1997)	soybean	10	Cp	RFLP-DAF	0.70	<0.01
Russell et al. (1997)	barley	18	Cr	RFLP-RAPD	0.20	<0.01
				RFLP-AFLP	0.71	<0.01
				RFLP-SSR	0.51	<0.01
				RAPD-AFLP	0.11	ns
				RAPD-SSR	0.24	<0.01
				AFLP-SSR	0.52	<0.01
Dos Santos et al. (1994)	Cabbage	45	Cr	RFLP-RAPD	0.75	<0.01
Engquist and Beeker (1994)	colza	17	Cr	RFLP-RAPD	0.76	<0.01
				RFLP-Iso	0.53	<0.01
				RAPD-Iso	0.67	<0.01
Thormann et al. (1994)	Brassica spp.	18	Tm	RFLP-RAPD	0.93	—

*Tm, Mantel test; Cp, Pearson coefficient of correlation; Cr, Spearman coefficient of rank correlation.
***Iso, isozymes; SSR, microsatellites; DAF, DNA amplification fingerprinting.

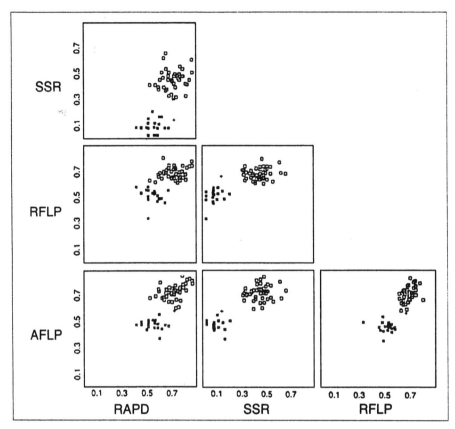

Fig. 1. Two-dimensional relations between the genetic similarities calculated from RFLP, AFLP, microsatellites (SSR), and RAPD in soybean. The white squares represent the similarities between accessions of *Glycine max*, the pluses represent similarities between accessions of *G. soja*, and the black squares represent similarities between accessions of the two species (Powell et al., 1996).

of loci analysed is most often less than 20. Messmer et al. (1991) calculated the type deviations of indexes of similarity by the jackknife method (Millier, 1974). They showed that the values of type deviations obtained with 22 isozymic loci were two to three times higher than values obtained with 144 enzyme-probe combinations of RFLP. This is also true for molecular techniques, which allow detection of more loci. Dos Santos et al. (1994) showed by the bootstrap method (Efron and Tibshirani, 1986) that the distribution of correlations between indexes calculated for 100 random samples of 56 RAPD bands taken two by two widely overlapped the distribution of correlations between indexes for 100 random samples of 56 RAPD bands and 56 RFLP bands taken two by two.

The technical quality of manipulations has a direct repercussion on the reliability of encoding of data and can thus also be a source of divergence in the estimation of similarities. The encoding may become laborious if technical quality is not high, particularly for AFLP, which reveals several bands simultaneously. Many studies have indicated problems of reproducibility with RAPD, which necessitates repeated analyses of each sample if serious errors are to be avoided in the encoding of data (Salimath et al., 1995; Yang et al., 1996). Moreover, the profiles are difficult to reproduce from one laboratory to another, which is a handicap for studies that are part of a network (Jones et al., 1997). Finally, the competition for primers may lead in certain cases to the non-amplification of a band present in the heterozygous state, because of its considerable dilution (Hallden et al., 1996).

More alleles can be detected at a single locus with microsatellites than with other techniques, which results in an average index of similarity between individuals that is generally much lower. This may explain the low correlation between this technique and the others, especially when the individuals are not closely related (the correlation is not significant in the study of Powell et al., 1996) (Table 2).

RAPD seems sometimes to give an estimation of similarity that differs from that of other techniques, particularly when the individuals are distant (of two different species). This divergence has been observed by Thormann et al. (1994) in the genus *Brassica* and by Powell et al. (1996) in *Glycine*. Thormann et al. (1994) have shown by hybridization that, in 3 cases out of 15, bands of the same mobility are not homologous in the different species studied. RAPD may be more sensitive to this type of confusion than other markers.

The influence of evolutionary mechanisms that underlie the variation observed cannot be completely ruled out in explaining the differences between markers. N'goran et al. (1994), for example, showed that in cacao the portion of bands amplified by RAPD corresponds to those repeat sequences and that the bands amplifying repeat sequences and unique sequences do not give the same structure of genetic material.

RELATIONSHIP WITH GENEALOGY AND MORPHOLOGICAL CHARACTERS

It may be asked whether the similarities calculated from markers can be used to organize the relations between individuals into a hierarchy in the same way as genealogy or morphological characters.

The relationship between the similarity evaluated with markers and the relation coefficient of Malecot has been studied in several plants for different markers (Table 3). Overall, the correlations between the two types of information are low but significant. The low value of the correlations can be explained by the fact that the calculation of coefficients of relation is based on unrealistic hypotheses, particularly the absence of relationship between distant ancestors whose genealogy is not known. Also, genetic drift and selection are not taken into account. This caused Graner et al. (1994) to suggest

Table 3. Correlation between the Malecot coefficient of relation f and the similarity
calculated from markers

Reference	Species	Technique	Test*	No.**	Correlation	Threshold (P)
Cox et al. (1985)	Soya	Isozymes	Cr	27	0.15	< 0.01
				32	0.40	< 0.01
				39	0.45	< 0.01
O'Donoughue et al. (1994)	Oat	RFLP	Tm	55	0.32	< 0.001
Graner et al. (1994)	Barley	RFLP	Cr	21	0.21	< 0.05
				17	0.42	< 0.01
Gerdes and Tracy (1994)	Maize	RFLP	Cp	42	0.54	< 0.01
		Isozymes	Cp	31	0.32	< 0.01
Tinker et al. (1993)	Barley	RAPD	Cr	27	0.51	< 0.01
Plaschke et al. (1995)	Wheat	Microsatellites	Cr	40	0.55	< 0.001

*Tm, Mantel test; Cp, Pearson coefficient of correlation; Cr, coefficient of rank correlation.
**Number of individuals used to calculate the correlation.

that the relationship calculated with molecular markers, which does not ignore these parameters, may be a more relevant tool than the f coefficient of Malecot.

Low but significant correlations have been observed between similarities calculated from molecular markers and distances calculated from morphological variables. For example, in *Avena sterilis*, a correlation of 0.24 ($P < 0.005$) was observed between RFLP and morphology and a correlation of 0.13 ($P < 0.005$) was found between isozymes and morphology (Beer et al., 1993). In maize, correlations of 0.167 and 0.124 ($P < 0.01$) were obtained in these two cases respectively (Gerdes and Tracy, 1994). More precisely, Dillmann et al. (1997) emphasized that the relationship between the two types of information is not linear in maize. A small distance calculated with RFLP markers corresponds always to a small morphological distance. On the other hand, a large distance calculated with markers may correspond to a large or small morphological distance (Fig. 2). This triangular relationship may be explained by the fact that, for a given morphological character, various combinations of genes may correspond to a single phenotype. Moreover, the linkage disequilibrium between the markers and the genes involved in the character can vary according to the sampling of loci. If it is low, the two types of information will be independent.

PRACTICAL IMPLEMENTATIONS OF MARKERS

Two major practical aspects of genetic marker techniques are the ease with which they can be used and their cost.

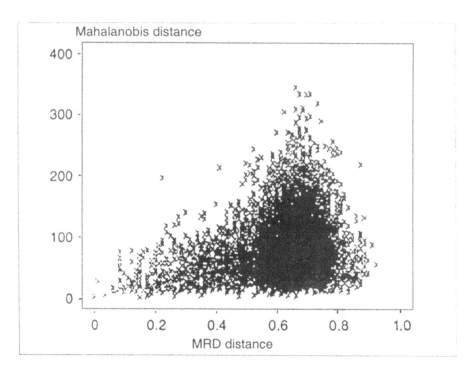

Fig. 2. Relation between the similarity (distance of Rogers, MRD) calculated from RFLP markers and the distance calculated from morphological data (Mahalanobis distance) in maize (Dillmann et al., 1997)

Use

Isozymes do not require large laboratory equipment and use only small quantities of plant material. They may be rather laborious to use on a new species. That depends on the difficulty of finding an equilibrium between the buffer for protein extraction and the electrophoresis buffer. This equilibrium is more difficult to reach for plants with organs rich in polyphenols. The main handicap of isozymes arises from the small number of loci that they can be used to detect. This number is always less than 50 and often less than 20.

Practical use of molecular markers depends on three criteria: the difficulty of using them during the start of a programme, the quantity of DNA needed to analyse each sample, and the essential laboratory equipment for routine implementation.

Genetic fingerprinting, RAPD, and AFLP can be used directly on any plant, without preliminary steps. On the other hand, RFLP and microsatellites may be very difficult to get underway unless a great deal of work has already been completed by other teams on the plant under study.

For RFLP, a source of probes must be arranged, which requires the construction and maintenance of a bank of genomic DNA or cDNA. However, in taxonomic groups that are widely studied, such as the Poaceae, it is possible to use probes that are said to be heterologous, coming from other species. A group of 180 probes of rice, barley, and oats, having a wide spectrum of hybridization over the group Poaceae, has been defined. It is distributed by Cornell University in the United States (van Deyne et al., 1998). These probes can, in principle, be used on all the species of this botanical family.

Microsatellites require a preliminary step involving screening of a bank and sequencing to define the primers in the flanking regions of the microsatellite sites, which is a difficult and costly process of molecular biology. In the most commonly studied plants, such as wheat, maize, and rice, the primer sequences of many microsatellite sites are published and this step is therefore no longer necessary, but that is not the case with most tropical plants. Moreover, sequence information often cannot be transferred from one species to another, even among taxonomically related species.

The quantity of DNA required may be a limiting factor when DNA is difficult to extract or when large samples are worked on. This quantity depends a great deal on whether the technique used relies on PCR. PCR requires low quantities of DNA. For RAPD and microsatellites, only a few nanograms of DNA are needed to complete an amplification. For AFLP, some hundreds of nanograms will do. On the other hand, 2 to 10 μg of DNA are required per track for RFLP. Moreover, microsatellites can be implemented with DNA of mediocre quality obtained with a basic extraction technique because of the high specificity of primer hybridization.

Microsatellites and RAPD can be used routinely in laboratories with rudimentary equipment. On the other hand, RFLP and AFLP rely on radioactive or biochemical marking of probes or primers and thus require greater laboratory infrastructure. AFLP may possibly be revealed with a DNA colorant such as silver nitrate (Cho et al., 1996). This technique is less constraining in terms of infrastructure but reading the profiles obtained often becomes a more delicate task.

Cost

Cost is inherently a determining factor in the choice of a molecular marker. The cost of a routine genotyping operation depends on the availability of certain molecular tools, such as probes for RFLP or primers for microsatellites, the perfection of which may lead to significant and variable costs (fabrication and maintenance of a probe bank, definition of microsatellite primers). If these costs are set aside, for a given operation, the cost of different markers can be calculated in two parts: the cost price at an elementary point (given one marker for an individual) and the time taken to obtain it. The cost price at one elementary point depends on how the laboratory gets its supplies and its geographic location in relation to the plant material and suppliers. The size of the laboratory evaluated in terms of consumption of products or volume of results obtained for a type of marker influences this component. The time taken to obtain results is related to salary costs and costs of material depreciation. The size of the laboratory in terms of equipment and critical mass of operators (technicity) influences this second component. The cost factor is thus very difficult to extrapolate from one laboratory to another.

In the laboratory of CIMMYT (Centro Internacional de Mejoramiento de Maíz y Trigo, Mexico), Ragot and Hoisington (1993) compared the cost of two types of markers, RAPD and RFLP, distinguishing for the latter the radioactive marking of probes (rRFLP) and biochemical markers (cRFLP). They concluded that the relative cost of different techniques depends on the size of samples treated. RAPD is less expensive for small samples and, even if its cost declines, it becomes costlier in relative terms for large samples. This is mainly due to the economy of scale from reuse of membranes for RFLP markers. Ragot and Hoisington (1993) estimated that the cost begins to drop significantly from three reuses onwards. For cRFLP, it is difficult to go beyond 10 uses, routinely. For rRFLP, the number of uses can vary from 15 to 20, according to our experience. According to Ragot and Hoisington (1993), the cost of DNA extraction may represent half the cost of RAPD. To reduce this cost, Mohan et al. (1997) suggested the use of the squash technique on plant tissues (Langridge et al., 1991). The concentration of polysaccharides and other 'contaminants' of DNA present in the tissues of a good number of tropical plants often makes this technique impracticable. There is thus no universal guideline, but the positive observations of Mohan et al. (1997) show that it is very difficult to understand all the possible situations.

Risterucci (1997) compared the routine cost price of different types of marker in the Biotrop programme of CIRAD. For example, for the genotyping of 100 individuals with 100 markers, the final cost was 0.17 euros for AFLP, 0.26 euros for RAPD, 0.48 euros for RFLP, 0.49 euros for microsatellites, and 0.61 euros for isozymes. It is clear from this comparison that the cost price of multiple allele techniques (RFLP, microsatellites, isozymes) is higher than that of genetic fingerprinting techniques. The high cost of microsatellites arises from the choice to use radioactivity to detect them. Moreover, when the final cost is considered as a function of the sample size, it appears that beyond 25 individuals studied the economies of scale are much more significant with the RFLP markers than with markers based on PCR, as was observed by Ragot and Hoisington (1993). RFLP costs almost the same as genetic fingerprinting techniques for a sample of about 200 individuals.

Even though certain trends can be inferred, it seems impractical to attempt to give a standard and universal cost for each type of marker. Rather than choose a technique on the basis of comparative cost in a necessarily highly specific context, it seems more realistic to choose a technique on the basis of the information required and then find a means of applying it as cheaply as possible. There are ways to reduce costs that may make the choice of type of marker totally independent of cost: the subcontracting of certain steps such as occasional sequencing to service providers, agreements with other competent research laboratories in a specific field and for limited periods, and the use of kits for short periods or for occasional projects.

CONCLUSION

With biochemical and molecular markers we can obtain an image of the non-selective diversity existing in a collection. Each type of marker has advantages and disadvantages and the choice of a particular technique must be reasoned on a case-by-case basis, taking into account the objectives, available means, and state of our knowledge of the species studied. As far as we know at present, the different techniques seem to give images sufficiently similar to the genetic structure. The differences observed can be attributed to the sampling of loci, to problems of repeatability and homoplasy for certain techniques such as RAPD, and the nature of polymorphism revealed and sequences targeted by each technique. It is probable that we lack distance enough to judge on this last point.

The relationship between neutral polymorphism revealed by markers and the polymorphism of useful morphoagronomic characters is not clear. Markers reveal more or less precisely the similarities of sequences between individuals for a sample of a locus. Morphoagronomic characters measure resemblances between individuals on the basis of variables whose level of expression depends on the number of potentially epistatic genes. Markers may, however, be a useful tool to better sample genes of agronomic interest. If the species is structured in genetic groups, that signifies that the entities (populations, groups of genotypes) have had independent evolutionary histories, characterized by limited gene exchange. It is thus possible that the role of alleles determining the expression of characters of agronomic interest is different from one group to the other in response to genetic drift or to various pressures of selection. In this case, the sources of variations for characters manipulated by the selector can be better sampled if the genetic structure revealed by the markers is taken into account.

APPENDIX

Description of Molecular Techniques

This section contains a succinct description of AFLP, RAPD, RFLP, and microsatellite techniques. For more detailed descriptions, the reader is referred to Karp et al. (1997), de Vienne and Santoni (1998), and the internet site of IPGRI (*http://www.cgie.org/ipgri/training*). Words marked by an asterisk are explained in the glossary at the end of this chapter. The internet site of the Institut Pasteur (*http://www.pasteur.fr/other/biology/francais/bio-docs-fr.html*) provides links to many dictionaries.

Isozymic techniques are not covered here, although they are competitive in terms of diversity analysis, because they are already fully described in the literature.

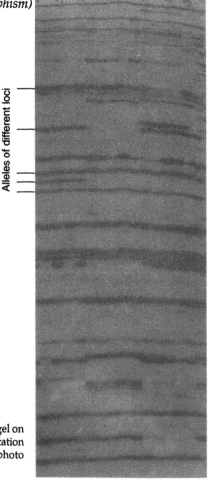

One individual per track

AFLP (amplified fragment length polymorphism) After extraction*, DNA is hydrolysed under the action of two restriction enzymes*. Specific adaptors of two types of restriction sites* are joined (ligation*) at the tips of the fragments obtained. The DNA thus prepared is used during a PCR amplification*, which uses two primers corresponding to adaptors* to which are added one to three arbitrary bases of 3´*. One of two primers is labelled. The fragments produced by this amplification are separated by electrophoresis* on a denaturant polyacrylamide gel*.

The gel is exposed* for some days to contact with an autoradiograph film*. For each individual, several tens of fragments, or bands, are materialized. Each band is interpreted as an allele of a particular locus.

Alleles of different loci

Autoradiograph of a denaturant polyacrylamide gel on which fragments from an AFLP radioactive amplification are separated. The total length of the gel is 40 cm (photo R. Purba, CIRAD).

Microsatellites

After extraction*, the DNA is used in the same way as during a PCR amplification*, which uses two primers* defined in the flanking regions* after sequencing* of a microsatellite site (simple sequence repetition, SSR). The fragments produced by this amplification are separated by electrophoresis* on a resolutive agarose gel* (concentrated) or, better still, on a denaturant polyacrylamide gel*.

It is the variation of number of simple sequence repetitions that is revealed. The monolocus nature of microsatellites allows some flexibility as to the mode of detection of products arising from PCR amplification*. The bands observed materialize the alleles of the target locus. Each allele corresponds to a particular number of repetitions of the base motif of the microsatellite.

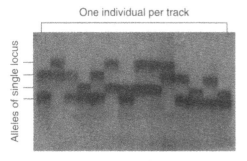

One individual per track

Alleles of single locus

Autoradiograph of a long gel (40 cm) of denaturant polyacrylamide after radioactive PCR amplification of a microsatellite locus of hevea (photo M. Rodier-Goud and M. Seguin, CIRAD).

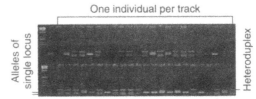

One individual per track

Alleles of single locus

Heteroduplex

UV photograph of agarose gel. The products of PCR amplification of a microsatellite locus of hevea are stained with ethidium bromide. This type of detection is useful because it is easy to implement. The disadvantages are the lower resolutive power than those of denaturant polyacrylamide gels, as well as the appearance of supplementary bands called heteroduplex, which are the result of an artificial reassociation of alleles of a heterozygous individual at the end of PCR (photo by M. Rodier-Goud and M. Seguin, CIRAD).

One individual per track

Alleles of single locus

Photograph of a short gel (12 cm) of denaturant polyacrylamide. The products of PCR amplification of a microsatellite locus of banana are stained with silver nitrate (photo by J.L. Noyer, CIRAD).

RAPD (random amplified polymorphic DNA)
After extraction*, the DNA is used as during a PCR amplification*, which results in a single primer* of 10 bases defined arbitrarily. For the amplification to occur, this primer must find two inverted homologous sequences*, located at a distance such that the amplification can take place. The fragments produced by this amplification are separated by electrophoresis* on a resolutive (concentrated) agarose gel*. The result is presented in the form of a photograph of gel stained with ethidium bromide. Each band is interpreted as an allele of a particular locus.

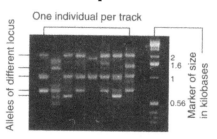

UV photograph of 4% agarose gel. The products of RAPD amplification of DNA of 8 cacao plants are stained with ethidium bromide (photo by J. N'goran, CIRAD).

RFLP (restriction fragment length polymorphism)
After extraction*, the DNA is hydrolysed under the action of a restriction enzyme*. Fragments of variable size are then separated by electrophoresis* on an agarose gel* and then transferred on a membrane (Southern*). A labelled probe is applied on this membrane. It hybridizes on the fragments that contain all or part of a sequence* that is homologous* to it. After several washes with increasing stringency, the membrane is exposed* for a few hours or up to a few days. The result is presented in the form of an autoradiograph.. The bands observed materialize the different alleles of a target locus. Each allele corresponds to a particular configuration of restriction sites around the homologous region of the probe.

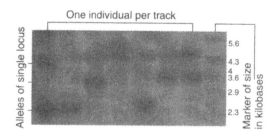

Autoradiograph of the DNA of seven diploid banana plants hybridized with a probe labelled with 32P (photo J.L. Noyer-CIRAD)

GLOSSARY

3´-5´ ends: The double strand of DNA is made up of two complementary antiparallel strands, oriented 3´OH-5´P. The synthesis of DNA occurs only from the 5´ end to the 3´ end. The bases of the primer located on the 3´ end, where the synthesis occurs, have greater significance in the accuracy of the required homology than the bases located at the other end of the primer during PCR amplifications.

Adaptor: Fragment of double-strand DNA, synthesized, the known sequence of which serves as a primer site for PCR amplifications. The adaptors are already linked to DNA fragments that are to be amplified.

Agarose gel: Support of migration in electrophoresis. The resolutive (separating) power increases with the concentration. All the same, one cannot go beyond a concentration of around 4%. The resolution does not routinely go beyond about 10 base pairs, even with high-performance agaroses.

Autoradiography: Sensitive film that can be imprinted by a radioactive or light ray. This radiation localized on a gel or on a membrane can provide a labelled primer at its tip or a marked DNA.

Denaturant polyacrylamide gel: This support of electrophoretic migration is a polyacrylamide gel in the composition of which is introduced a chaotropic (generally of urea) that keeps the denatured complementary strands dissociated. The migration technique allows the separation of DNA fragments that otherwise would not be able to penetrate the gel beyond a certain size.

Electrophoresis: A technique that allows the migration of a polarized molecule under the action of an electric field. At a constant charge density, the smallest molecules migrate further from their origin than the largest molecules. Since DNA has a negative charge in solution, one can separate fragments of different sizes.

Endonuclease: Enzyme capable of precisely catalysing the hydrolysis of DNA, which causes cleavage of two strands within the chain (as opposed to Exonucleases, which cause cleavage at the extremities of the chain).

Ethidium bromide: Intercalary agent of DNA that is fixed inside the double helix. Once fixed, it becomes fluorescent under the UV light. This characteristic is used to visualize DNA on a gel. (Ethidium bromide is a toxic product.)

Exposure: When DNA fragments have been labelled, they emit a rather luminous or radioactive glow. The radiation is detected on contact with a sensitive radiographic film (Autoradiography). According to the intensity of the signal, the detection requires a reasonably long period of contact with a substrate (such as gel or membrane) that contains DNA.

Extraction: There are various techniques to extract DNA from plants. The high proportion of 'contaminants' that accompany DNA in this phase obliges technicians to carry out long purification steps in addition to the extraction itself. This difficulty of extraction, specific to the plant kingdom, has long delayed the development of techniques of molecular biology in comparison with those used in the animal kingdom. Even today, some techniques are difficult to use because of the sometimes mediocre quality of the DNA obtained.

Flanking regions: DNA sequences located on either side of a target sequence.

Homologue: See Homologous sequence.

Homologous sequence: Sequences identical in the series of bases that constitute them. The notion of homology can be relative and arbitrary thresholds can be fixed on it. Sequences can thus be considered homologous even if they are not totally identical, in which case one speaks of a percentage of homology (see also Hybridization).

Hybridize/Hybridization: Pairing of two complementary strands of DNA. The stringency can be controlled by temperature for a given saline charge. For a PCR primer of 10 to 30 bases, the hybridization temperature (Tm) can be calculated with precision, even if the calculation formulae have been obtained empirically. A drop of some degrees in the temperature calculated can bring the primer to hybridize on sites that are not totally homologous to it, particularly at its 5′ end. For probes used in RFLP, which are larger, the control of homology is not absolute (see also Homologous sequence).

Hydrolysis: see Endonuclease.

Ligation: Linkage of two double-strand fragments of DNA at their tips. For this, ligase is used, an enzyme that can effect this linkage no matter what kind of cut (plain or sticky ends) has produced the strands.

Marker/Marking (also *Labelling*): Technique that allows the tracking and location of a DNA molecule by addition or replacement of one or several nucleotides by nucleotides modified by an artificial radioisotope of the ^{32}P, ^{33}P, or ^{35}S type or a chemoluminescent molecule. A large fragment (more than 100-200 nucleotides) is marked by replacing a normal nucleotide by a modified nucleotide (nick translation or random priming labelling). A small fragment or oligonucleotide is marked by the addition of a base modified in terminal position (end labelling).

PCR amplification: Reaction of synthesis of DNA fragments following a series of cycles of denaturation, hybridization, and synthesis, which allows the amplification in quantity of a target DNA sequence. In theory, an amplification of n cycles gives 2n times the number of copies of fragment synthesized in the course of the first cycle. In practice, it always gives less. The discovery of heat-stable DNA polymerases, resistant to several cycles of denaturation of DNA at 95°C, has led to the spectacular development of this technique.

Primer: Fragment of single-strand DNA, synthesized, which after hybridization on a homologous sequence of a denatured strand of DNA serves as a point of anchorage to trigger the synthesis of a complementary strand. During the hybridization on genomic DNA, it is the length of the primers that defines the specificity of the target locus. In the case of RAPD, one primer of about 10 bases will hybridize on several hundreds, even thousands, of sites in each direction and allow the amplification of a small number of fragments (about 10). In the case of microsatellites, two primers of 20 bases will statistically allow the amplification of a single locus (apart from duplications).

Probe: Marked DNA fragment the sequence and function of which are not necessarily known (but may be) and that is used to locate homologous sequences among other sequences.

Restriction enzyme: Endonuclease that cuts DNA at a specific site. This site is defined by a sequence of bases. The most frequently used enzymes recognize sequences of 4 or 6 bases.

Restriction site: A sequence recognized by a restriction enzyme.

Sequence: A determined series of bases that compose a DNA strand.

Sequencing: A process that determines the sequence of a DNA fragment.

Southern (blot): After T. Southern, inventor of the technique. A technique of transfer of DNA on a membrane (blot) following a hybridization, which then allows research by means of a probe labelled with target DNA fragments. By analogy between the name of the inventor and the cardinal points, the Western blot technique was developed for the proteins and the Northern blot for RNA. The East remains to be conquered.

Stringency: Quality of a medium of hybridization or reassociation of nucleic acids. Stringency depends mainly on conditions of temperature, pH, and saline concentration. If stringency is high, all the base pairs of two nucleic acids that are to reassociate must be complementary. When the stringency drops, the reassociation is made by tolerating a growing number of base pairs with a lower percentage of homology.

Washing: In RFLP, a technique that allows the elimination of non-specific hybridization of probes from a membrane, that is, probes that do not have a sufficiently high degree of homology. As for hybridization, it is the saline charge and the temperature that determine the stringency.

REFERENCES

Beer, S.C., Goffreda, J., Phillips, T.D., Murphy, J.P., and Sorrells, M.E. 1993. Assessment of genetic variation in *Avena sterilis* using morphological traits, isozymes and RFLPs. *Crop Science*, 33: 1386-1393.

Bennett, M.D. and Leitch, I. 1995. Nuclear DNA amounts in angiosperms. *Annals of Botany*, 76: 113-176.

Bennett, M.D. and Smith, J.B. 1991. Nuclear DNA amounts in angiosperms. *Philosophical Transactions of the Royal Society of London*, Series B, 334: 309-345.

Caetano-Anolles, G.B. and Gresshoff, P.M. 1991. DNA amplification fingerprinting using very short arbitrary oligonucleotides primers. *Bio/Technology*, 9: 553-557.

Cho, Y.G., Blair, M.W., Panaud O., and McCouch, S. 1996. Cloning and mapping of variety-specific rice genomic DNA sequences : amplified fragment length polymorphisms (AFLP) from silver-stained polyacrylamide gels. *Genome*, 39: 373-378.

Cox, T.S., Kiang, Y.T., Gorman, M.B., and Rodgers, D.M. 1985. Relationships between coefficient of parentage and genetic similarity indices in the soybean. *Crop Science*, 25: 529-532.

de Vienne D. and Santoni, S. 1998. Les principales sources de marqueurs moléculaires. In: *Les Marqueurs Moléculaires en Génétique et Biotechnologies Végétales*. D. de Vienne, ed., paris, France, Inra, 200 p.

Dillmann, C., Bar-Hen, A., Guerin, D., Charcosset, A., and Murigneux, A. 1997. Comparison of RFLP and morphological distances between maize, *Zea mays* L., inbred lines: consequences for germplasm protection purposes. *Theoretical and Applied Genetics*, 95: 92-102.

Dolezel, J. 1991. Flow cytometric analysis of nuclear DNA content in higher plants. *Phytochemical Analysis*, 2: 143-154.

Dos Santos, J.B., Nienhuis, J., Skroch, P., Tivang, J., and Slocum, M.K. 1994. Comparison of RAPD and RFLP genetic markers in determining genetic similarity among *Brassica oleracea* L. genotypes. *Theoretical and Applied Genetics*, 87: 909-915.

Dubreuil, P. and Charcosset, A. 1998. Genetic diversity within and among maize populations: a comparison between isozyme and nuclear RFLP loci. *Theoretical and Applied Genetics*, 96: 577-587.

Efron, B. and Tibshirani, R. 1986. Bootstrap methods for standard errors, confidence intervals, and other measures of statistical accuracy. *Statistical Science*, 1: 54-77.

Engqvist, G.M. and Becker, H.C. 1994. Genetic diversity for allozymes, RFLPs and RAPDs in resynthesized repe. In: *Biometrics in Plant Breeding:*

Applications of Molecular Markers; IXth meetimg of the EUCARPIA section biometrics in plant breeding, J. Jansen, ed., Wageningen, Netherlands, EUCARPIA.

Gerdes, J.T. and Tracy, W.F. 1994. Diversity of historically important sweet corn inbreds as estimated by RFLPs, morphology, isozymes and pedigree. *Crop Science,* 34: 26-33.

Graner, A., Ludwig, W.F., and Melchinger, A.E. 1994. Relationships among European barley germplasm. 2. Comparison of RFLP and pedigree data. *Crop Science,* 34: 1199-1205.

Hallden, C., Hansen, M., Nilsson, N.O., Hjerdin, A., and Sall, T. 1996. Competition as a source of errors in RAPD analysis. *Theoretical and Applied Genetics,* 93: 1185-1192.

Henry, A.M. and Damerval, C. 1997. High rates of polymorphism and recombination at the *Opaque-2* locus in cultivated maize. *Molecular and General Genetics,* 256: 147-157.

Heun, M., Murpy, J.P., and Philips, T.D. 1994. A comparison of RAPD and isozyme analyses for determining the genetic relationships among *Avena sterilis* L. accessions. *Theoretical and Applied Genetics,* 87: 689-696.

Inomata, N., Shibata, H., Okuyama, E., and Yamazaki, T. 1995. Evolutionary relationships and sequence variation of α-amylase variants encoded by duplicated genes in the *Amy* locus of *Drosophila melanogaster. Genetics,* 141: 237-244.

Jarne, P. and Lagoda, J.L. 1996. Microsatellites, from molecules to populations and back. *Tree,* 11: 424-429.

Jones, C.J., Edwards, K.J., Castaglione, S., Winfield, M.O., Sala, F., van de Wile, C., Bredemeijer, G., Vosman, B., Matthes, M., Daly, A., Brettschneider, R., Bettini, P., Buiatti, M., Maestri, E., Malcevschi, A., Marmiroli, N., Aert, R., Volckaert, G., Rueda, J., Linacero, R., Vazquez, A., and Karp, A. 1997. Reproducibility testing of RAPD, AFLP and SSR markers in plants by a network of European laboratories. *Molecular Breeding,* 3: 381-390.

Karp, A., Isaac, P.G., and Ingram, D.S. 1998. *Molecular Tools for Screening Biodiversity.* Londres, Royaume-Uni, Chapman and Hall, 498 p.

Karp, A., Kresovich, S., Bhat, K.V., Ayad, W.G., and Hodgkin, T. 1997. Molecular tools in plant genetic resources conservation: a guide to the technologies. Rome, IPGRI, Technical Bulletin no. 2, 47 p.

Karp, A., Seberg, O., and Buiatti, M. 1996. Molecular techiques in the assessment of botanical diversity. *Annals of Botany,* 78: 143-149.

Kremer, A. 1998. Les marqueurs moléculaires en génétique des populations. In: *Les Marqueurs Moléculaires en Génétique et Biotechnologies Végétales.* D. de Vienne ed., Paris, Inra, 200 p.

Langridge, U., Schwall, M., and Langridge, P. 1991. Squashes of plant tissue as substrate for PCR. *Nucleic Acids Research*, 19: 6954.

Liu, Z. and Furnier, G.R. 1993. Comparison of allozyme, RFLP, and RAPD markers for revealing genetic variation within and between trembling aspen and bigtooth aspen. *Theoretical and Applied Genetics*, 87: 97-105.

Lu, J., Knox, M.R., Ambrose, M.J., Brown, J.K.M., and Ellis, T.H.N. 1996. Comparative analysis of genetic diversity in pea assessed by RFLP- and PCR-based methods. *Theoretical and Applied Genetics*, 93: 1103-1111.

Lynch, M. and Milligan, B.G. 1994. Analysis of population genetic structure with RAPD markers. *Molecular Ecology*, 3: 91-99.

Messmer, M.M., Melchinger, A.E., Lee, M., Woodman, W.L., Lee, E.A. and Lamkey, K.R. 1991. Genetic diversity among progenitors and elite lines from the Iowa stiff stalk synthetic (BSSS) maize population: Comparison of allozyme and RFLP data. *Theoretical and Applied Genetics*, 83: 97-107.

Miller, R.G. 1974. The jackknife: a review. *Biometrika*, 61: 1-15.

Mitchell, M.J. and Quiros, C. 1992. Genetic diversity at isozyme and RFLP loci in *Brassica campestris* as related to crop type and geographical origin. *Theoretical and Applied Genetics*, 83: 783-790.

Mohan, M., Nair, S., Bhagwat, A., Krishna, T.G., Yano, M., Bhatia, C.R., and Sasaki, T. 1997. Genome mapping, molecular markers and marker-assisted selection in crop plants. *Molecular Breeding*, 3: 87-103.

Morgante, M., Rafalski, A., Biddle, P., Tingey, S., and Olivieri, A.M. 1994. Genetic mapping and variability of seven soybean simple sequence repeat loci. *Genome*, 37: 763-769.

N'goran, J.A.K., Laurent, V., Risterucci, A.M., and Lanaud, C. 1994. Comparative genetic diversity studies of *Theobroma cacao* L. using RFLP and RAPD markers. *Heredity*, 73: 589-597.

O'Donoughue, L.S., Souza, E., Tanksley, S.D., and Sorrells, M.E. 1994. Relationships among North American oat cultivars based on restriction fragment length polymorphisms. *Crop Science*, 34: 1251-1258.

Olufowote, J.O., Xu, Y., Chen, X., Park, W.D., Beachell, H.M., Dilday, R.H., Goto, M., and McCouch, S.R. 1997. Comparative evaluation of within-cultivar variation of rice (*Oryza sativa* L.) using microsatellite and RFLP markers. *Genome*, 40: 370-378.

Orti, G., Pearse, D.E., and Avise, J.C. 1997. Phylogenetic assessment of length variation at a microsatellite locus. *Proceedings of the National Academy of Sciences of the United States of America*, 94: 10745-10749.

Peakall, R., Smouse, P.E., and Huff, D.R. 1995. Evolutionary implications of allozyme and RAPD variation in diploid populations of dioecious buffalograss *Buchloe dactyloides*. *Molecular Ecology*, 4: 135-147.

Perrier, X., Flori, A., and Bonnot, F. 1999. Les méthodes d'analyse des données. In: *Diversité Génétique des Plantes Tropicales Cultivées*. P. Hamon et al. eds., Montpellier, France, Cirad, collection Repéres, pp. 43-76.

Plaschke, J., Ganal, M.W., and Röder, M.S. 1995. Detection of genetic diversity in closely related bread wheat using microsatellites markers. *Theoretical and Applied Genetics*, 91: 1001-1007.

Powell, W., Morgante, M., Andre, C., Hanafey, M., Vogel, J., Tingey, S., and Rafalski, A. 1996. The comparison of RFLP, RAPD, AFLP and SSR (microsatellite) markers for germplasm analysis. *Molecular Breeding*, 2: 225-238.

Prabhu, R.R., Webb, D., Jessen, H., Luk, S., Smith, S., and Gresshoff, P.M., 1997. Genetic relatedness among soybean genotypes using DNA amplification fingerprinting (DAF), RFLP, and pedigree. *Crop Science*, 37: 1590-1595.

Ragot, M. and Hoisington, D.A. 1993. Molecular markers for plant breeding: comparisons of RFLP and RAPD genotyping costs. *Theoretical and Applied Genetics*, 86: 975-984.

Risterucci, A.M. 1997. Coût des techniques de marquage moléculaire. In: *Utilisation des Marqueurs Moléculaies en Génétique et en Amélioration des Plantes*. Training seminar. Montpellier, France, Cirad-Biotrop.

Rus-Kortekaas, W., Smulders, M.J.M., Arens, P., and Vosman, B. 1994. Direct comparison of levels of genetic variation in tomato detected by a gaca-containing microsatellite probe and by random amplified polymorphic DNA. *Genome*, 37: 375-381.

Russell, J.R., Fuller, J.D., MaCaulay, M. Hatz, B.G., Jahoor, A., Powell, W., and Waugh, R., 1997. Direct comparison of levels of genetic variation among barley accessions detected by RFLPs, AFLPs, SSRs, and RAPDs. *Theoretical and Applied Genetics*, 95: 714-722.

Saghai-Maroof, M.A., Biyashev, R.M., Yang, G.P., Zhang, Q., and Allard, R.W. 1994. Extraordinarily polymorphic microsatellite DNA in barley: species diversity, chromosomal locations, and population dynamics. *Proceedings of the National Academy of Sciences of the United States of America*, 91: 5466-5470.

Salimath, S.S., de Oliveira, O.A.C., Godwin, I.D., and Bennetzen, J.L. 1995. Assessment of genome origins and genetic deversity in the genus *Eleusine* with DNA markers. *Genome*, 38: 757-763.

Santoni, S. 1996. Les marqueurs moléculaires utilisables en amélioration des plantes. *Le Sélectionneur Français*, 46: 3-18.

Taramino, G. and Tingey, S. 1996. Simple sequence repeats for germplasm analysis and mapping in maize. *Genome*, 39: 277-287.

Thormann, C.E., Ferreira, M.E., Camargo, L.E.A., Tivang, J.G., and Osborn, T.C. 1994. Comparison of RFLP and RAPD markers to estimating genetic relationships within and among cruciferous species. *Theoretical and Applied Genetics*, 88: 973-980.

Tinker, N.A., Fortin, M.G., and Mather, D.E. 1993. Random amplified polymorphic DNA and pedigree relationships in spring barley. *Theoretical and Applied Genetics*, 85: 976-984.

van Deyne, A.E., Sorello, M.E., Park, W.D., Ayres, N.M., Fu, H., Cartinhour, S.W., Paul, E., and McCouch S.R. 1998. Anchor probes for comparative mapping of grass genera. *Theoretical and Applied Genetics*, 97: 356-369.

Vos, P., Hogers, R., Bleeker, M., Reijans, M., van de Lee, T., Hornes, M., Frijters, A., Pot, J., Peleman, J., Kuiper, M., and Zabeau, M. 1995. AFLP: a new technique for DNA fingerprinting. *Nucleic Acids Research*, 23: 4407-4414.

Weir, B.S. 1990. *Genetics Data Analysis, Methods for Discrete Population Genetic Data*. Sunderland, USA, Sinaur Associates.

Weising, K., Nybom, H., Wolff, K., and Meyer, W. 1995. DNA fingerprinting in plants and fungi. London, CRC Press.

Williams, J.G.K., Kubelik, A.R., Livak, K.J., Rafalski, J.A., and Tingey, S.V. 1990. DNA polymorphisms amplified by arbitrary primers are useful as genetic markers. *Nucleic Acids Research*, 18: 6531-6535.

Wu, K.S. and Tanksley, S.D. 1993. Abundance, polymorphism and genetic mapping of microsatellites in rice. *Molecular and General Genetics*, 241: 225-235.

Yang, W., de Oliveira, A.C., Godwin, I., Schertz, K., and Bennetzen, J.L. 1996. Comparison of DNA marker technologies in characterizing plant genome diversity: variability in Chinese sorghums. *Crop Science*, 36: 1669-1676.

Zhang, Q., Sanghi-Maroof, M.A., and Kleinhofs, A. 1993. Comparative diversity analysis of RFLPs and isozymes within and among populations of *Hordeum vulgare* subsp. *spontaneum*. *Genetics*, 134: 909-916.

Zietkiewicz, E., Rafalski, A., and Labuda, D. 1994. Genome fingerprinting by simple-sequence repeat (SSR)-anchored polymerase chain reaction amplification. *Genomics*, 20: 176-183.

Methods of
Data Analysis

Xavier Perrier, Albert Flori and François Bonnot

Analysis of data on diversity consists mostly of analysis of crossing tables of taxonomic units—species, populations, cultivars, etc.—and of variables that characterize the diversity. These variables are quantitative agronomic characters, morphological descriptors, mostly qualitative, biochemical markers (isozymes), or molecular markers (RFLP, RAPD), often coded as presence or absence. The taxonomic units are generally observed individually, but they can also correspond to populations; the observations are thus of means of characters, of allele frequencies, and so on. In any case, the analysis of these data aims to discover a possible structure of taxonomic units by analysis of resemblances or dissimilarities between these units.

This objective comes under classification, a vague term that has two meanings in current usage. The first involves the separation of objects into homogeneous groups. This ambition has been that of all the great taxonomists, from Aristotle to Linnaeus. The second meaning comes under the action of logical ordering. This is essentially to do with relations of order between taxonomic units, that is, with relations of filiation between units in the sense of evolution, which is phylogenetics.

Both these tasks may result in graphic representations in the form of a tree. The informative content of each is, however, fundamentally different. In the first case, the nodes of the tree correspond to concepts of grouping. In the second, the nodes are the parental units that are not observed but are supposed to exist, for example, ancestral forms that have disappeared. The links thus include the temporal dimension of evolution.

The problem of diversity treated here comes partly from these two processes, since we wish to define groups of comparable units, as well as to describe the relations of parentage between these groups.

Two approaches can be taken to treat evolutionary organization. The first, called phenetic, describes an organization from objective measures of dissimilarities between the units and refers the introduction of genetic hypothesis to the interpretation. These dissimilarities are estimated overall for all the characters observed and the number of characters gives relevance to the measures. It is often extremely difficult to detect the organization of units by direct examination of their dissimilarities. The object of the analysis is thus to find a close but easily readable representation of them: a factorial plane or a tree structure. The advantage of readability is gained at the cost of a certain loss of information, a loss that one of course seeks to minimize.

The second approach, called cladistic, directly looks at observed characters and attempts to distinguish identical characters inherited from a common ancestor, which are only informative, from phenomena of accidental convergence. It relies on a model of genetic transformation and was established initially on morphological characters that were known to be monogenic and for which the order of different modalities in terms of evolution was known. This approach is now most often used for genome sequencing data provided with an explicit genetic model, of the type developed by Jukes and Cantor (1969) or derived from that type.

The cladistic or related methods such as the methods of compatibility or likelihood are not described and are not within the scope of this work. The reader is referred, for example, to Darlu and Tassy (1993). The nature of markers used often allows only the use of phenetic approaches relevant to the analysis of dissimilarities.

The object of this chapter is to present relevant methodological aspects of these approaches. The base data will be a table of taxonomic units described by markers of diversity. There may also be a stacking up of tables when the same set of units is characterized by several series of markers of various kinds that cannot be grouped together. These tables can be considered separately at first, but researchers ultimately try to analyse them simultaneously.

This chapter is made up of three major parts.

The first involves the initial step of any analysis, the definition of a measure of the resemblance or dissimilarity between individuals. The choice of this measure is of primary importance and corresponds to a deliberate choice of perspective on the data. When the individuals are populations and the observations are of allele frequencies, various measures of dissimilarity, called genetic, can be used—e.g., Nei distances, Gregorius distances. Their properties have often been studied, for example in Lefort-Busson and de Vienne (1985). When the data are of individuals, described by a series of characters, the nature of variables—quantitative, qualitative, binary— determines the usable types of dissimilarity. It will be shown, on the example of molecular markers of RFLP or RAPD type, that knowledge of the biological characteristics of markers helps us choose an index of dissimilarity.

The second part describes factorial methods. Once a two-by-two dissimilarity matrix between individuals is established, it is analysed so that a simplified but faithful representation can be found. If the dissimilarity is Euclidean, factorial methods of analysis can be used. The bases of principal methods are summarized, particularly techniques of simultaneous analysis of several tables. If the dissimilarity is not Euclidean, methods of multidimensional scaling can be used to find a decomposition on a fixed number of axes. These iterative methods require considerable time for calculation when there are a large number of units to be treated. They have not been covered in this work, and the reader is referred, for example, to Escoufier (1975).

The third part concerns tree representations. An evolutionary process, by accumulation of transmissible mutations, will produce an organization that can be represented in the form of a tree. This representation, because of its biological foundation, is particularly relevant for the analysis of diversity, even if the modalities of evolution are generally more complex. Various methods for inferring trees will be presented, as well as techniques that allow the construction of synthetic trees from several types of markers.

Factorial methods and tree methods constitute two very different approaches to the representation of diversity. They must be considered complementary rather than concurrent. Factorial methods aim above all to make an overall representation of diversity that is as far as possible free of individual effects. On the other hand, the more commonly used tree methods tend to represent individual relations faithfully. They are thus two different ways of viewing the data.

The main statistical software proposes various methods of factorial analysis. The tools linked to analysis of dissimilarities and to their tree representation are less frequent. The Ntsys software is for instance a complete and easily accessible software (Rohlf, 1987). Various non-available functions or partly modified methods have been grouped in a specialized software available from the authors (Darwin, dissimilarity analysis and representation; Perrier, 1998).

CHOOSING A DISSIMILARITY INDEX

The aim of a dissimilarity index is to define a means of measuring the resemblance (similarity) between two individuals or, on the other hand, the difference (dissimilarity), since one can go from one to the other simply by a linear transformation.

There are, in fact, a large number of measures of dissimilarity and, depending on the data, various choices are available. Some of these dissimilarities have mathematical properties useful for the analysis or are stable in the face of data errors, which are relevant arguments for their use. The final choice depends on the real informative content of the markers used and to the point of view one wishes to take towards this information.

Finally, certain dissimilarities, such as ultrametric or additive distances, have special properties and can be represented in the form of a tree. There is, of course, no reason for the measure established on the data to have these properties. The object of tree construction methods is to transform the measure established on the data into a tree-representable dissimilarity, with the least possible deformation.

Definitions and Properties

The most general mathematical object representing the difference is a dissimilarity; this is a function d of the set of pairs (i,j) of individuals in the set of positive or null reals, symmetrical $(d(i,j) = d(j,i))$ and such that $d(i,i) = 0$ for any i. This definition is relatively simple and covers a large number of possible measures. On the other hand, it opens up only a few mathematical properties and it is necessary to add other constraints to acquire certain useful properties.

A distance, sometimes called a metric, is a particular dissimilarity obtained by adding the condition $d(i,j) = 0 \Rightarrow i = j$ and above all the triangular inequality between three individuals, $d(i, j) \leq d(i,k) + d(j,k)$. This natural condition translates simply into the possibility of representing any triplet of points by a two-dimensional triangle. This very useful property allows us especially to avoid the problem of negative edge lengths in the construction of a tree.

CITY BLOCK AND EUCLIDEAN DISTANCE

An important group of distances is made up of Minkowski distances of order $p(p \geq 1)$. A distance belongs to this group if there exists an integer K and a series of K values x_{ik} applicable to each individual i, the distance thus being written as $d(i,j) = (\Sigma_k \mid x_{ik} - x_{jk} \mid^p)^{1/p}$.

The only cases used in practice correspond to the values 1 and 2 of p. When $p = 1$, the index is known as the city block, $d(i,j) = \Sigma_k \mid x_{ik} - x_{jk} \mid$. For $p = 2$, the usual Euclidean distance is found as follows: $d(i,j) = (\Sigma_k \mid x_{ik} - x_{jk} \mid^2)^{1/2}$. Of course, by definition, city block or Euclidean distances are obtained when we calculate a distance from a table of individuals $\times$ variables by summing, the absolute values or the squares of the deviations between i and j. However, certain indexes, for example those calculated on data of presence or absence, can be of either of these two types without that following evidently from the mode of construction.

The city block distance and Euclidean distance belong to the same group and are linked by certain relationships. It can be shown, for example, that any Euclidean distance is a city block distance. It has also been shown that the square root of a city block distance is a Euclidean distance.

POWER TRANSFORMATION

Power transformations of a dissimilarity, for certain powers, give it properties of a distance or even of a Euclidean distance. First, we must recall that two dissimilarities d and d are equivalent in order if $d(i,j) \leq d(k,l) \Leftrightarrow d(i,j) \leq d(k,l)$. These two dissimilarities arrange the pairs of individuals in the same fashion and are thus of comparable interpretation. In particular, if d' can be written as an increasing monotone function of d, then d' and d are equivalent in order. This is the case of power functions $\alpha(\alpha \geq 0)$; d^α and d are equivalent in order.

It can be shown that, if d is a dissimilarity, one can always find a value α between 0 and 1 such that then for any λ ($0 \leq \lambda \leq \alpha$), d^λ is a distance that, by the preceding property, is equivalent in order to d. Thus, if the informative content of markers results in the choice of a measurement that is only a dissimilarity, a power transformation can give it the properties of a distance that is useful for methods of classification.

Similarly, it can be demonstrated that if d is a distance, one can always find a value α between 0 and 1 such that for any $\lambda(0 \leq \lambda \leq \alpha)$, d^λ is a Euclidean distance equivalent in order to d. A power transformation can thus allow the passing from a distance to a Euclidean distance just as one passes from a dissimilarity to a distance. A useful application involves factorial analysis. Although they apply only to Euclidean distances, these methods can be used for any dissimilarity after an appropriate power transformation. This transformation seems preferable to the usual technique of adding a constant, possibly very large, to each pair distance, a technique less meaningful with respect to the initial data.

In some cases, the α values of this transformation are known; for example, it has already been indicated that the square root ($\alpha = \frac{1}{2}$) of a city block distance is a Euclidean distance. Often it is not known how to establish this value, which must be estimated numerically on the data.

ULTRAMETRIC AND ADDITIVE TREE DISTANCE

In adding supplementary constraints to the definition of a dissimilarity, we can give it the property of being a distance that can be represented as a tree. The classification methods all aim to approach as closely as possible, in terms of some criterion, the dissimilarity observed by one of these representable distances.

The best known of these dissimilarities is the ultrametric, a distance verifying the inequality $d(i,j) \leq \max[d(i,k), d(j,k)]$ for any three individuals i, j and k. This property expresses that, among the three distances, the two largest are equal: if $d(i,j)$ is the smallest, then $d(i,j) \leq d(i,k) = d(j,k)$. Any triplet of points thus forms a sharp isosceles triangle. This property allows a representation in the usual form of a dendrogram (Fig. 1a), which is a tree that has a particular point, the root, located at an equal distance from all the leaves of the tree. It is shown that an ultrametric is a Euclidean distance.

In the 1970s, the distance called the additive tree distance was proposed. It verifies the inequality $d(i,j) + d(k,l) \leq \max[d(i,k) + d(j,l), d(i,l) + d(j,k)]$ for any four individuals i, j, k and l. This property, called the four points property, is an extension of the ultrametric condition to four points. It expresses that, among the three sums of distances two by two, the two largest are equal. This property allows a tree representation. It is verified on an example of four points (Fig. 2) that the condition is effectively respected since $d(i,j) + d(k,l) = a_i + a_j + a_k + a_l$ and $d(i,k) + d(j,l) = d(i,l) + d(j,k) = a_i + a_j + a_k + a_l + 2a_c$. This tree can be represented in a hierarchical form such as a dendrogram (Fig. 1b). However, in the absence of an objective root, a radial representation is often preferred (Fig. 1c).

The ultrametric condition is shown to be only a particular case of the four points condition. The ultrametric appears thus as an additive distance of a particular tree subjected to a more narrow constraint. This supplementary constraint is that, among three individuals, the two largest distances are equal, a constraint absent in additive tree distance, which allows for unequal branch lengths. This lesser constraint thus enables a more faithful representation of initial dissimilarities.

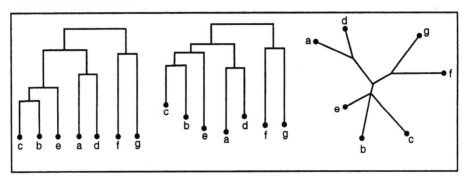

Fig. 1. Ultrametric (a) and additive distance in hierarchical representation (b) or radial representation (c).

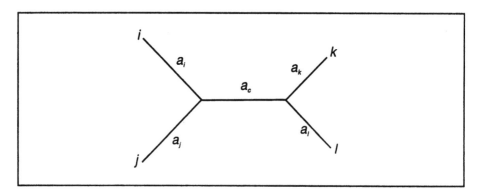

Fig. 2. Tree with four leaves and edge lengths.

INCLUSION RELATION BETWEEN FAMILIES OF DISSIMILARITIES

Different families of dissimilarities have been established by adding supplementary constraints from the initial definition of a dissimilarity. This induces relations of inclusion between the different families (Fig. 3). The ultrametrics, for example, are Euclidean distances and particular additive distances. They are themselves city block distances, which in turn are distances, a particular case of dissimilarities. When we look at Fig. 3 from top to bottom, we find indexes supporting conditions that are increasingly strong and thus less and less apt to describe accurately the relations between individuals. Conversely, from the bottom to the top, the indexes lose their properties of representability. The ultrametrics and the additive distances can be represented in the form of a tree in two dimensions, and the Euclidean distances and city block distances can be represented in spaces of higher dimension. On the contrary, distances and dissimilarities have no useful properties for this purpose. Thus, among the representable distances, the city block distance appears to be the least constrained. In the absence of a contrary indication, the choice must logically include such an index.

In practice, a factorial analysis on a Euclidean distance and a tree representation are often used in parallel. The level equivalent to the Euclidean distance and the additive distance can be noted on the graph. The coherence would thus require that the tree representation be founded on an additive distance rather than on an ultrametric, as is often the case.

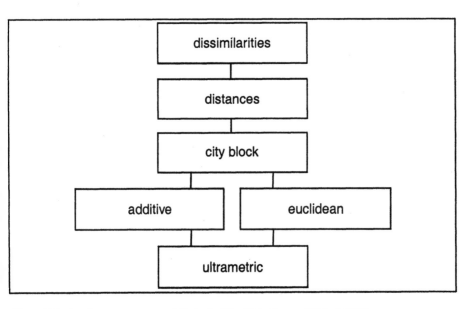

Fig. 3. Relations between groups of dissimilarities (Critchley and Fichet, 1994).

Measures of Dissimilarities on Quantitative Variables

The dissimilarity most often used on quantitative data is the Euclidean distance $d(i,j) = [\Sigma_k (x_{ik} - x_{jk})^2]^{1/2}$ (x_{ik} is the value of the k^e variable for the individual i). It is classical in statistics and can be treated by factorial methods. If the variables are of different nature, it is often useful to reduce them and, to render the value of the distance independent of the number P of variables, one often balances it with $1/P$. From this we get the general expression $d^2(i,j) = (\Sigma_k [(x_{ik} - x_{jk})/\sigma_k]^2)/P$.

The mathematical advantage of the city block distance, $d(i,j) = \Sigma_k \mid x_{ik} - x_{jk} \mid$, has been emphasized. It has been observed that the Euclidean distance, when the differences are squared, gives a heavy weight to significant deviations. This is not always inherently justified; the city block distance, which gives the same weight to all the differences, thus seems preferable. This distance is generally balanced by the number P of variables and, if the order of magnitude of variables is very different, it is necessary to bring them to the same scale, for example by relating them to their amplitude.

Measures of Dissimilarities on Presence/Absence Variables

The descriptors of genetic diversity are often binary variables, which represent the absence or presence of a character. For biochemical or molecular descriptors, these characters in general code the presence or absence of a band on an electrophoresis gel. We conventionally use 0 to denote absence and 1 to denote presence.

Several measures of resemblance between individuals, or association coefficients (Sneath and Sokal, 1973), have been defined on binary data. They have most often been proposed by researchers in a particular discipline—botany, zoology, palaeontology—and are justified mainly because they accurately translate the idea these researchers have of the resemblance between their subjects of study. We will limit ourselves here to discussing indexes for which we can explain the logic of construction and which present a reasonable behaviour (Beaulieu, 1989). For example, if a new marker is added, the dissimilarity between two individuals must increase (or decrease) if these two individuals take different (or identical) values for this marker.

For two individuals i and j, a is the number of markers that are simultaneously present in i and j. Similarly, d is the number of absences in common, b is the number of presences in i and absences in j, and c is the number of absences in i and presences in j. Table 1 lists 13 of these indexes. They are customarily expressed in the form of the similarity S, dissimilarity being obtained for these indexes by $D = 1 - S$.

The general principle of construction of all these indexes is the same, the similitude is estimated by the number of agreements. However, this value does not have an absolute meaning and must be reported with a basis of

Table 1. Major indexes of similarity on presence/absence variables

Authors	Expression	Properties*			
		(1)	(2)	(3)	(4)
S1 Russel and Rao	a/P	y	y	y	
S2 Simpson	$a/\min[(a + b), (a + c)]$	n		n	
S3 Braun-Blanquet	$a/\max[(a + b), (a + c)]$				
S4 Dice	$a/[a + (b + c)/2]$				
S5 Ochiai	$a/[(a + b), (a + c)]^{1/2}$	n		n	S7, S8
S6 Kulczynsky 1	$(a/2)([1/(a + b)] + [1/(a + c)])$	n		y	
S7 Jaccard	$a/(a + b + c)$	n		n	
S8 Sokal and Sneath un2	$a/[a + 2(b + c)]$	y	y	y	S4, S8
S9 Kulczynski 2	$a/(b + c)]$	y	y	y	S4, S7
S10 Sokal and Michener	$(a + d)/P$				
S11 Rogers and Tanimoto	$(a + d)/[a + d + 2(b + c)]$	y	y	y	S11, S12
S12 Sokal and Sneath un1	$(a + d)/[a + d + (b + c)/2]$	y	y	y	S11, S12
S13 Sokal and Sneath un3	$(a + d)/(b + c)$	n		n	S10, S11

*(1) The associated dissimilarity is (y) or is not (n) a distance. (2) It is (y) or is not (n) a city block distance. (3) Its square root is (y) or is not (n) Euclidean. (4) It is equivalent in order to the indexes indicated (absence of mention indicates that the response is not presently known).

comparison. The indexes differ in their mode of estimating the number of agreements and in the choice of the basis of comparison.

The estimation of agreements depends on the meaning assigned to the absence modality. If only modality 1 is considered informative, modality 0 expresses mainly an absence of information; then the number of agreements is a, the number of presences in common (indexes S1 to S9). If 0 and 1 are informative and can be considered as two modalities of a qualitative variable, then the number of agreements is $a + d$, the number of presences and absences in common (indexes S10 to S13). The choice between these two attitudes depends entirely on the nature of characters analysed and is a prerequisite to any reflection on the choice of a dissimilarity index. However, it is not always easy to separate these two points of view. Regarding genetic markers of diversity, it is clear that biological knowledge of the markers being considered will enable us to choose the most appropriate model.

AGREEMENTS ESTIMATED BY PRESENCES IN COMMON: INDEXES S1 TO S9

The numerator of these indexes is always a. The denominator should be an estimation of the number of agreements that two identical individuals will present. For index S1 this number is P, the number of variables. This choice is not judicious since two individuals can thus be identical only if they have the value 1 for all the variables. It is better to estimate the denominator from numbers of presences in i, $a + b$, and in j, $a + c$. Of course, these two values are not equal, and a consensual expression must be defined for them. We can take the minimum (S2), the maximum (S3), or a series of values between these two extremes: the arithmetic mean (S4), the geometric mean (S5), and the harmonic mean (S6). The knowledge of characters can sometimes guide the choice of one of these indexes. However, in the absence of a clear justification, we prefer to retain the neutral behaviour of the Dice index (S4) based on the arithmetic mean.

Another approach is to consider that the basis of comparison is the number of presences found in i or in j, $a + b + c$, whence the Jaccard index (S7). With the S7 index, as with S1, a is compared to the number of variables, but the double absences are treated as missing data. This point of view is, in a good number of cases, highly reasonable and explains the success of the S7 index, which is certainly one of the most widely used.

Index S8, which is used often, and index S9 seem difficult to justify. Note that indexes S4, S7 and S8 can be written in the form $S = a/[a + \beta(b + c)]$ with, respectively, $\beta = \frac{1}{2}$, $\beta = 1$, and $\beta = 2$. We could construct new indexes by choosing other values for β.

AGREEMENTS ESTIMATED BY PRESENCES AND ABSENCES IN COMMON: INDEXES S10 TO S13

Indexes S10 to S13 are obtained by extension of indexes S1, S8, S4 and S9, simply replacing a by $a + d$. On the evidence, these indexes must be symmetrical in a and d, the notation 0 and 1 being purely arbitrary. The extension of other indexes leads to indexes that are non-symmetrical in a and d, or sometimes non-symmetrical in b and c, and thus unacceptable. The most appropriate in many cases are index S10, an extension of S1 but also of S7, the Jaccard index, and S12, an extension of S4 based on the arithmetic mean. Indexes S12, S10 and S11 can be written in the form $S = (a + d)/[(a + d) + \beta(b + c)]$ with, respectively, $\beta = \frac{1}{2}$, $\beta = 1$, and $\beta = 2$.

PROPERTIES OF INDEXES S1 TO S13

Complementary to the logic of construction, the possession of particular properties can guide the choice of an index. Table 1 summarizes certain properties of dissimilarities associated with the proposed indexes. A primary characteristic is the fact of being a true distance. It can also be shown that some of these indexes are city block distances. Although none of them is a Euclidean distance, it is known that city block distances have Euclidean square roots. Index S5, without being a city block distance, also has a Euclidean square root. These indexes, after transformation, can thus be treated by factorial analysis.

Finally, several of these indexes, which have comparable modes of construction, are equivalent in order. Between several indexes giving identical orders, those having useful properties are preferred, S10 or S11, rather than S12, for example.

The algorithms of tree construction are highly sensitive to small variations of dissimilarities. The behaviour of these different indexes faced with data errors is thus an important criterion for selection. This behaviour can be addressed by studying the properties of a matrix subjected to a random noise exchanging the values 1 with 0 and inversely, with a probability t. It is thus possible to estimate the expectation D' of the index considered for a noise of rate t of data.

Figure 4 presents, for indexes S10, S11, and S12, symmetrical in 0 and 1, the value of D' as a function of D. The rate of error used is 0.10, but that value does not change the meaning of conclusions that may be drawn. For index S11, data errors may lead to serious overestimations of small values of dissimilarity, while large values are relatively not modified. Index S12 has an inverse behaviour in leaving the small values relatively not modified and underestimating the large values. Index S10 has a more neutral behaviour; the bias is null for $D = 0.5$ and increases symmetrically. It is thus a useful compromise. However, the most widely used algorithms of tree construction are agglomerative; the smallest dissimilarities determine the first groups, which then determine the entire tree. It is thus preferable to minimize the noise on the small values, as with index S12.

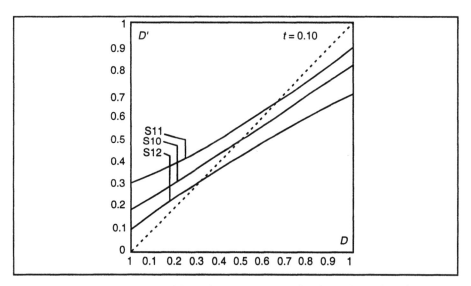

Fig. 4. Estimation by simulation of dissimilarities D' associated with S10, S11 and S12 for an error rate of 0.10 as a function of initial dissimilarity D.

The study of indexes that estimate the agreements by presences in common is more complicated. The distortions observed are greater than for the symmetrical indexes. They can be very high (up to 50% for the small values of D, even with rates of error of 0.05), when the frequencies of 0 are high (Fig. 5). This situation is frequent with highly polymorphic molecular

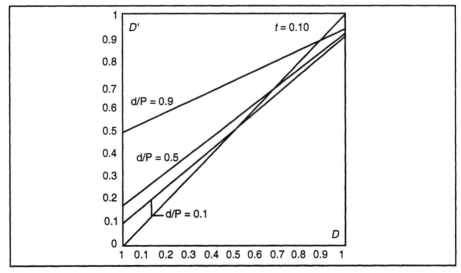

Fig. 5. Estimation by simulation of dissimilarities D' associated with index S7 (Jaccard) for an error rate of 0.10 as a function of initial dissimilarity D.

markers, for which the number of 1 for each individual is very low. This type of matrix must therefore be used with some caution, and the stability of structures must be verified by comparing results obtained on the complete matrix and free of particularly rare markers. Such markers lead to an increase in the number of absences in common for most of the pairs of individuals. Indexes S4, S5 and S6, corresponding to three types of mean for the denominator, have a closely comparable behaviour. They tend to give smaller distortions of small values of dissimilarity than the Jaccard index, S7, which in turn is more favourable than S8. These two indexes are thus preferred.

Measures of Dissimilarities on Qualitative Variables

Here the characters are of qualitative variables that present a finite number of modalities: petal colour, form of stigmata, and so on. Intuitively, two individuals are closely related if they have the same modality for a large number of variables.

The usable indexes correspond to the generalization of indexes presented for the binary values symmetrical in 0 and 1, which are in fact qualitative variables of two modalities.

The number of variables in agreement is m, the number of variables in disagreement is u, and P is always the number of variables. The corresponding indexes of similarity are written as follows:

$m/(m + u) = m/P$ (Sokal and Michener);

$m/(m + 2u) = m/(P + u)$ (Rogers and Tanimoto);

$2m/(2m + u) = 2m/(P + m)$ (Sokal and Sneath).

The dissimilarity is obtained directly by $d = 1 - s$.

The χ^2 distance is often used on a binary table. For one variable, and for each of these modalities, a binary is created that takes the value 1 if the individual presents this modality, 0 if not, and the classic χ^2 is calculated on this new data matrix. This distance does not give the same weight to all the modalities, particularly to rare modalities, which are weighted heavily. Although such an effect can be sometimes desirable, it often is not.

Choice of Index of Similarity on Biochemical and Molecular Markers

The genotype resemblance between two individuals can be measured by a genetic similarity defined from allele forms of genes observed. Certain types of markers (e.g., isozymes, microsatellites) generally give direct genetic information and allow coding in alleles by identification of the allele composition of each locus. The genetic similarity, as defined here, can thus be directly estimated. Other markers do not allow access to all of the genetic information and only one phenotype is observed. The genetic similarity cannot be calculated directly, but the characteristics of these markers be used to define the most relevant similarity index. Moreover, it is useful to evaluate the order of magnitude of the error related to this loss of information.

MULTIPLE ALLELE MARKERS

The genetic similarity T_{ij} between two individuals i and j is defined *a priori* as the mean on L loci of the ratio of the number n_{ls} of alleles for the locus l present simultaneously in the two individuals and of the number n_{lc} of alleles compared: $T_{ij} = (\Sigma_l \, n_{ls}/n_{lc})/L$. If π is the ploidy, n_{lc} would be π for all the loci. When $n_s = \Sigma_l \, n_{ls}$ and $n_c = \pi L$, then $T_{ij} = n_s/n_c = n_s/(\pi L)$.

For a diploid species, each allele of a locus can be coded as a variable taking the values 2, 1 or 0. Following this coding for two individuals i and j, the combination of genotypes of each locus belongs to one of seven groups, named A to G (Table 2). The number of loci of each group is designated I_A to I_G. We can thus write T_{ij} from $n_s = 2I_A + I_B + 2I_E + I_F$ and $n_c = 2(I_A + I_B + I_C + I_D + I_E + I_F + I_G)$.

The parameters $n_{r,s}$ ($r \geq s$) are defined as the number of alleles present r times in one individual and s times in the other. Each of the groups A to G contributes to these parameters. For example, for a locus with a_l alleles, a

Table 2. The seven possible combinations of genotypes of two diploid individuals

	A	B	C	D	E	F	G
Individual i	20..	200..	200..	2000..	110..	110..	1100..
Individual j	20..	110..	020..	0110..	110..	101..	0011..

pair of group D participates one time at $n_{2,0}$, two times at $n_{1,0}$, and a_l — 3 times at $n_{0,0}$.

The numbers I_A to I_G can be expressed from these parameters $n_{r,s}$, from which we can write T_{ij} as the ratio of $n_s = 2n_{2,2} + n_{2,1} + n_{1,1}$ and $n_c = (2n_{2,2} + n_{2,1} + n_{1,1}) + (n_{2,1} + 2n_{2,0} + n_{1,0})/2$.

In the expression of T_{ij}, therefore, a generalization of S4, the Dice index, can be recognized. The number of agreements is effectively 1 for the genotypes contributing to $n_{2,1}$ and $n_{1,1}$ and 2 for those contributing to $n_{2,2}$. Similarly, the disagreements are 1 for $n_{2,1}$ and $n_{1,0}$ and 2 for $n_{2,0}$. It can be noted that for haploid species, some microorganisms, for example, we can directly find the Dice index.

The Dice index belongs to the family of indexes that do not take into account the information carried by double absences. This point of view is logical since the number of double absences does not carry information on the proximity of individuals and depends only on the number of alleles of the loci.

CODOMINANT MARKERS CODED IN BANDS

For markers of the RFLP type or the isozymes, it is sometimes impossible to identify alleles belonging to the same locus and there is only coding in bands. For a diploid species, the presence of a band, coded 1, can thus correspond to a homozygous locus, which must be coded 2, or to one of the alleles of a heterozygous locus, normally coded 1. The different groups of genotypes defined for multiple allele markers are not identifiable and the phenotype observed differs from the real genotype (Table 3).

Table 3. The seven possible combinations of genotypes and phenotypes observed for codominant markers

	A	B	C	D	E	F	G
Genotype i	20..	200..	200..	2000.	110..	110..	1100..
Genotype j	20..	110..	020..	0110.	110..	101..	0011..
Phenotype i	10..	100..	100..	1000.	110..	110..	1100..
Phenotype j	10..	110..	010..	0110.	110..	101..	0011..

As in the preceding case, the information of double absences is not relevant and, by analogy with the Dice index, one can retain, as a measure of resemblance, an index of the group $S_{ij} = n_{1,1}/(n_{1,1} + \beta n_{1,0})$. The value of β is chosen such that, under certain hypotheses, S_{ij} more closely approaches T_{ij}.

For an autogamous diploid species that will be homozygous for all the loci, $S_{ij} = T_{ij}$ for $\beta = \frac{1}{2}$; S_{ij} is thus directly the Dice index.

For a given species, a value of t, the rate of average heterozygosity, is generally known. Considering that this value may be applied at each locus, we can define the value of β, function of t, that annuls $S_{ij} - T_{ij}$. The expression of this difference is complex and the relations between β and t have been researched numerically for different distributions of allele number per locus, the maximum being fixed at 7 (Fig. 6). The distributions I, II and IV correspond to extreme distributions, and the distribution III reproduces a distribution observed on hevea (M. Seguin, personal communication).

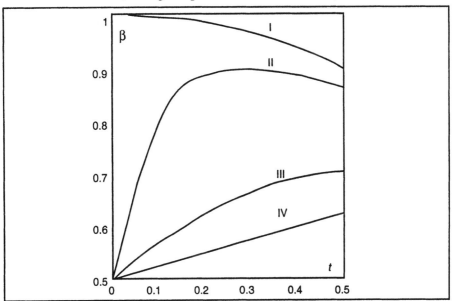

Fig. 6. Values of the index β of the dissimilarity $D = a/[a + β(c + d)]$ such that D_{ij} is equal to the genetic dissimilarity T_{ij} calculated from the complete genetic information, as a function of rates of heterozygosity t and for various distributions of number of alleles:
I   :   all the loci have 2 alleles;
II   :   loci with 2 or 3 alleles in proportions 0.9 and 0.1;
III   :   loci with 2 to 7 alleles in proportions 0.27, 0.32, 0.19, 0.11, 0.07 and 0.03;
IV   :   all the loci have 7 alleles.

For loci with a small number of alleles (I or II), the genetic similarity is approached, for low heterozygosity, only for values of β close to 1.

For a more realistic distribution as distribution III, one can propose to fix the value of β at 0.5 for $t ≤ 0.1$, at 0.6 for $0.1 < t ≤ 0.3$, and at 0.7 for $t > 0.3$.

The order of magnitude of the difference has been evaluated by simulation for different rates of heterozygosity and a series of values of b between ½ and 1, from a population of 40 individuals described by 200 loci. The deviation $S_{ij} - T_{ij}$ is composed of a systematic bias, without consequence on the methods of analysis, and of a true error that is the only relevant point here. This error is expressed by the half-deviation between the lowest and the highest of the values observed in the simulations; it is expressed as a percentage of the mean genetic similarity. For a type III distribution, this maximal error is of the order of 3% for t close to 0.2, but it reaches 5% for highly heterozygous species.

It can be demonstrated, moreover, that this error depends mostly on t and very little on β. In practice, we can most often retain $\beta = 0.5$, corresponding to the Dice index.

DOMINANT MARKERS CODED IN BANDS

Dominant markers, RAPD or AFLP, can be considered loci with two alleles, since a single allele can be located, the other being a null allele that is not materialized on the gel. These markers are said to be dominant, because it is impossible to know the number of copies of an allele for a particular locus and thus to distinguish the homozygotes from heterozygotes.

Among the types A to G defined previously for a diploid species, only A, B, C and E exist, since the loci have only two alleles. The two alleles do not play an identical role. Genotypes A and B must be subdivided according to the phenotypes, distinguishing the located allele from the null allele. The impossibility of coding into alleles leads, as for RFLP markers, to a phenotype designated theoretical in Table 4. As the null allele, noted by a point, cannot be located, only one of these allele states is thus readable, from which we get the phenotype that is actually read.

Table 4. Genotypes, theoretical phenotypes, and phenotypes observed for dominant markers

	A1	A2	B1	B2	C	D
Genotype i	20	02	20	02	20	11
Genotype j	20	02	11	11	02	11
Theoretical phenotype i	10	0.	10	0.	10	1.
Theoretical phenotype j	10	0.	1.	1.	0.	1.
Read phenotype i	1	0	1	0	1	1
Read phenotype j	1	0	1	1	0	1

The number L of loci is the number P of different bands observed in the population. We must note immediately that $n_{0,0}$, the number of double absences, here has a clear meaning. It corresponds to the number of homozygous loci for the null allele in the two individuals and carries as much information as the number of double presences. It will be logical to retain an index symmetrical in 0 and 1. The denominator of the genetic similarity T_{ij} is equal, in this particular case, to P. This value leads us to retain the Sokal and Michener index, here expressed by $S_{ij} = (n_{1,1} + n_{0,0})/P$.

The difference $S_{ij} - T_{ij}$, which measures the error of estimation of genetic similarity, is expressed simply by $(I_{B1} - I_{B2})/P$. This difference is nil if the frequencies of genotypes (2,0) and (0,2) are equal, that is, if the frequencies of the two alleles are identical for all the loci; it remains low if they tend to be identical, on average, for all the loci. On the other hand, if the species is highly homozygous, the proportions of genotypes B are low and $I_{B1} - I_{B2}$ remains low.

For more heterozygous species, this difference can be high since the genotypes B can represent up to 50% of the population. As before, numerical simulations are used to estimate the order of magnitude for different levels of heterozygosity and for different proportions of the two alleles of a locus. The true error is, as before, the demi-amplitude of extremes observed related to the mean genetic similarity. It increases with heterozygosity and reaches 13% for $t = 0.5$ and no allele deficit, goes beyond 18% if the non-identifiable alleles are in a majority, and falls to 10% in the opposite case. In any case, the imprecision induced by the loss of genetic information is clearly greater than for the codominant markers.

In conclusion, it must be emphasized that the informative content of markers must determine the choice of an index of similarity, a criterion that is often missing in most publications in this field.

METHODS OF FACTORIAL ANALYSIS

Methods of factorial analysis aim to discover the strongest structures in populations that are being studied and to eliminate the occasional peculiarities that hamper the general perception of phenomena. They can therefore be very useful in a study of the diversity of a species or a population.

The aim of these methods is to produce a geometric representation of measures of differences between units. Such a representation allows us to manipulate specifically the notion of diversity. We thus get a hierarchical composition of the diversity, which allows us to distinguish the basic tendencies of various characteristics.

The use of these techniques requires a knowledge of the principles they are based on and the conventions they use. After a general presentation of factorial analysis emphasizing its conventional aspects, we will see how it integrates specially in three widely known applications: principal coordinates analysis (PCoA), principal components analysis (PCA), and multiple correspondence analysis (MCA).

Then we will discuss the treatment of multiple tables, a problem found particularly in the study of a population described by several types of markers. When there is information coming from different points of view about the individuals, we can use multiple factorial analysis (MFA) to reveal the major characters, which are manifested in a consensual manner, and those that on the other hand are specific to one type of measure.

Factorial Analysis

CLOUD AND POINTS OF DISPERSAL

If the distance chosen to measure the differences between the units studied is Euclidean, the group of units can be represented in the form of a cloud of points in a vast space. In this space one can mark out a point B, called the mean point, or barycentre, which is defined as being as close as possible to all the points of the cloud. It corresponds to a neutral 'mean type' of which each real unit differs by its own characteristics.

In these conditions, the total diversity of the population is represented by the dispersal of the cloud around B. Factorial analysis quantifies this dispersal by the inertia of the cloud in relation to B, that is, by the sum of squares of distances from each point to B.

PRINCIPAL AXES OF INERTIA AND COORDINATES OF INDIVIDUALS

In general, the space in which the cloud is found is a space of high dimension. Practically, this means that in leaving B one can go in a multitude of directions. In other words, there are many ways for an individual to be distinguished from the mean type and all the units observed are different from it, each in its own way. The aim of factorial analysis is to bring any peculiarity back approximately to the composition of a small number of directions that are very commonly used and independent from each other.

To determine the 'frequency of use' of a direction U, one constructs a criterion by summing for all the individuals a value that is large enough so that the angle that U makes with the specific direction of the individual is acute and so that the individual is at a great distance from the mean point B. This criterion is called the inertia of the cloud along U. The direction that maximizes this criterion is called the principal axis of the inertia of the cloud. Factorial analysis successively produces other axes by choosing for each the direction that maximizes the criterion of inertia among the orthogonal directions of the axes already identified. The orthogonality of axes ensures the independence of variations summarized by them.

The result of this process is a system of independent axes successively explaining the maximum of variation. The coordinates of each unit can be calculated on these axes and they can thus be placed in the orthogonal system. The set of coordinates of all the individuals on an axis is called a *factor*.

From the factorial coordinates, the total dispersal of the population can be reconstituted. However, and this is the essential advantage of factorial analysis, the selection of just the primary factors will enable generally the reconstruction of a large part of the total dispersal. Several relatively empirical criteria have been proposed to determine the number of axes to be conserved (Saporta, 1990). However, it is advisable to remain circumspect and the choice of axes must be reasoned case by case.

In the absence of other information, the coordinates of individuals on the axes have little significance. For a factorial analysis to be interpretable, there must be an 'external' source of information available that one can link to other factorials. This external information may not be explicit and may appear only when one observes that such an axis resembles or, on the contrary, opposes certain recognizable individuals. Most often, however, an axis will be interpreted through observation that it is linked to variables measured on the individuals, and that they have participated or not participated in the calculation of the distance matrix.

Only when an axis is interpreted and when the dispersed part attached to it is comprehensible do we consider that there is a factor that must be taken again into account in the elaboration of the diversity.

GRAPHIC REPRESENTATIONS

To provide a visual representation of diversity, but also to facilitate the interpretation of axes, the methods of factorial analysis propose graphic representations of individuals and, if necessary, of variables. The graphs are critical inputs of these methods. They allow us to see, literally, the oppositions, groupings, and trends that are difficult to perceive from enumerated statistics. However, each type of graph must be read according to particular rules, and it is advisable to master those rules to avoid any chance misreading.

Classic Methods of Factorial Analysis

In practice, factorial analysis of the kind discussed here is only one step in a process that starts with a table of data, operates possible coding, calculates distances if necessary, and finds factorial axes produced with interpretative aids adapted to the nature of the data. It is the entire process that forms a 'method' of factorial analysis. According to the type of data available, the most commonly used methods are: PCoA for a table of distances, PCA for a table of quantitative variables, and MCA for a table of qualitative variables.

PRINCIPAL COORDINATES ANALYSIS

Principal coordinates analysis is used to treat a matrix of distances calculated with an original index, chosen in a manner adapted to its own data. It is, however, necessary that the distance obtained be a Euclidean distance or a distance rendered Euclidean by transformation.

In this case, two individuals i and j can be represented by two points e_i and e_j of a space of unknown dimension K such that the square of the distance d_{ij} can be written as follows: $d_{ij}^2 = (e_i - e_j)'(e_i - e_j)$.

In terms of calculation, it is demonstrated that the successive axes of the factorial analysis and the inertia of the cloud on each axis correspond to vectors and values of the matrix W of scalar products between elements. If the origin of the coordinates is placed at the midpoint of the cloud, the scalar products $w_{ij} = <e_i, e_j>$ are entirely determined by the d_{ij} according the formula of Torgerson: $w_{ij} = -(d_{ij}^2 - d_{i.}^2 - d_{.j}^2 + d^2)/2$.

The diagonalization of this matrix W allows extraction of the eigenvectors and eigenvalues associated with them.

The number of eigenvectors associated with a non-null eigenvalue is the dimension K of the space containing a cloud of points.

The PCoA is the simplest use of factorial analysis. In these conditions, the only aid to interpretation available is the graph of individuals on any choice of two factorial axes.

PRINCIPAL COMPONENTS ANALYSIS

Principal components analysis treats quantitative variables and imposes a Euclidean distance between individuals. The Euclidean distances that can be calculated differ only in the weight assigned to the variables. The most common weightage system consists of bringing all the variables to the same scale and weighing them by the inverse of their standard deviation. The PCA is thus said to be normal, or centred-reduced.

As with PCoA, the coordinates of individuals on the principal components can be calculated from the matrix of scalar products. This can be obtained immediately from a table of centred data.

The interpretation of axes provided by a PCA is more rich because correlations can be calculated between the factors and the variables of the table. From examination of these correlations for each axis, we can reveal the variable or variables most closely linked to the axis. The bundle of variables that vary conjointly can thus be revealed; the conjoint variation allows a distinct discrimination of individuals.

MULTIPLE CORRESPONDENCE ANALYSIS

Multiple correspondence analysis (MCA) allows the study of individuals described by qualitative variables with modalities. This method calculates distances between individuals by the χ^2 distance on the binary table (see above). That is, two individuals are distant from each other to the extent that they present numerous disagreements and that their disagreements cause one or another to adopt rare modalities.

This method is also known as multiple correspondence analysis (MCA), correspondence analysis on a binary table (CA on BT), or simply correspondence analysis (CA).

The interpretation of axes provided by factorial analysis of this matrix of distances is deduced from relations that they have with three types of objects: individuals, variables, and modalities. The individuals provide the same interpretations as in other methods of factorial analysis.

As with PCA, one can evaluate the link between the variables of the table and each axis. For a qualitative variable, this link is increased by the ratio η^2, which is equal to the part of the total inertia of the axis reconstituted by the sum of squares of the intermodality deviations. One factor can thus be perceived as a numerical synthesis of a certain number of variables that vary conjointly.

A qualitative variable is a more complex object than a quantitative variable and the diversity that it induces manifests itself on as many axes as the variable has modalities. A more refined interpretation of axes and of liaisons between variables must thus be looked for at the level of modalities. However, the choice of the χ^2 distance determines that the considerable contribution of a modality to an axis can only be due to the rarity of this modality and thus to the eccentricity of the small number of individuals that possess it. When this is not the case, one axis reveals a set of modalities in mutual association, that is, present or absent simultaneously in a large number of individuals.

Regarding graphic representations, the χ^2 distance means that when the individuals and the modalities are plotted on a single graph, an individual is close to modalities that it possesses and a modality is close to the individuals that possess it.

Treatment of Multiple Tables and Multiple Factorial Analysis

CONJOINT ANALYSIS OF SEVERAL TABLES

Faced with data of various kinds, we may wish to combine the information contributed by the different tables in such a way as to clarify the finer structures in the population.

An organization of individuals that appears during the analysis of a certain table and that is clearly due to variations of certain important characters of this table may sometimes manifest itself only during analysis of another type of data. One approach could be to make a conjoint analysis of a supertable juxtaposing the different data. All the possible oppositions can thus be made apparent; those that are common to all the types of data and those that, on the contrary, are specific to a given table can be revealed.

There are, however, two major obstacles to the immediate analysis of such a supertable:

(1) The type of variable—qualitative or quantitative—is not necessarily the same for all the tables. It is advisable to analyse the qualitative tables beforehand by MCA and the quantitative tables by PCA.

(2) All the tables do not have necessarily the same inertia and do not clearly express the same organization of the diversity. The most highly structured tables may thus influence the results excessively, and the input of the more loosely structured tables may go unnoticed.

A useful result allows us to overcome the first obstacle: it can be demonstrated that one can obtain the same factors as those produced by MCA by making a PCA, weighted appropriately, of the table formed by the indicators of the qualitative variables, that is, of the table coded in a binary form. In the case of molecular data, this means that each marker must be represented by two columns: one indicating its presence and the other its absence.

The second obstacle is overcome by adopting a weighting of tables that reduces the influence of tables that are highly structured. For that purpose, we must quantify beforehand the part of the total inertia of the table that corresponds to its interpretable structure. Such a quantification is relatively arbitrary and this is one of the points that distinguishes the different methods of analysis of multiple tables. The Statis method (Lavit, 1988) uses as a measure the sum of squares of eigenvalues of the separate analysis of the table. The MFA (Escofier and Pages, 1993) prefers the largest of these eigenvalues. Without going into detail, we can say that the first method can reduce the significance of a table when, by chance, the last eigenvalues are high. The second may on the other hand give more weight than should be given to a table in which the first eigenvalues are not very different from each other.

MULTIPLE FACTORIAL ANALYSIS

Multiple factorial analysis is thus a particular PCA, in which all the variables in a table are weighted by the inverse of the variance of the principal axis of inertia of the separate analysis of the table.

As does PCA, MFA can be used to locate individuals that resemble each other for the set of variables and thus to make a typology of individuals. It also enables us to attribute the passage from one class of this typology to another to the conjoint variation of some variables. It is these common directions of variation that MFA produces as axes of inertia, under the name of axes of global analysis.

However, the organization of data in the form of juxtaposition of tables enriches the interpretation of MFA. On the one hand, MFA provides indexes for each axis of overall analysis that allow us to determine whether it is due only to variables of a single table or whether it is common to several types of data. On the other hand, the MFA calculates the correlations between the axes of global analysis and the axes that will produce separate analysis of each table. When the preceding criterion indicates that a certain axis of the global analysis is common to several tables, this allows us to determine at what character precisely within each type of measure the common axis is linked.

Finally, from these correlations, we can find out, for a certain axis of a separate analysis whose meaning is known, the number of the axis of the global analysis to which it essentially contributes. If this global axis is common to several tables, that signifies that the partial axis manifests itself in a consensual manner. On the other hand, if the global axis is specific, that signifies that the partial axis corresponds to a character that is proper to the type of observations of separate tables.

MFA also has properties useful in terms of representation of individuals. In particular, it is shown to offer an interesting compromise between the quality of representation of clouds corresponding to each separate table and the proximity of these partial representations for a single individual.

TREE REPRESENTATION OF DISSIMILARITIES

The principle of any tree representation is to approach as closely as possible the dissimilarity δ, chosen for its relevance in describing the relationships between individuals, by a distance d that can be represented as a tree, that is, an ultrametric or an additive tree distance (Barthelemy and Guenoche, 1988). To find the exact solution, we must enumerate all the possible tree configurations possible for n individuals. For each possible tree structure, the edge length of the tree is estimated in terms of least squares. Finally, we retain the tree that allows us to minimize the sum of squares of deviations between initial dissimilarity and distance reconstituted in the tree. It is shown, by recurrence, that the number of different binary trees constructed over n individuals is $\Pi_{i=3,n}(2i-5)$. For $n = 10$, there are more than 2×10^6 different trees and for $n = 20$ there are more than 2×10^{20}. It is thus impossible to enumerate all the trees when n goes beyond some tens, even with the most powerful computers.

In such situations, the only possibility is to construct, from reasonable heuristics, solutions that will be the best possible but that can never be guaranteed to be optimal. Various heuristics have been proposed, permitting the construction of algorithms of more or less great complexity, the complexity of an algorithm measuring the increase in calculation time when n increases. In the scope of this work, we have discussed only the methods of grouping that are of sufficiently low complexity to treat some hundreds of individuals.

The individuals studied are often described by several types of variables that cannot be combined in the calculation of a single dissimilarity. Each set of variables is thus treated separately and the presence of common structures in the different trees is examined. Two methods of constructing synthetic trees, consensus trees and the common sub-trees, will be presented.

Methods of Grouping

Methods of grouping are iterative methods that proceed by successive ascending agglomerations, constructing the tree step by step. Initially, the matrix treated has as many elements as individuals in the population studied and the tree has a star structure. At each iteration, two elements, individuals or groups already formed and defined as neighbours, are joined to form one group. They are chosen so that the tree that traces their grouping optimizes a fixed criterion. This group becomes a new fictive element that replaces the two combined elements; the matrix is updated and thus reduced by one unit. The process is reiterated until all the individuals are united in a single group.

The various methods of this type are characterized by different choices at three key points of each iteration: the selection of elements to be joined, which depends on the definition of 'neighbourhood' used, the updating of the dissimilarity matrix by calculation of a dissimilarity between the group formed and the other elements, and the construction of lengths of the two edges derived from the two combined elements.

DEFINITION OF NEIGHBOURHOOD

The most natural definition of neighbours is the two individuals or groups that have the least dissimilarity. The elements i and j are defined as neighbours if $\delta(i,j)$ is the smallest dissimilarity.

This criterion allows us to find the tree solution in a theoretical case in which the initial distance is already an ultrametric. But it does not necessarily allow us to find the right structure if the initial dissimilarity is an additive tree distance. In the example in Fig. 2, on just four points, we can imagine that the distance $d(i,k) = a_i + a_c + a_k$ is the smallest of distances even though i and k are in two pairs opposed by the central edge. This criterion necessarily groups i and k and does not allow us to find the true tree. For that, other criteria must be used that take all of the distances into account to judge a neighbourhood.

In the context of genetic diversity, Saitou and Nei (1987) proposed a criterion of neighbourhood based on the principle of parsimony, which is at the basis of the phylogenetic approach. The objective is to create a tree that will be of minimal total length. It can be considered that an edge represents a number of mutational events and, by virtue of the basic principle that evolution proceeds always by the simplest genetic modification, the number of events, and thus the total length, must be minimized. These considerations lead to a definition of relative neighbourhood, defined by minimizing criterion $Q(i,j)$, function of $\delta(i,j)$ and of the average of dissimilarities of i and j at the $n - 2$ other elements k:

$$Q(i,j) = \delta(i,j) - (\Sigma_k [\delta(i,k) + \delta(j,k)]) / (n - 2)$$

It can be demonstrated that this criterion has properties of optimality in terms of least squares and that its domain of application goes beyond the scope of the phylogenetic reconstruction. It can be interpreted, very generally, as a weighting of the dissimilarity between two individuals by their dissimilarities to other individuals. Two related individuals that differ considerably from other individuals resemble each other more than two related individuals that are equally related to other individuals. This attitude is justified in many cases and explains the success of the method.

Sattath and Tversky (1977) adopt a very different approach. They start from the characterization of an additive tree distance by the four points condition. For any four individuals i, j, k and l, if $d(i,j) + d(k,l)$ is the smallest of three sums of distances two by two, then the two largest are equal: $d(i,k) + d(j,l) = d(i,l) + d(j,k)$. The initial dissimilarity δ is not an additive tree distance but it is always possible to form the sums of dissimilarities two by two. Among these three sums, one is the smallest, $\delta(i,j) + \delta(k,l)$, for example; the pairs of points (i,j) and (k,l) are thus considered good candidates to be neighbours. They are assigned a score of 1, while other pairs of points, (i,k), (j,l), (i,l) and (j,k), are attributed a null score. All the quadruplets of points that can be formed on the n points are scanned and the individuals of the pair that has the largest total score are considered neighbours. This definition of topological neighbourhood is ordinal in nature, since only the order of the three sums is important, and not directly the dissimilarity values that constitute them. It seems particularly appropriate when the dissimilarity values are somewhat marred by errors.

UPDATING DISSIMILARITIES

When two groups, each possibly composed of a single individual, are combined to form a new group, it is necessary to define a dissimilarity between the new fictive element created and the other elements present at this iteration. Given i and j the groups combined to form a new element s, c_i and c_j the numbers of these groups, and k another element, the most natural definition of $\delta(s,k)$ is the arithmetic mean of dissimilarities between k and the individuals constituting i and j: $\delta(s,k) = [c_i\delta(i,k) + c_j\delta(j,k)]/(c_i + c_j)$. It corresponds to an unweighted criterion since all the individuals play the same role.

Often, an arithmetic mean is used that is said to be weighted, in the sense that a different weight must be attributed to individuals of the groups in order that the mean does not depend on the numbers of the groups and is written: $\delta(s,k) = [\delta(i,k) + \delta(j,k)]/2$. The choice between these two criteria depends on the nature of the population studied. If the whole arises from a real process of sampling on a given structure of diversity, then the number of individuals in each element has a significance and must be taken into account in the calculation of dissimilarity. If the whole is only a circumstantial group of units that is not representative, the number of each group is only the result of chance and is not to be considered in the calculation of diversity.

In the case of adjustment to an additive distance, the formula $\delta(s,k) = [\delta(i,k) + \delta(j,k) - \delta(i,j)]/2$ is used, or its weighted equivalent. In the reconstruction of edges, this formula can be used to find the correct edge lengths when the initial tree is already an additive distance. This modification has no influence on the choice of subsequent steps.

It has sometimes been proposed, for adjustment to an ultrametric, that criteria called simple link and complete link be used that correspond respectively to $\delta(s,k) = \min[\delta(i,k), \delta(j,k)]$ and $\delta(s,k) = \max[\delta(i,k), \delta(j,k)]$. Except in very particular cases, it is difficult to justify these criteria in problems of diversity. If the simple link has been studied a great deal, it is essentially because it induces particular mathematical properties in the ultrametric produced.

RECONSTRUCTION OF EDGES

The elements i and j are combined to form the node s of the tree, and edge lengths $l(i,s)$ and $l(j,s)$ remain to be fixed. The dissimilarity $\delta(i,j)$ can simply be divided equally between the two edges: $l(i,s) = l(j,s) = \delta(i,j)/2$. This mode of calculation corresponds to an adjustment of the initial dissimilarity by an ultrametric.

If one does not impose an ultrametric condition, one accepts that the two edges can have different lengths for, on average, the dissimilarities of $n - 2$ elements k to i and to j to be represented as well as possible. The difference of edge lengths, designated e, is the sum on all the elements k of $[\delta(i,k) - \delta(j,k)]/$ $(n - 2)$. From this we get the formulae to calculate the edge lengths $l(i,s) = [\delta(i,j) + e]/2$ and $l(j,s) = [\delta(i,j) - e]/2$. This mode of calculation corresponds to the adjustment of the initial dissimilarity by an additive tree distance.

These formulae of edge reconstruction do not guarantee that the estimations will be the best. Fortunately, the edge lengths are not involved in the subsequent steps. It is thus preferable to reestimate them afterwards, when the tree topology has been established. We know how to estimate the length of each edge of a given tree topology overall, in the sense of least squares. For this, we express the distance in the tree between two individuals i and j as the sum of edge lengths belonging on the way from i to j and we expect that this distance will be the closest, in the sense of least squares, to the initial dissimilarity. We thus resolve a system of $n(n - 1)/2$ equations with as many unknowns as edges. This reestimation will cause loss of the ultrametric property.

SOME KNOWN ALGORITHMS

The most widely known algorithms correspond to certain compatible combinations among the different possible combinations of modes of definition of neighbourhood, various formulae for updating dissimilarities, and modes of estimating the edge lengths.

The set of methods often described as hierarchical clustering correspond to the definition of neighbourhood according to the minimal dissimilarity, with an adjustment to an ultrametric, and various formulae are proposed for updating. Among these, formulae of type average or weighted average are the most commonly used and the corresponding methods are frequently referred to as UPGMA and WPGMA, for unweighted or weighted pair group method using average.

The NJtree method (for neighbour-joining) proposed by Saitou and Nei (1987) is often used in genetic diversity. It uses the criterion of relative neighbourhood, the criterion of updating the dissimilarities is that of the unweighted average, and the dissimilarities are adjusted to an additive tree distance. The calculation complexity is of the order of n^3 and allows us to treat matrices of some hundreds of individuals. Several studies by simulation have shown the effectiveness of this method in finding the true tree. Gascuel (1997) proposes a version, UNJ (unweighted neighbour-joining), which uses a criterion of weighted average. The choice between these two methods depends, as has already been mentioned, on the sampling process.

Sattath and Tversky (1977) propose the Addtree algorithm (or method of scores), which uses the average as a criterion of aggregation and makes an adjustment to an additive tree distance. The notion of neighbourhood is that of topological neighbourhood. The complexity of n^4 is higher than that of NJtree. This method, which also has a proven effectiveness, can be preferred for its topological point of view.

Consensus Trees and Common Sub-trees

The same set of individuals is often described by several types of markers of diversity. To each type of marker corresponds a matrix of dissimilarity and an adjustment to a tree. Each type of marker contributes different information; the trees produced are thus different. However, one can hope to build from them a strong structure, a consensus tree, which will be common to different trees.

The problem of comparison of trees is also found in the resampling approaches (bootstrap or jackknife). The object of these approaches is to test the stability of tree representations obtained. The principle is to compare the trees established from several random samples in the set of markers observed (Felsenstein, 1985). If the markers are numerous and of equivalent informative content, molecular markers, for example, each of these samples is recorded to give a single image of diversity. If all the trees obtained are closely related, we conclude the stability of the structure revealed. On the other hand, if the different random samples of markers give very different trees, then the structure obtained must be regarded with the greatest caution. A measure of dispersion around a synthetic consensus tree quantifies the deviations between these trees.

Considering that the principal information on diversity involves much more of the tree structure than the edge length, the methods presented here construct synthetic trees in terms of structure, without the edge lengths being taken into account.

CONSENSUS TREES

Various methods of consensus have been proposed on rooted trees. A direct generalization to non-rooted trees does not pose a major problem. The strict consensus only retains the edges that are present in all the trees compared. Since this method may be considered too severe, the majority rule has been proposed, which retains the edges present in more than 50% of the trees. If one compares two trees only, the two types of consensus are equivalent.

A measure of the difference between two trees can be defined as the sum of differences of each to their consensus tree. The consensus tree is always simpler in organization than the initial trees. This organization is quantified by an index of complexity v; the dissimilarity between two trees H and H', of consensus H_c, is thus:

$$d(H,H') = v(H) + v(H') - 2v(H_c).$$

Various definitions of complexity can be considered. The most common way is to measure the complexity of a tree is by its edge number. For the binary trees that have only nodes of the third degree, the complexity of H and H' will be $2n - 3$; the complexity of H_c is the sum of the number of its external edges, n, and of the number of its internal edges, I_c, from which $d(H,H') = 2n - 6 - 2I_c$.

The distance, called the edge distance, thus varies according to the number of internal edges conserved in the consensus tree. If I_c is 0, the consensus is a star, and the trees compared have no common structure. It is the opposite if I_c is $n - 3$; the initial trees are identical. This distance is often standardized by $2n - 6$ to keep the variation between 0 and 1.

The value of a distance can only be interpreted by reference to the value that will be obtained on the trees drawn randomly, thus without any common structure other than a random one. The distribution of this index, for random trees, is known only asymptotically, that is, when the number of individuals tends towards infinity. We can, however, estimate these distributions for finite numbers by simulation and thus construct an approximate statistical test. Table 5 gives the threshold values for different values of n and different thresholds of probability. For example, for a tree of 40 individuals, there is only a 5% chance that trees of distance less than 0.973 will be constructed. This distance seems hardly discriminating and the random hypothesis is often refused even if the trees have only a very small number of internal edges in common.

The edge distance carries the same judgement on all the edges no matter what their position on the tree. The complexity of a tree depends only on its edge number independent of its internal organization. An intuitive idea will

Table 5. Edge distance between trees. Threshold values for different levels of probability. Empirical test established from simulation of 20,000 pairs of random trees with n leaves

n	Probability (%)			
	1	5	10	20
20	0.882	0.941	0.941	1
40	0.973	0.973	0.973	1
60	0.982	0.982	0.982	1
80	0.987	0.987	0.987	1
100	0.990	0.990	0.990	1

be to give a different weight to edges according to their aptitude to 'structure' the tree. For example, an external edge that involves a single element is less structuring, and an edge that separates the group into two sub-groups of number $n/2$ is more structuring. One associates at an edge a weight that is the product of numbers of each of the sub-groups defined by this edge. The complexity is the sum of these weights on all the edges. A distance known as bipartition distance is thus expressed, as previously, as a function of the complexity of trees and of their strict consensus. The maximum reached is not simply expressed but can be calculated from n; the distance is standardized by this maximum. As for the edge distance, the distribution under a random model is not known but can be approached by simulation. Table 6 gives the

Table 6. Bipartition distance between trees. Threshold values for different levels of probability. Empirical test established from simulation of 20,000 pairs of random trees with n leaves

	Probability (%)			
n	1	5	10	20
20	0.727	0.753	0.766	0.782
40	0.580	0.600	0.611	0.626
60	0.492	0.511	0.522	0.537
80	0.435	0.453	0.464	0.477
100	0.394	0.412	0.421	0.434

threshold values for different probabilities of an approximate statistical test. It is clearly more discriminating.

COMMON SUB-TREES

It is not unusual, in certain applications, to obtain consensus trees that look like a star, thus without marked structure, while direct examination of the initial trees reveals evident structural analogies.

The consensus tree makes the hypothesis that all the individuals are correctly represented; the exchange of an individual from one group to another is a strong indication that there is no real separation of the groups. For several applications this hypothesis is not entirely acceptable. For example, it has been demonstrated that the dissimilarities estimated from molecular markers can be compromised by a certain error. If the population studied combines units that are genetically closely related, on the infraspecific scale, for example, the imprecision of the measure of dissimilarities may suffice to explain the erratic character of certain units.

Considering that just a few individuals may mask a common structure, it is useful to try to identify these 'fluctuating' individuals rather than allow them to have the same weight as others in the determination of the common structure. The research of structures common to several trees thus needs to be reformulated. The problem becomes that of determining the smallest subgroup that needs to be pruned in each of these trees to obtain identical trees or, inversely, the largest subgroup of individuals having the same structure in all the trees. These individuals form the common sub-tree. In Fig. 7, this common sub-tree and the consensus tree are compared on a simple example.

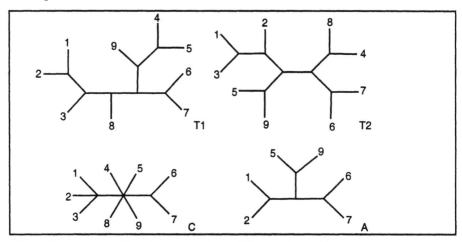

Fig. 7. Consensus tree (C) and common sub-tree (A) of trees T1 and T2.

This approach is much less widespread than the consensus approach and has been the subject of very few studies. The simple statement of the problem masks a relatively complex algorithmic problem. Kubicka et al. (1995) published the basis of an algorithm that gives an exact solution while retaining a sufficiently low complexity to be used in practice. The algorithm relies on the enumeration of all the solutions possible while reducing complexity by limiting the depth of exploration of the branches on a stop criterion.

The order o of common sub-tree, that is, the number of individuals conserved, can be considered as a measure of the resemblance between trees. The maximum order is n, and it is obtained for two identical trees. On the other hand, the minimal value of the order is 3, since it has only a single possible typology of three distinct points, and it is thus common to two trees. From this arises a definition of dissimilarity between trees $D = n - o$, which one can standardize as $n - 3$ to keep the variations between 0 and 1. The practical use of this criterion requires knowledge of the distribution of o under the hypothesis of independence of trees. This distribution can be approached by simulation and has been calculated for binary trees of 20 to 100 leaves (Table 7). It can be used to fix a rule of interpretation. By fixing a threshold of 5%, for example, it can be considered that chance gives less than 5 out of 100 trees of order superior or equal to 10, 13, 16, 18 and 19 when n varies from 20 to 100 by steps of 20.

Table 7. Distribution of number of leaves of common sub-trees of 1000 pairs of random
binary trees with 20, 40, 60, 80 and 100 leaves

	6	7	8	9	10	11	12	13	14	15	16	17	18	19	Mean
20	1	40	50	9											7.67
40			5		34	43	17	0	1						10.76
60				2		30	48	17	3						12.89
80						3	25	39	24	7	2				15.13
100							2	19	36	28	11	4			16.39

It is necessary to emphasize that the consensus tree and common sub-tree do not have the same point of view on data. Each is justified according to the interpretation that one can make on the shift of individuals in the initial trees. If it is a matter of accidental exchange of some individuals within a strong common structure, then the common sub-tree appears more adapted. If the exchange really expresses fundamental differences of structure, then the consensus tree is logically more adapted. In practice it is rarely possible to opt boldly for one of these two hypotheses and the two approaches can be implemented in a complementary manner.

CONCLUSION

The dynamism of the field of numerical taxonomy must be emphasized. It was considered, with the work of Sneath and Sokal (1973), that everything, or almost everything, had been said on the subject. Recent theoretical developments, the considerable increase in information technology, and the use of new types of markers have again thrown open this discipline, which is presently highly active at the interface between mathematics and biology.

On reading this chapter, the reader will understand the necessarily joint intervention of these two fields. The question that naturally arises from such a presentation—are some methods better than others?—does not have a real meaning in this field. There is no universal mathematical solution that is valid in all circumstances. Among the various approaches available we can choose the most relevant only on the basis of our knowledge of the plant species studied, its level of diversity, mode of reproduction, heterozygosity, and so on, the nature of the sample analysed, and the characteristics of the markers used.

REFERENCES

Barthelemy, J.P. and Guenoche, A. 1988. *Les arbres et les représentations des proximités*. Paris, Masson, 239 p.

Beaulieu, F.B. 1989. A classification of presence/absence based dissimilarity coefficients. *Journal of Classification*, 6: 233-246.

Darlu, P. and Tassy P. 1993. *Reconstruction phylogénétique*. Paris, Masson, 245 p.

Critchley, F. and Fichet, B., 1994. The partial order by inclusion of the principal classes of dissimilarity on a finite set and some of their basic properties. In: *Classification and Dissimilarity Analysis*. B. van Custem ed, New York, Springer, Lecture Notes in Statistics, pp. 5-65.

Escofier, B. and Pages, J. 1993. *Analyses Factorielles Simples et Multiples: Objectifs, Méthodes et Interprétation*. Paris, Dunod.

Escoufier, Y. 1975. Le positionnement multidimensionnel. *Revue de statistique appliquée*, 23(4): 5-14.

Felsenstein, J. 1985. Confidence limits on phylogenies: an approach using the bootstrap. *Evolution*, 39(4): 783-791.

Gascuel, O. 1997. Concerning the NJ algorithm and its unweighted version, UNJ. In: *Mathematical Hierarchies and Biology*, DIMACS workshop. American Mathematical Society, Series in Discrete Mathematics and Theoretical Computer Science 37, pp. 149-170.

Jukes, T.H., and Cantor, C.R. 1969. Evolution of protein molecules. In: *Mammalian Protein Metabolism*, H.N. Munro, ed., New York, Academic Press, pp. 21-132.

Kubicka, E., Kubicki, G., and McMorris, F.R. 1995. An algorithm to find agreement subtrees. *Journal of Classification*, 12: 91-99.

Lavit, C., 1988. *Analyse Conjointe de Tableaux Quantitatifs*. Paris, Masson.

Lefort-Busson, M. and de Vienne, D. 1985. *Les distances génétiques, estimations et applications*. Paris, Inra, 181 p.

Perrier, X. 1998. Analyse de la diversité génétique: mesures de dissimilarité et représentation arborée. Doct. thesis, Université Montpellier II, Montpellier, France, 192 p.

Rohlf, F.J. 1987. *Ntsys-pc Numerical Taxonomy and Multivariate Analysis System*. New York, Applied Biostatistics Inc., Setauket.

Saitou, N. and Nei, M. 1987. The neighbor-joining method: a new method for reconstructing phylogenetic trees. *Molecular Biology and Evolution*, 4(4): 406-425.

Saporta, G. 1990. *Probabilités, Analyse de Données et Statistiques*. Paris, Technip, 493 p.

Sattath, S. and Tversky, A. 1977. Additive similarity trees. *Psychometrika*, 42(3): 319-345.

Sneath, P.H.A. and Sokal, R.R. 1973. *Numerical Taxonomy*. San Francisco, Freeman, 573 p.

A Method for Building Core Collections

Michel Noirot, Francois Anthony, Stephane Dussert and Serge Hamon

About 10,000 years ago, our ancestors discovered agriculture and thereby exerted new selection pressures on wild plants: the process of domestication began. Over the course of millennia, this process led to what are generally called 'primitive varieties'. From the 16th century onwards, with the development of intercontinental migrations, the cultivation of plants went far beyond their zone of origin and diversification. Finally, the 20th century has been marked by the rapid development of selection techniques and modes of cultivation and by the production of new idiotypes. Thus, a considerable diversity of species and forms has been created.

Vavilov (1935) was among the first to demonstrate the importance of gene banks in plant breeding. However, some decades passed before Harlan (1970), Frankel and Bennett (1970) and subsequently Pernes (1984) encouraged the sampling and evaluation of the diversity of natural populations.

The major cultivated plants, their related wild species, and the minor cultivated plants have been sampled on the initiative of the International Board for Plant Genetic Resources (IBPGR). Routine prospections rapidly led to difficulties in managing and conserving the collections thus assembled.

By the late 1980s, collections became enormous and difficult to regenerate and maintain. This considerable growth in collection size, as well as the inadequate documentation available for the samples, have often been cited as limitations on the effective use of genetic resources (Holden, 1984). Potential users require either populations representative of the diversity or accessions that present particular agronomic characters (e.g., disease resistance, drought resistance). In either case the managers of collections find it difficult to meet such needs.

Frankel and Brown (1984) were the first to emphasize the need to constitute small collections with maximal diversity: the core collections. For most users, a core collection helps them avoid the redundancy of genotypes (doubles), often linked to the mode of reproduction or overrepresentation of cultivated varieties. This redundancy is rare in allogamous species, very frequent in autogamous species, and common in apomictic species or in plants with vegetative propagation.

In terms of practical use, the three major objectives of the core collection are to set up as wide a representation as possible of the genetic diversity, to be able to conduct intensive studies on a reduced set of genotypes, and to attempt to extrapolate the results thus obtained to facilitate research on appropriate genotypes in the base collections.

Presently, in the constitution of a core collection, most researchers agree on the need for a stratification prior to the sampling. In other words, the organization of the variability in groups and subgroups must be taken into account (Frankel and Brown, 1984; van Hintum, 1995; Yonezawa et al., 1995).

On the other hand, several processes are proposed for sampling within groups and subgroups. One such process is to take a random sampling in each previously defined group (or from the whole base collection if the organization into groups is unknown or does not exist). This type of sampling has the advantage of creating a core collection that is statistically representative of the base collection.

In this chapter, we propose a new method of sampling, principal component scoring (PCS), the aim of which is to maximize the diversity sampled. This diversity is measured by using quantitative or qualitative variables, the choice of which is discussed. We also examine the effects of PCS on the stratification of the sampling and on the size of the sample. Finally, we present some examples of presently developed core collections.

PRINCIPLES AND METHODS

The within-population diversity is determined by differences between individuals for one or several characters. These differences can be estimated by a distance that, for our purposes, must be a metric distance. The choice of this distance depends on the characters observed, quantitative or qualitative.

Quantitative Variables

THE CHOICE OF DISTANCE, COLINEARITY AND WEIGHTING

Quantitative characteristics are generally heterogeneous. They correspond to lengths (plant height, stem diameter), areas (leaf area, area of stigmata), weights (aerial biomass, reproductive biomass), or time (date of flowering, duration of fructification). They have, moreover, different forms of variability. In order to give the same weight to each character j, the Euclidean distance is weighted by the inverse of the standard deviation σ_j. The distance d_{ik} between two individuals i and k for J quantitative characters is defined by the following formula:

$$d_{ik} = \sqrt{\sum_{j=1}^{j} \left[\left(x_{ij} - x_{kj} \right) \sigma_j^{-1} \right]^2}$$

where x_{ij} is the value of the character j observed on the individual i and x_{kj} is the value of the character j for the individual k.

The distance between individuals is directly linked to the differences. If the differences come from characters that are strongly correlated, positively or negatively, the distance between certain individuals is greatly overestimated. Thus, if we measure the diameter of a tree trunk at different heights from the ground (1 m, 1.10 m, 1.20 m, etc.), the Euclidean distance between two trees will be greatly influenced by differences in diameter. This example is obvious, but such an effect, called the colinearity effect, is found for all the correlated characters.

To eliminate the colinearity effects, the principal components analysis has been applied to standardized variables to give J new centred variables, which are statistically independent: the factors. The distance between two individuals i and k for J factors is calculated by using a similar formula:

$$d_{ik} = \sqrt{\sum_{j=1}^{j} \left[\left(z_{ij} - z_{kj} \right) \sqrt{\lambda_j^{-1}} \right]^2}$$

where the square root of the eigenvalue λ_j allows the weighting, and where z_{ij} and z_{kj} are respectively the coordinates of the individuals i and k on the factor j.

Such a procedure gives the same weight to all the factors in the estimation of distance, including the residual components resulting from noise or observation errors. Factors for which the eigenvalue is less than 1 (Kaiser criterion) are eliminated in order to prevent their intervention in the calculation of distance.

THE CHOICE OF INDIVIDUALS MAXIMIZING DIVERSITY

The sum of generalized squares (SGS) of a lot of N individuals in the factorial space of K standardized variables (mean = 0, variance = 1) and independent variables (correlation coefficient = 0) is equal to the product NK (Lebart et al., 1977). The contribution P_i of the individual i to the SGS is equal to the sum of squares of these K new coordinates:

$$P_i = \sum_{j=1}^{k} x_{ij}^2$$

The relative contribution CR_i of the individual i to the SGS of the whole is given by:

$$CR_i = P_i(NK)$$

Conserving the greatest variability is equivalent to maximizing the score of the subgroup of individuals sampled by using an estimator SGS. The first step is to sample the individual furthest from the barycentre of the group, that is, the individual that makes the greatest relative contribution. The iterative selection of individuals maximizing the diversity of the core collection increases the core collection size. At each iteration, the cumulative SGS of the core collection, expressed as a percentage of the total SGS, is known. The procedure may be concluded either according to the size of the core collection or according to the percentage of diversity retained. The two criteria can be taken into account simultaneously. In that case, when the first criterion is met, the sampling comes to an end.

Qualitative Variables

The method described above is designed for quantitative data. The modifications required for qualitative data concern the first steps of the PCS. As with the quantitative data, there are relationships between the variables. For example, two molecular markers may be linked genetically. In order to eliminate the effects of this type of relationship on the distance and in order to give the same weight to independent variables, a method of multivariate analysis has been used to transform the initial data into factorial coordinates: this involves correspondence analysis (Benzecri, 1972).

The χ^2 distance is retained in place of the Euclidean distance and the analysis uses a binary table. In this table, the presence and absence of an allele are considered two different variables taking the values 1 and 0. With p molecular markers observed on N individuals, we obtain a table $2pN$. In consequence, all the individuals have the same marginal frequency equal to p. Moreover, the term $p\lambda_i$ ($p\lambda_i$ is the eigenvalue of the factor i) is equal to the sum of r^2 of this factor with p variables. This term is equivalent to the eigenvalue observed in the principal components analysis. The sum of $p\lambda_i$ is equal to the number of markers (for analysis of the principal components on quantitative data, the sum of eigenvalues is equal to the number of variables). As for quantitative data, the factorial coordinates are weighted. In our case, the weights are square roots of the corresponding $p\lambda_i$ values. The Kaiser criterion for the choice of number of factors is applied to the term $p\lambda_i$. The subsequent steps of the PCS are the same as for the quantitative characters.

DISCUSSION

Stratification: Conditions and Consequences

The simplest method for creating a core collection is random sampling over the entire base collection. When the genetic structure of the base collection is not known, such sampling is the best solution (Brown, 1989a). Nevertheless, it is less effective for alleles that are common locally but rare over the entire collection. This is why Brown (1989b) has suggested a stratification of the sampling.

Peeters and Martinelli (1989), Holbrook and Anderson (1995), and van Hintum (1995) are of a divided opinion and suggest that, as a basis of stratification, the country of origin of the plant material should be taken into account. Peeters et al. (1993) advise the use of precise ecogeographical data. These have been taken into consideration in establishing the soybean core collection (Perry et al., 1991). The major agronomic or biological characters—mode of reproduction, duration of cycle—are also used for the stratification (Spagnoletti-Zeulli and Qualset, 1987; Hamon and van Sloten, 1989; Diwan et al., 1994; Hamon et al., 1995). These characters are quantitative (height, diameter) as well as qualitative (colour, appearance), so they are difficult to treat simultaneously to obtain matrices of distances, except when the quantitative data are recorded so as to obtain classes of equal numbers or equal amplitudes. However, Cole-Rogers et al. (1997) propose an original method, the normed binary scale, which allows calculation of matrices of distances that integrate these two types of variables.

Molecular markers, used by Lux and Hammer (1994) and strongly recommended by Gepts (1995), have only begun to be taken into account in stratification. The percentage of accessions evaluated with the help of this type of marker is still low. Breeders are often not interested in it and the structuring of groups determined by the molecular markers does not always coincide with that which follows from morphoagronomic diversity.

However, random sampling, even within groups, does not enable us to reach the main aim of the core collection, which is to sample the maximum range of diversity. For example, the production of hydrocyanic acid in white clover (*Trifolium repens*) is controlled by two independent loci. This character confers a resistance to several species of insects and molluscs, and its expression is regulated by climatic parameters (temperature, day length, humidity). The base collection of the National Plant Germplasm System (NPGS) in the United States contains 602 accessions of white clover. A core collection of 91 accessions has moreover been established on the basis of a geographic stratification and according to a random selection within the groups. Pederson et al. (1996) determined the proportion of cyanogenic plants in the base collection and compared it to that of the core collection. No significant difference of frequency was found between the two collections, which proves that the core collection did not 'maximize' the variability: it is simply a reduced version of the base collection.

Conversely, PCS modifies the mode of sampling so that it is no longer random, maximizes the diversity, and in most cases prevents doubles. It thus meets the objectives defined for a core collection.

By preferentially sampling the most distant individuals, the PCS method requires three conditions to be functional and effective. The first, and most important, is that all the individuals of the core collection can cross with each other and give intermediate types. This implies a prior knowledge of the genetic structure of the species complex (Pernes, 1984) and a good estimation of the level of reproductive barriers between compartments. The second condition concerns the efficiency of the PCS, which supposes a generalized additivity and high heritability (in the wide sense) for the quantitative characters. In cases where this hypothesis is not valid (dominance, superdominance, high plasticity, etc.), selection on phenotype diversity will not necessarily lead to selection on genetic diversity. The sampling can thus be considered random in relation to the hidden variability. The third condition is the absence of a polymodal structure of the base collection. Indeed, the existence of several groups may lead to a sampling of individuals alternatively in the most distant groups. In this case, maximizing the diversity increases redundancy. The stratification of the sampling is here a determining preliminary step, as in the case of random sampling.

Thus, for PCS, the stratification of sampling must depend on the genetic structure of populations and the limits of recombination. When such data are lost or absent, taxonomy must be taken into account. The bioclimatic and biogeographic information must then modify the structure by establishing subgroups corresponding to the genetic differentiation, in subspecies and in ecotypes. The PCS is thus applied within each group and subgroup.

Size of the Core Collection and its Strata

Discussion on the size of the core collection is always topical. Brown (1989a), using the theory of neutral alleles (Kimura and Crow, 1964) and that of sampling, demonstrated that a sample of 10% of the base collection contains at least 80% of the alleles, with a statistical risk of error of 5%. According to this author, the results are reliable in relation to the type of frequency distribution of alleles at each locus. This value of 10% is not modified by our mode of sampling.

The size of each subgroup within the core collection was studied by Brown (1989b). Three methods were compared to determine this size: choosing the same number of accessions per group, defining a number of accessions proportionate to the size of the group, or opting for a number proportionate to the logarithm of the group size. The author demonstrated that the third solution is a good compromise. Nevertheless, the choice of the size of the subsample according to the size of the group supposes a relationship between the diversity of the group and its size, which is far from being always the case in the base collections. The ratio between the diversity and the size of the group depends, among other things, on the mode of reproduction and the economic importance of the plant (cultivated plant or related species). For example, in the base collections of the species complex of the genus *Coffea*, the cultivated and autogamous species *C. arabica* is overrepresented in relation to the wild and allogamous species *C. sessiliflora* from Africa and the East (Noirot et al., 1993).

The PCS method of sampling allows management of the diversity of the core collection. The sampling of individuals to increase the core collection can be halted according to the percentage of diversity already achieved. Taking into account the diversity already achieved is particularly useful in the case of species with high natural redundancy (agamic complexes, autogamous plants, plants that reproduce vegetatively).

The Primary Uses of Core Collections

Whatever the strategy used, the core collections are conceived to help managers to conserve and use genetic resources. The two examples recorded here show that users can find them helpful.

The US Department of Agriculture had in 1990 a base collection of perennial alfalfa of 2400 accessions. In order to extract a core collection of 200 accessions, Basigalup et al. (1995) decided on directed selection of genotypes after a geographic stratification, among the eight methods tested. Jung et al. (1997) subsequently used this core collection for research on proteic composition, biodegradation of leaves, digestibility, and lignin composition. Thus, this core collection was useful for characters other than those that determined its constitution.

Holbrook et al. (1993) established a core collection of 831 accessions of peanut from the base collection, which numbered 7432. This collection comprises 70% of samples evaluated on morphoagronomic characters and 30% of non-evaluated samples. For the evaluated accessions, multivariate analyses having revealed a structure in groups, 10% of accessions of each group are taken randomly. For the accessions that are not evaluated, 10% are taken randomly after stratification by country. Holbrook and Anderson (1995) tested the relevance of this core collection in relation to the base collection for resistance to cercosporiosis due to *Cercosporidium personatum*. This involved determining the number of resistant accessions that would allow us to detect the core collection in relation to the base collection. The process comprised two steps: In the first, the entire core collection was tested for this character, then the groups that seemed to indicate resistance in the core collection were examined in detail in the base collection. The rates of efficiency, in terms of proportion of resistant accessions identified, increased from 1/64 in the base collection to 1/8 in the core collection. This result demonstrated the utility of a core collection not only in developing the genetic material, but also in improving the efficiency of the search for particular characteristics.

The PCS method of constituting the core collections enables us to achieve a greater efficiency. Of course, as with peanut, the implementation of this process depends on the availability of data required for the selections. Hamon et al. (1998) demonstrated on four plants—rice, coffee, sorghum and hevea—that the variability of quantitative characters in the core collections is only slightly or is not modified when the selection is qualitative. On the other hand, the means and variances of morphoagronomic characters are greatly modified by a quantitative selection. Qualitative selection seems the most effective to conserve the rare alleles and increase the overall diversity with limited numbers at the quantitative level. Quantitative selection leads to the loss of 6% of rare alleles (from initial frequency in the base collection of less than 5%) in hevea, 11% in sorghum, 12% in rice, and 33% in coffee. When qualitative selection is used, the losses are reduced to 2% for rice, sorghum, and hevea and 6% for coffee. In other words, this approach demonstrates that it is possible to maximize the allele richness—also known as the neutral variability—in the core collection while preserving the representativity of the morphoagronomic variability.

Core collections have now been set up for many plants. A consensus seems to have been reached on their size (around 10% of the base collection) and on the need for stratification. Random sampling leads to conservation of the variability contained in the base collection while retaining its faults (overrepresentation, redundancy, sampling bias). The main advantage of PCS is that it allows an increase in the neutral allelic diversity of the core collection without modifying the agronomic representativity or changing the relative intensity of the sampling (10% of the base collection). The present state of progress of molecular biology will certainly facilitate the use of molecular markers in the estimation and structuring of genetic diversity. The constitution of core collections must thus take into account the relations between different levels of variability.

REFERENCES

Basigalup, D.H., Barnes, D.K., and Stucker, R.E. 1995. Development of a core collection for perennial *Medicago* plant introductions. *Crop Science*, 35: 1163-1168.

Benzecri, J.P. 1972. Pratique de l'analyse des données: analyse des correspondances. Paris, Dunod, 424 p.

Brown, A.D.H. 1989a. Size and structure of collection: the case for core collection. In: *The Use of Plant Genetic Resources*. T. Hodgkin et al., eds., Chichester, Wiley, pp. 136-156.

Brown, A.D.H. 1989b. Core collections: a practical approach to genetic resources management. *Genome*, 31: 818-824.

Cole-Rogers, P., Smith, D.W., and Bosland, P.W. 1997. A novel statistical approach to analyze genetic evaluations using *Capsicum* as an example. *Crop Science*, 37: 1000-1002.

Diwan, N., Bauchan, G.R., and McIntosh, M.S. 1994. A core collection for the United States annual *Medicago* germplasm collection. *Crop Science*, 34: 279-285.

Frankel, O.H. and Bennett, E. 1970. Genetic resources. In: *Genetic Resources in Plants, their Exploration and Conservation*. O.H. Frankel and E. Bennett, eds., Oxford, Blackwell, 547 p.

Frankel, O.H. and Brown, A.H.D. 1984. Current plant genetic resources: a critical appraisal. In: *Genetics, New Frontiers* (vol. IV). New Delhi, Oxford and IBH.

Gepts, P. 1995. Genetic markers and core collections. In: *Core Collections of Plant Genetic Resources*. T. Hodgkin et al., eds., Chichester, Wiley, pp. 127-146.

Hamon, S., Dussert, J., Deu, M., Hamon, P., Seguin, M., Glaszmann, J.C., Grivet, L., Chantereau, J., Chevallier, M.H., Flori, A., Lashermes, P., Legnate, H., and Noirot, M. 1998. Effects of quantitative and qualitative principal component score strategies on the structure of coffee, rice, rubber tree and sorghum core collections. *Genetics, Selection, Evolution*, 30 (suppl. 1): 237-258.

Hamon, S., Noirot, M., and Anthony, F. 1995. Developing a coffee core collection using the principal components score strategy with quantitative data. In: *Core Collections of Plant Genetic Resources*. T. Hodgkin et al. eds., Chichester, Wiley, pp. 117-126.

Hamon, S. and van Sloten, D.H. 1989. Characterization and evaluation of okra. In: *The Use of Plant Genetic Resources*. A.D.H. Brown and O. Frankel, eds., Cambridge, Cambridge University Press, pp. 173-196.

Harlan, J.R. 1970. The evolution of cultivated plants. In: *Genetic Resources in Plants*, Their Exploration and Conservation. O.H. Frankel and E. Bennett, eds., Oxford, Blackwell, 547 p.

Holbrook, C.C. and Anderson, W.F. 1995. Evaluation of a core collection to identify resistance to late leafspot peanut. *Crop Science*, 35: 1700-1702.

Holbrook, C.C., Anderson, W.F., and Pittman, R.N. 1993. Selection of a core collection from the US germplasm collection of peanut. *Crop Science*, 33: 859-861.

Holden, J.H.W. 1984. The second ten years. In: *Crop Genetic Resources: Conservation and Evaluation*. J.H.W. Holden and J.T. Williams, eds. London, George Allen and Unwin, 296 p.

Jung, H.G., Sheaffer, C.C., Barnes, D.K., and Halgerson, J.L. 1997. Forage quality variation in the US alfalfa core collection. *Crop Science*, 37: 1361-1366.

Kimura, M. and Crow, J.F. 1964. The number of alleles that can be maintained in a finite population. *Genetics*, 49: 725-738.

Lebart, L., Morineau, A., and Tabard, N. 1977. *Techniques de la Description Statistique: Méthodes et Logiciels pour l'Analyse des Grands Tableaux*. Paris, Dunod.

Lux, H. and Hammer, K. 1994. Molecular markers and genetic diversity: some experience from the genebank. In: EUCARPIA meeting on evaluation and exploitation of genetic resources pre-breeding. Clermont-Ferrand, France, EUCARPIA, pp. 49-53.

Noirot, M., Hamon, S., and Anthony, F. 1993. L'obtention d'une core collection de caféiers: définition des groupes d'échantillonnage et méthodologie. In: XVIe Colloque scientifique international sur le café. Paris, ASIC.

Pederson, G.A., Fairbrother, T.E., and Greene, S.L. 1996. Cyanogenesis and climatic relationships in the US white clover germplasm and core subset. *Crop Science*, 36: 427-433.

Peeters, J.P. and Martinelli, J.A. 1989. Hierarchical cluster analysis as a tool to manage variation in germplasm collections. *Theoretical and Applied Genetics*, 78: 42-48.

Peeters, J.P., Wilkes, H.G., and Galwey, N.W. 1993. The use of ecogeographical data in the exploitation of variation from gene bank. *Theoretical and Applied Genetics*, 80: 110-112.

Pernes, J. 1984. Gestion des Ressources Génétiques des Plantes. Paris, ACCT, 212 p.

Perry, M.C., McIntosh, M.S., and Stoner, A.K. 1991. Geographical patterns of variation in the USDA soybean germplasm collection. 2. Allozyme frequencies. *Crop Science*, 31: 1356-1360.

Spagnoletti-Zeuli, P.L. and Qualset, C.O. 1987. Geographical diversity for quantitative spike characters in a world collection of durum wheat. *Crop Science,* 27: 235-241.

van Hintum, T.J.L. 1995. Hierarchical approaches to the analysis of genetic diversity of crop plants. In: *Core Collections of Plant Genetic Resources.* T. Hodgkin et al. eds., Chichester, Wiley, pp. 23-34.

Vavilov, N.I. 1935. The origin, variation, immunity and breeding of cultivated plants. *Chronica Botanica,* 13 (6 volumes).

Yonezawa, K., Nomura, T., and Morishima, H.1995. Sampling strategies for use in stratified germplasm collections. In: *Core Collections of Plant Genetic Resources.* T. Hodgkin et al., eds., Chichester, Wiley, pp. 35-53.

Asian Rice

Jean–Christophe Glaszmann, Laurent Grivet,
Brigitte Courtois, Jean-Louis Noyer, Claude
Luce, Michel Jacquot, Laurence Albar, Alain
Ghesquière, and Gérard Second

Rice is part of the heritage of the most ancient civilizations—in China, close to the Yangtze river, traces of rice cultivation date to about seven millennia ago. Today, in the developing countries, rice represents nearly 50% of the cereal production, far more than wheat and maize. The world production of paddy, or dehulled rice, is close to 540 million t. It is cultivated on about 150 million ha, of which 90% is in Asia. In this region, the per capita annual consumption of milled rice is often higher than 100 kg, and can reach up to 200 kg in Myanmar or Laos. On other continents, the consumption is more variable but can reach 100 kg, particularly in Madagascar, Sierra Leone, or Surinam. In the European Union, the per capita consumption of rice is about 5 kg. Most rice is consumed locally, international commercial trading amounting to less than 5% of the production. However, with economic progress, the populations that are the highest consumers of rice are gradually diversifying their diet, while populations that consume small amounts are progressively integrating more rice into their diet. According to demographic forecasts, by the year 2025 the demand for rice is likely to increase by 70%.

Rice is a semi-aquatic plant: it tolerates aquatic cultivation conditions but does not absolutely depend on them. Since the beginning, development of rice cultivation has been dependent on a sufficient natural fertility of the soil and on a low weed competition. The most favourable situations are found in flooded lands and recently cleared forest lands.

Cultivation on flooded lands has developed most, particularly because of improvements in irrigation and drainage and the possibility of growing rice continuously. Rice is direct-seeded or transplanted from nurseries. The wet lowlands of valleys and wide plains are systematically exploited. Deep-water rice cultivation is practised in areas subject to floods of 4 to 5 m in Bangladesh and Mali, using varieties that can elongate their stems by several centimetres a day. Close to the coastal areas, soil desalination techniques using fresh water have been developed to grow one rice crop, for example in

Indonesia and Guinea. Inland, in hilly areas, terrace cultivation is practised, as in Nepal, the Philippines, and Bali. Aquatic rice cultivation is found thus from the equator to latitudes of 40°N, even 50°N in China, and in the tropical regions from sea level to over 2000 m altitude in Nepal and Madagascar notably.

In the humid tropical regions, apart from conditions of aquatic cultivation, upland rice cultivation depending exclusively on rainfall for its water requirement is also practised. It is first developed on cleared forest land. After slashing and burning of the forest, rice is broadcast or dibbled. The cultivation requires little input in the first two or three years, then the natural fertility of the soil decreases while weeds begin to invade the area. Then the cultivation area is shifted. This practice gradually disappears with increasing demographic pressure. Other cropping systems are found in Brazil, where upland rice is cultivated on millions of hectares before the installation of pastures, as well as in several countries where upland rice is intercropped with young plantations of rubber tree, coffee, or other perennial crops.

TAXONOMY AND GENETIC RESOURCES

Cultivated Rice Within the Genus *Oryza*

Cultivated rice belongs to the genus *Oryza*, which is highly diverse, and comprises several genomes recognized on a cytogenetic basis, as well as diploid and allotetraploid forms. More than 20 wild species are recognized, of which 9 are tetraploid. They are now distributed on all the continents but the origin of the genus is probably Asia (Second, 1985).

Two species are cultivated: one is of African origin and found almost exclusively in West Africa, and the other is of Asian origin. Both are diploid and autogamous. The African cultivated species, *O. glaberrima*, results from domestication of *O. breviligulata* in Africa. The Asian cultivated species, *O. sativa*, has an original genetic structure, which is the focus of the work described in this chapter. Its domestication, from the species *O. rufipogon*, is ancient: remains of grains associated with human activities almost 4500 years old have been found in Pakistan, some dating from over 8500 years ago in northern India, some dating from at least 6000 years ago in Thailand, and some dating from 7000 years ago in China. There have been constant migrations in a larger circle towards the west to Pakistan and Iran, towards the east—rice was introduced in Japan from the beginning of the first century—and towards the southern islands of Sri Lanka, Malaysia, the Philippines, Indonesia, and Taiwan. Malay navigators probably introduced rice from Indonesia to Madagascar in the 5th or 6th century. Much later, Europeans introduced Asian rice in tropical Africa and then in America.

Cultivated rices have a wide diversity of forms. Geneticists, notably in Japan and India, attempted very early to classify the varieties to summarize the global variability into a few subgroups. To do so, they used morphological and physiological characters and observed the crossing behaviour. Two subspecies were soon identified, *indica* and *japonica*, with 'temperate' and 'tropical' types for the latter (Oka, 1958). Some breeders distinguished three morphological types, Indica, Japonica, and Javanica (Chang, 1976), this last type being variously positioned.

Genetic Resources

The genetic resources of rice have received a great deal of attention, notably because of the existence of the International Rice Research Institute (IRRI), focussing strictly on rice (Chang et al., 1984). From the 1930s to the 1950s, the major rice-producing countries in Asia built up important national germplasm banks: the germplasm bank in India comprised over 20,000 accessions, and the one in China close to 40,000 accessions. The IRRI, established in 1961, rapidly assumed a leading role, serving as a centre for international seed exchange and for medium-term conservation thanks to high-quality facilities. The spread of semi-dwarf, high-yielding varieties subsequently stimulated international collaboration in the task of collection. Financial support from another international organization, the IBPGR (International Board for Plant Genetic Resources), facilitated this collaboration. Between 1971 and 1982, more than 25,000 samples were collected. French institutions participated by supporting collections in Africa. Since 1994, a programme of complementary collection in the less represented regions has been financed by the Agence Suisse pour le Développement et la Coopération. Today the international germplasm bank has more than 80,000 accessions, 95% of which belong to the *O. sativa* species (Jackson, 1997).

The international germplasm bank comprises all the samples conserved in some but not all the national germplasm banks. A duplicate of the samples is systematically deposited at the conservation centre at Fort Collins, in the United States, after rejuvenation of the accessions. The African accessions are duplicated at IITA (International Institute for Tropical Agriculture) in Nigeria and at ADRAO (Association pour le Développement de la Riziculture en Afrique de l'Ouest), in Côte d'Ivoire. Besides research on the physiology and conservation of seeds, many studies have been undertaken to characterize and evaluate the accessions for botanical descriptors and characters of agronomic interest (Jackson, 1997).

ORGANIZATION OF GENETIC DIVERSITY: THE CONTRIBUTION OF MOLECULAR MARKERS

Molecular markers contribute a great deal to our comprehension of the genetic diversity of rice. Isozymes, despite the limited number of loci and alleles available, provided the primary elements of quantification of intravarietal differentiation. In Asia, the cultivated species seems fundamentally bipolar, which is remarkably consistent with the work of Oka (1958), but with a continuum of intermediate forms (Second, 1982). Most of the traditional varieties can be divided into two major groups, corresponding to the *indica* and *japonica* subspecies. Only 20% of the varieties are not clearly classified into either group (Glaszmann, 1987, 1988). Isozymes have also shown a triangular equidistance between the African cultivated species, the typical *indica* forms, and the typical *japonica* forms (Second, 1982). The concept of a molecular clock was applied to the scenario of paleoenvironmental modifications (Second, 1985) and has enabled an overall interpretation of the genetic structure of the genus *Oryza*, with a scenario of triple domestication from wild species appearing around two million years ago, following the isolation of tropical Africa by the cooling of the climate and the emergence of the Himalayas as a geographic barrier in Asia. A particular form was also discovered in western India; it presents traces of introgression of *O. glaberrima* (Lolo and Second, 1988). The inventory of forms that are not classified into the *indica* and *japonica* subspecies cannot be considered complete.

The classification established on the basis of isozymes globally agrees with the local, empirical classifications as well as with synthetic classifications such as those of Oka (1958) and Cheng et al. (1984), established on multiple criteria associating morphological components such as grain width or hairiness with physiological components such as cold tolerance of seedlings (Zhou et al., 1988).

With the advent of new molecular techniques, studies based on different types of molecular markers have been undertaken, from which different infraspecific classifications can be constructed. Each newly obtained scheme has generally been comparable directly to one, sometimes two or three, of the existing schemes using the material common to the various studies, the more global comparisons being done step by step. Remarkably similar images, which indicate a very marked bipolar structure, have been obtained with the different tools used: isozymes (Second, 1982, 1985; Glaszmann et al., 1984; Glaszmann, 1985, 1987, 1988), RFLP (Wang and Tanksley, 1989; Zhang et al., 1992; Ishii et al., 1995; Second and Ghesquiere, 1995), RAPD (Virk et al., 1995; Mackill, 1995; Parsons et al., 1997), ISSR-PCR (Parsons et al., 1997; Blair et al., 1999), AFLP (Zhu et al., 1998), and microsatellites (Yang et al., 1994), the last giving a hazier image because of its greater polymorphism and numerous rare allelic forms.

Moreover, Virk et al. (1996) indicated the value of molecular markers in predicting characters of agronomic interest with quantitative expression when a sample of highly diverse varieties is considered. This observation carries in itself the question of the organization of the genetic diversity and its control: What structure is associated with what characters? With what mechanisms of maintenance or evolution? And with what uses?

We propose here to compare several classifications based on different types of characters, concentrating the analyses on the common samples (see Appendix). Then, we examine some specific situations that enable us to tackle some aspects of the dynamics that conditions the genetic diversity of cultivated rice.

Diversity Revealed by Different Types of Markers

ISOZYMES

The largest study was done on a sample of 1688 traditional Asian varieties characterized for 15 polymorphic loci (Glaszmann, 1987). Six groups were distinguished: two major groups (I and VI) with pan-Asian distribution, two minor groups (II and VI) found essentially on the Himalayan foothills of South Asia, and two particular groups (II and IV) restricted to deep-water zones in Bangladesh. Group I represents the typical *indica* forms and group VI the typical *japonica* forms. Group II contains notably the rices of the 'aus' ecotype of northern India and Bangladesh. Group V is represented by high-quality rices (aroma, elongation on cooking) of the 'basmati' group of India and Pakistan. Groups II and IV correspond to floating rices, among which some, such as the 'rayada' of group IV, are capable of stem elongation to more than 5 m.

A sample A of 270 varieties, constituted to represent the geographic, ecotypic, and enzymatic diversity of cultivated rice of Asian origin (see Appendix), was used for a similar analysis. From the 15 loci, 57 allelic forms could be revealed. These data were subjected to a correspondence analysis (CA). The percentage of variation explained by the CA axes decreased very rapidly: it was 18.7%, 9.2%, 7.9%, 6.3%, and 5.1% for axes 1 to 5 respectively. This rapid decrease illustrates a very marked structure. Figure 1 shows the distribution of accessions in the 1-2 plane of the CA. The six groups, I to VI, are identified in agreement with the results obtained on larger samples.

RFLP

A sample B of 147 varieties was analysed using 202 probes (see Appendix). Among these, 148, or 74%, revealed a polymorphism and produced a total of 482 polymorphic bands. These data were subjected to a CA. The percentage of variation explained by the CA axes decreased very rapidly: they were 16.5%, 6.2%, 4.4%, 3.8%, and 3.1% for axes 1 to 5, indicating, as for the isozymes, a marked general structure.

Figure 2a shows the distribution of the 147 varieties studied in the 1-2 plane. The position of the varieties on the primary axis is by far the one that characterizes them the best. It is particularly interesting to remark that this axis represents the variation between the reference *indica* varieties (isozymic group I) and the reference *japonica* varieties (isozymic group VI). Groups II,

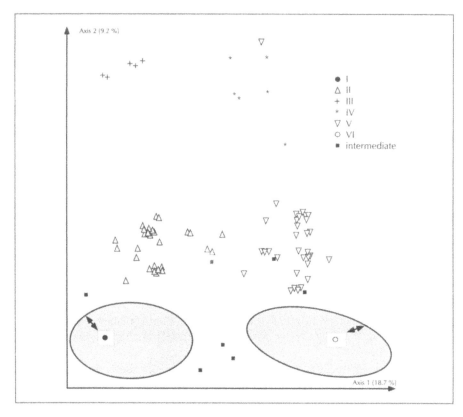

Fig. 1. Distribution on the 1-2 plane of the CA on isozyme data (15 loci and 57 alleles) characterizing sample A.

III, IV, V, and 'intermediates' appear overall in an intermediate position between the reference *indica* and *japonica* varieties, with, however, a low percentage of specific alleles. With this restriction, *O. sativa* appears organized according to a principal axis of *indica-japonica* variation.

In the refined analysis of *O. sativa* diversity, a new result is seen on axis 3: the differentiation of the subspecies *japonica* into two groups, a 'temperate' one comprising varieties of the temperate regions and high-altitude tropical regions and a 'tropical' one including the upland rice varieties. This differentiation is based on only a few loci and matches the morphological differentiation of Oka (1958).

Finally, two additional RFLP probes characterize the cytoplasmic diversity as a function of earlier data that enabled the identification of two major types, at the chloroplast level (Dally and Second, 1990) as well as the mitochondrial level (Second and Wang, 1992). Figure 2b shows the distribution of two cytoplasmic types among 141 of the 147 varieties. It is clear that one of the types characterizes the subspecies *indica* and some intermediates, while the other type is found in all varietal groups.

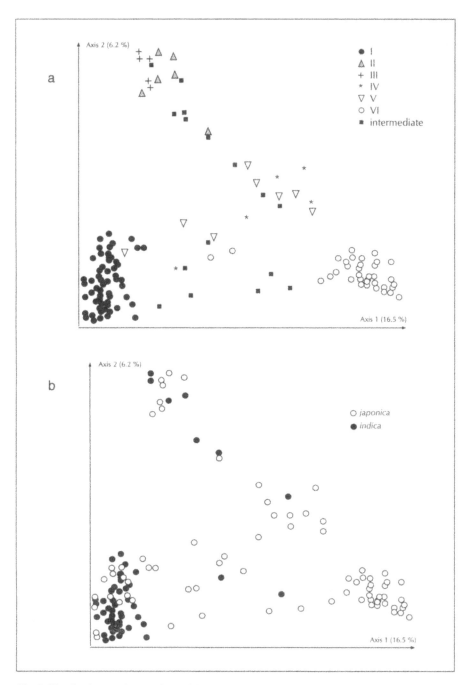

Fig. 2. Distribution on the 1-2 plane of the CA on RFLP data (202 nuclear probes and 482 bands) characterizing sample B, with (a) indication of enzymatic classification and (b) indication of cytoplasmic type (Dally and Second, 1990).

MICROSATELLITES

The diversity of microsatellites was studied for 12 loci (*RM7, RM12, RM13, RM19, RM122, RM148, RM164, RM167, RM168, OSR4, OSR7, OSR35*) within a subgroup of 54 varieties representative of known elements of structure (see Appendix). The loci revealed 4 to 19 alleles per locus, with an average of 10.8. A CA was done on the 130 alleles obtained. The primary axes explained 6.0%, 5.8%, 4.8%, 4.7%, and 3.9% of the variation; this gradual decrease revealed a relatively weak structure. The 1-2 plane of the CA (Fig. 3) shows three groups: the first corresponds to group 1 of the enzymatic classification, the second to groups II and III, and the third to groups IV, V, and VI. Axes 3 and 4 separate the groups IV, V, and VI. On the other hand, groups II and III overlap. A great within-group diversity is observed.

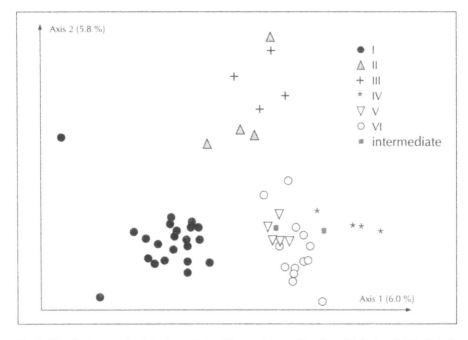

Fig. 3. Distribution on the 1-2 plane of the CA on microsatellite data (12 loci and 130 alleles) characterizing a sample of 54 varieties common to A and B.

COMPARISON OF DIFFERENT MARKER TYPES

The broadest-based studies (isozymes) and the most intensive ones (RFLPs) give the same image of the species: a highly bipolar structure with nevertheless some peculiar, partly intermediate forms. With the microsatellites, the discrimination between the most differentiated groups is conserved, but a high polymorphism is revealed within the groups. Figure 4 shows the

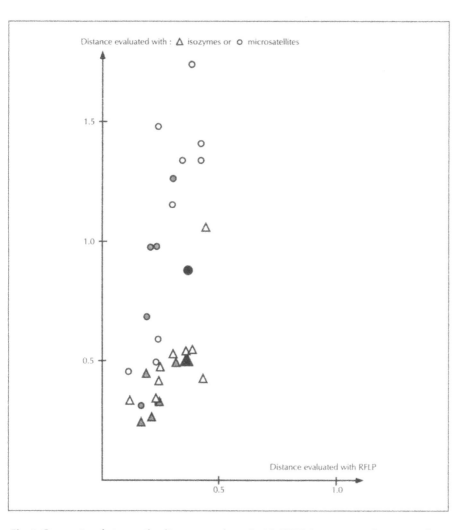

Fig. 4. Comparison between the distances evaluated with RFLP, isozymes, and microsatellites. The symbols in black characterize the distances between the two major groups I and VI; the symbols in grey characterize those between the large groups I, II, V, and VI; the symbols in white characterize those involving a small group, III or IV.

relationships between the intergroup genetic distances evaluated from different types of markers. The disparities between the numbers of accessions and loci taken into account for each of these types limit the conclusions that one can draw from this comparison. However, it is noted that, in comparison with RFLPs and isozymes, microsatellites considerably amplify the genetic distances between groups and that, among the comparisons between the largest groups (I, II, V, VI), the distance between groups I and VI is the largest with RFLPs and with isozymes, while it is far from being so with the microsatellites.

Thus, microsatellites reveal a greater polymorphism, which refines the resolution in the groups that are generally the least polymorphic. This polymorphism preserves the general imprint of the structure of the species but blurs the relations between groups.

Molecular Classification and Diversity for Characters of Agronomic Interest

MORPHOLOGICAL DIVERSITY

Data characterizing sample A were recovered from IRRI. A total of 248 varieties presented complete data for 11 variables; among these varieties all the enzymatic groups were represented except group IV, which flowered with difficulty in IRRI conditions. These data were subjected to a principal components analysis (PCA). The percentage of variation explained by the axes was 31.3%, 19.1%, 12.0%, 7.9%, and 7.0% for axes 1 to 5. Figure 5 shows the distribution of accessions in the 1-2 plane. Axis 1 separates the varieties with long organs (ligules, leaves, stems, and, to a lesser extent, panicles) and a rather thin grain, which are found close to negative values, from

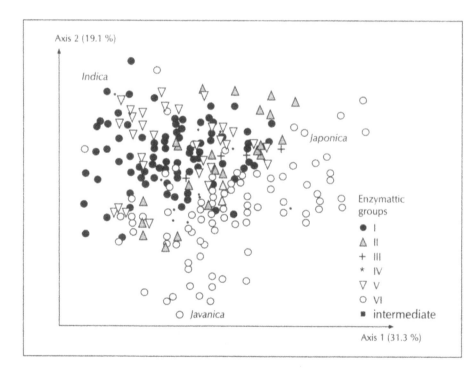

Fig. 5. Distribution in the 1-2 plane of the PCA of the variation among 248 varieties for 11 morphological characters

complementary varieties, which are on the side of positive values. Axis 2 draws towards the negative values the varieties with low tillering, large leaves, rather thick stem, and a wide and heavy grain. Thus, a tripolar structure is recognized with Indica types in the upper left corner of the plane, Japonica types in the extreme right corner of the plane, and Javanica types in the lowest part of the plane.

The morphological type Indica covers highly diversified groups in terms of molecular markers, with mostly representatives of groups I and V, but also of groups II and VI. The typical morphological types of Japonica and Javanica correspond only to group VI. While groups I and VI occupy nearly separate areas of the plane, the accessions of groups II to V are more frequent at the edge of these two major zones.

Within group VI, highly dispersed over the plane, there is a differentiation between the geographic origins that opposes the temperate forms (Japonica group) to the tropical forms (Javanica group). Between these two extremes, varieties of temperate, subtropical, and tropical origin are found, but mostly high-elevation tropical varieties from mountainous regions of Southeast Asia and South Asia. This distribution illustrates the observations made by Glaszmann and Arraudeau (1986): with the morphological types Japonica and Javanica, two forms were observed that were differentiated on a narrow genetic basis and represent the extremes of a continuous variation that follows a geographic cline according to the latitude and altitude. The classic distinction between these two types is probably due to the fact that the first studies of morphological diversity were done on samples in which the original types of Japan and Java, areas more accessible to collectors than the mountains of Southeast or South Asia, were overrepresented. The RFLP data show a similar cline with the dispersion of varieties on axis 3 of the CA.

DIVERSITY OF REACTION TO VARIOUS PATHOGENS

The 270 varieties of sample A were tested for their reaction to the major fungal, bacterial, and viral diseases (Bonman et al., 1990; Glaszmann et al., 1995). This sample revealed a great diversity of responses. Two examples are given here.

Blast, caused by the fungus *Magnaporthe grisea*, is the principal fungal disease of rice. There is a great variability in the fungus and several resistance genes are known in the plant. The location of these resistance genes is a major task. Accessions of sample A were inoculated with 13 fungal strains in a collaborative study. The symptoms were recorded on the basis of the size of the lesions detected on several plants for each accession, according to a scale of 0 to 5. The average score for an accession was used to characterize its resistance to the strain tested. A PCA was done on the complete data set obtained for 257 accessions (Fig. 6a). Axis 1 separates the varieties that are generally resistant (negative values) from the varieties that are generally

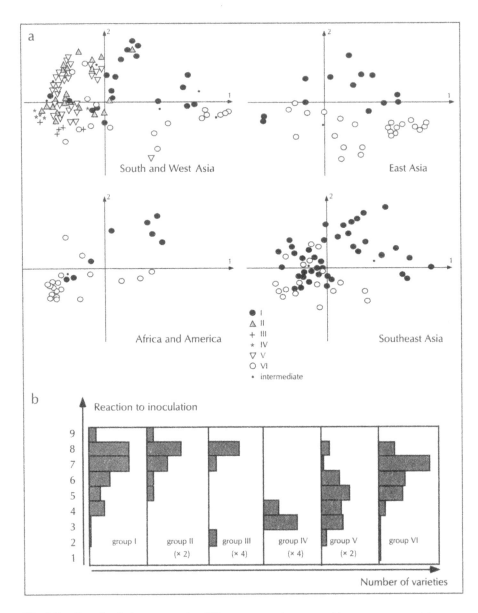

Fig. 6. Reaction of varieties representing different enzymatic groups with respect to several diseases. (a) Blast resistance of varieties representing different enzymatic groups: distribution of 257 varieties on the 1-2 plane of the PCA on the size of lesions produced by inoculation of 13 strains of *Magnaporthe grisea*. (b) Relative tolerance of 6 enzymatic groups to tungro virus: distribution of reactions to inoculation by vector (scored 1 for complete absence of symptoms to 9 for maximum susceptibility) on a set of 261 varieties representing different enzymatic groups.

susceptible. In South Asia, where all the enzymatic groups are present, groups II to V are differentiated from major groups I and VI by a higher level of resistance. Among the varieties most often susceptible, axis 2 tends to differentiate the reference *indica* varieties (group I) from the reference *japonica* varieties (group VI), because of the susceptibility of each group to several differential strains. Within group VI, a differentiation is found between the temperate forms (South and East Asia) and the tropical forms (Southeast Asia, Africa, America). It is not possible to say whether these differences are due to a general differentiation that marks the ancient isolation between these forms or to a coevolution of host and pathogen. The clearest fact to be retained pertains to the diversity of responses obtained from varieties of neighbouring geographic origin, which are *a priori* exposed to pathogen populations that are not differentiated or only weakly differentiated, as a function of their belonging to different molecular groups. Thus, the groups of the molecular classification present marked differences in behaviour, with differences in frequencies of resistance, and high specialization of the reference strains.

Tungro is a viral disease of primary importance. To evaluate the resistance of each genotype, plants were inoculated by a vector and the reactions were recorded on a scale from 1 for complete absence of symptoms to 9 for maximal susceptibility. Figure 6b shows the distribution of reactions on a set of 261 varieties according to their molecular groups. Tolerance is clearly more frequent in groups IV and V.

A pioneering study showed that blast strains are classified into two groups as a function of their differential reaction against *indica* and *japonica* varieties (Morishima, 1969). A later study on bacterial blight showed that certain genes of complete resistance are distributed largely according to the groups of the enzymatic classification (Busto et al., 1990). Our results confirm these observations and allow us to conclude that the structuration of rice into varietal groups is essential to an understanding of the host-pathogen interactions and to finding new sources of resistance.

Association between Characters and the Dynamics of the Genome

DIVERSITY AT MADAGASCAR, INSULAR SITE OF EVOLUTION

Madagascar represents a peculiar case. Asian rice was introduced there in ancient times. The fact that it is an island made it an evolutionary laboratory for cultivated rice by limiting the influences, apart from a few major introductions of, most likely, Indian, Malaysian, or Indonesian origin. The observation by breeders of varietal types particular to Madagascar justified a detailed analysis of the diversity found on the island (Rabary et al., 1989; Ahmadi et al., 1991).

A collection of Madagascar varieties was studied from morphological and isozymic perspectives. Some peculiar morphological types intermediate between the Indica and Javanica types were found with high frequency (Fig. 7). The enzymatic diversity revealed was high, but much lower than that of the cultivated rices of Asia—the H index of diversity of Nei was 0.29 in Madagascar against 0.62 in Asia. The genotype structure is clearly bipolar and recalls the *indica-japonica* opposition. However, some particular types carry an allele coding for an aminopeptidase, Amp_1^2, which is very rarely observed in enzymatic groups I and VI in Asia. Among these types, some classic associations between isozymes and morphological characters are weak, even nonexistent.

These peculiarities suggest that several recombinations between varietal groups occurred since rice was first introduced in Madagascar. Despite these recombinations, the associations between the isozyme markers of the *indica-japonica* differentiation were maintained, while they involved genes located on several chromosomes. This situation demonstrates that it is possible to recombine genes involved in control of plant architecture, but it also indicates selection pressures, natural and human, that tend to maintain the gametic disequilibrium implied in the *indica-japonica* differentiation. Numerous enzymatic markers seem located in zones of the genome subject to these

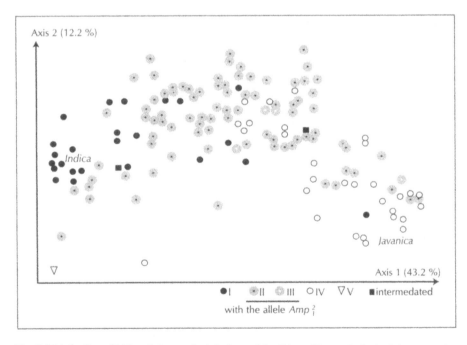

Fig. 7. Distribution of 144 varieties on the 1-2 plane of the CA on 39 morphological characters in the Madagascar material. The same symbols as in the preceding figures indicate the affiliation to enzymatic groups; a supplementary distinction is made for varieties possessing the allele Amp_1^2.

pressures, which confirms the value of classification based on enzymatic analysis.

THE CYTOPLASMIC GENOME AND ORIENTATION OF GENE FLOWS

From the particular situation of Madagascar, a phenomenon probably generalized in Asia has been documented, that of introgression between *indica* and *japonica*, source of diversity of the cultivated rices (Second, 1982).

Second and Ghesquiere (1995) compared nuclear and cytoplasmic characteristics within sample B. Two cytoplasmic types were distinguished, which was consistent with the global interpretation of the analysis of the genus *Oryza* for molecular markers: an '*indica*' type, I, and a '*japonica*' type, J (Dally and Second, 1990; Second and Wang, 1992). The *japonica* cytoplasmic type is found only in varieties of group VI, but together with the *indica* type (distribution J:I around 3:1) in the varieties of intermediate groups II to V. It is also present with a significant frequency among the varieties of group I (distribution J:I around 1:3). The japonica cytoplasm would thus have been introgressed by a number of varieties of group I, which indicates, given the maternal heredity of cytoplasm, that the hybrids at the origin of the introgressions of nuclear genes between the *indica* and *japonica* subspecies were all formed by hybridization of a female *japonica* parent pollinated by an *indica*.

DISTRIBUTION OF MOLECULAR DIVERSITY ALONG THE GENOME

The distribution of polymorphism along the genome presents a bias in contrast with what could be a random distribution (Second and Ghesquiere, 1995; Zhuang et al., 1998). It has been observed that monomorphic loci have a significant tendency to group in the species. Similarly, there is a tendency for *indica* markers to cluster together as *japonica* markers do (Second et al., 1995). This indicates clearly that the chromosomes of rice could be interpreted as a mosaic of regions that arise either from the ancestral type *indica* or from the ancestral type *japonica* and of monomorphic regions, which genetic drift could have fixed within the species in its present state from either one or the other of the ancestral types. The case of chromosome 12 is a good example of this phenomenon (Albar, 1998). There are strong associations, verified on several enzyme-probe combinations, between alleles of the same origin, *indica* or *japonica* (Fig. 8). This linkage disequilibrium is observed between loci that are very close (called a), located in the centromeric zone of chromosome 12, or between these loci and loci that are clearly further apart (called b). In this framework, monomorphic or nearly monomorphic loci and the loci for which the polymorphism is independent of the *indica-japonica* structure (called c, d, and e) could be intercalated. These loci represent probably areas subject to selection pressures of a different kind, as has been observed around certain isozymic loci in the Madagascar varieties.

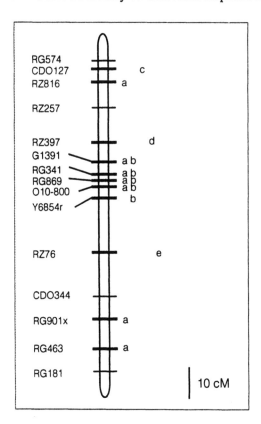

Fig. 8. Linkage disequilibrium along chromosome 12. On the basis of analysis of polymorphism within sample B, 11 loci out of the 15 tested were found to be polymorphic and are indicated by a thick horizontal bar. Linkage disequilibrium between loci was tested on the basis of the two most frequent alleles. Loci with significant disequilibrium ($p < 0.01$) between them are indicated by the presence of a single letter next to the horizontal bar.

CONCLUSION

Cultivated rices from Asia display a very wide diversity. The *indica* and *japonica* groups represent almost two different species if one accepts the hypothesis of a double origin from wild populations that were already differentiated (Second, 1982, 1984). Thus, there is a very marked structure, which is found whatever the type of marker used. The complementary inputs of markers such as isozymes and RFLPs on the one hand and microsatellites on the other are clearly seen. The isozymes and RFLPs give similar images and comparable indexes of differentiation. The RFLPs, because they are more numerous, allow an additional element to be seen: the ecogeographic differentiation between the temperate and tropical Japonica forms. Microsatellites less effectively record an ancient structure (*indica-japonica*), but they reveal an extended variability within each component.

Molecular markers bring to light important elements that are inaccessible through the study of morphological diversity alone: several groups (I to V) are identified within a group considered homogeneous in morphological terms. On the other hand, a clear continuity appears between two different morphological types, Japonica and Javanica, which corroborates the conclusions drawn from the behaviour of progenies resulting from hybridization between the two types: they are generally fertile and are easy to fix through successive selfing generations.

As compared to earlier works on the species structure, molecular markers draw attention to particular forms, represented by groups II to V of the enzymatic classification. These results confirm various observations made by breeders, such as the particular behaviour of the 'aus' types (group II) in crosses with typical *indica* and *japonica* varieties (Wan and Ikehashi, 1997) or that of 'basmati' varieties (group V), of Indica morphology but very difficult to exploit in combination with high-yielding varieties of group I (G.S. Khush, personal communication). Our results on the behaviour of varieties exposed to the major diseases also illustrate the specificities of these groups, which can help to broaden the sources of resistance present in the major groups. These groups were identified through an approach of structured sampling. A preliminary set of 1688 Asian varieties was put together by diversifying the geographic origin and ecosystem as much as possible and taking local classifications into account (Glaszmann, 1987). This sample was subsequently extended to other continents. The structuration into enzymatic groups was used to overrepresent the subgroups rich in particular diversity. This process has helped to constitute a 'nucleus' collection of a manageable size for refined evaluations.

The results of these evaluations confirm the general view. It has been observed, for example, that the minor groups harbour original sources of resistance to blast or tungro and that the japonica varieties of Africa are generally more resistant to blast than those of Asia. These associations within the sample indicate simple parameters—geographic origin, type of cultivation, classification established using a few selected markers—that it is advisable to diversify in order to enrich the collection. On the contrary, one could use the results to quickly find varieties that are exceptions to the associations or to the general correlations in order to move towards a precise selection. For example, to produce varieties with few tillers and long, thick panicle (general characteristics of group VI), but also fine grain (characteristic of other groups), one could select varieties that have the finest grain among the varieties of group VI, which generally have bold grains. The process would be more effective than if these characters were recombined between the groups. The complications linked to intergroup recombinations, all of which indicate that they disturb the association of coadapted factors, would be avoided. Then one would benefit from the accumulation of factors of grain fineness in certain varieties that have the desired genetic background, whether this accumulation be the result of a juxtaposition of factors dispersed in the populations of group VI or the result of the successful introgression of factors specific to other groups.

Our observations indicate a multitude of associations between molecular and quantitative characters, because of the existence of varietal groups differentiated for the two types of characters, as observed by Virk et al. (1996). Such associations have schematically two principal determinants.

The first is a foundation effect at the time of domestication, which possibly reinforces the joint action of natural selection and genetic drift that have differentiated the founding populations. The probably separate origin of the *indica* and *japonica* types derived from forms that diverged 2 to 3 million years ago (Second, 1985) lends a particular importance to this phenomenon in Asian rice and explains the omnipresence of associations. The origin of minor groups remains to be established. It is nevertheless clear in the light of our results that, even in the absence of an overall satisfying interpretation, the molecular data can constitute a valuable source of information to diversify the sources of useful genes. This has already proved of practical use (Kameswaro-Rao and Jackson, 1997).

The second determinant is the introgression between ancestral types. Several studies, including ours on rice from Madagascar, show that gene exchanges between varieties of different types are frequent. Nevertheless, if there was a generalized recombination, the associations would disappear rapidly over time. The possibility of evaluating the molecular diversity with markers mapped on the genome is of great interest in this context. The results available suggest that molecular polymorphism is distributed on the genome in non-random segments (Second and Ghesquiere, 1995; Zhuang et al., 1998). It could have a discrimination power differentiated, particularly for the varietal group II, according to the position of these markers in relation to the centromeres (Parsons et al., 1997). These phenomena merit more detailed study, as has been the case for chromosome 12 (Albar, 1998). In the search for the QTL for partial resistance to rice yellow mosaic virus, the indication of interactions between markers —or between markers and QTL—dispersed on the genome but belonging each time to zones that have a high *indica-japonica* linkage disequilibrium may illustrate the importance of phenomena of coadaptation in a crop such as cultivated rice of Asian origin (Pressoir et al., 1998). In the present state of our understanding, we hold that the apparent associations are the result of two phenomena: linkage disequilibrium that involves tightly linked genes and markers and associations between coadapted genes that could be dispersed on the genome.

The establishment of a database of molecular data and detailed evaluation for agronomic characters on pure lines that are results of controlled crosses as well as on unrelated lines will allow us to accumulate results and better understand and use the dynamics of evolution of genetic diversity of rice.

APPENDIX

Plant Material

The results presented are an overall perspective on earlier data that have already been partly published and new data.

On the basis of information compiled before 1985, a preliminary sample of 270 varieties, sample A, was put together to represent the geographic, ecotypic, and enzymatic diversity of cultivated rices of Asian origin. The samples were purified—through one plant selection and controlled selfings—to serve as a core collection to study the diversity for useful characters (Bonman et al., 1990; Glaszmann et al., 1995). These varieties are here used to compare the classification obtained from isozymes with the morphological diversity and the diversity of response to several diseases. The morphological data are extracted from the database maintained by the IRRI. The data on reactions to various pathogens are those reported by Glaszmann et al. (1995).

A second sample of 147 accessions, sample B, served in analysing the average diversity of RFLP for 202 mapped nuclear probes covering the entire rice genome. Among these accessions, 141 were analysed using two probes of cytoplasmic DNA: a probe located at the mutation point 28 of chloroplastic DNA (Dally and Second, 1990) and the mitochondrial probe Col (Second and Wang, 1992).

The subgroup common to A and B, which comprises 54 varieties, was used to study the diversity revealed by microsatellites: 12 loci were taken into account.

Another sample of 144 Madagascar varieties was studied to compare the molecular and morphological classifications in an island environment isolated from the area of origin of rice (Ahmadi et al., 1991). Twenty-four quantitative characters coded according to a distribution into large classes as well as 15 qualitative characters were used.

Data Analysis

A principal component analysis was done on quantitative data and correspondence analysis on qualitative data. Various marker types were compared by quantifying the genetic diversity between varietal groups using the Nei index of distance (1978).

REFERENCES

Ahmadi, N., Glaszmann, J.C., and Rabary, E. 1991. Traditional highland rices originating from intersubspecific recombination in Madagascar. In: *Rice Genetics* II. Manila, Philippines, IRRI, pp. 67-79.

Albar, L. 1989. Etude des bases génétiques de la résistance partielle du riz au virus de la panachure jaune (RYMV): cartographie moléculaire et relation avec la diversité génétique. Doct. thèsis, Université Paris XI, Paris, 187 p.

Blair, N.W., Panaud, O., and McCouch, S.R. 1999. Inter-simple sequence repeat (ISSR) amplification for analysis of microsatellite motif frequency and fingerprinting in rice (*Oryza sativa* L.). *Theoretical and Applied Genetics*, 98: 780-792.

Bonman, J.M., Mackill, A.O. and Glaszmann, J.C. 1990. Resistance to *Gerlachia oryzae* in rice. *Plant Disease*, 74: 306-309.

Busto, G.A., Ogawa, T., Endo, N., Tabien, R.E., and Ikeda, R. 1990. Distribution of genes for resistance to bacterial blight of rice in Asian countries. *Rice Genetics Newsletter*, 7: 127-128.

Chang, T.T. 1976. The origin, evolution, cultivation, dissemination, and diversification of Asian and African rices. *Euphytica*, 25: 435-441.

Chang, T.T., Adair, C.R., and Johnston, T.H. 1984. The conservation and use of rice genetic resources. *Advances in Agromony*, 35: 37-90.

Cheng, K.S., Wang, X.K., Zhou, J.W., Lu, Y.X., Lou, J., and Huang, H.W. 1984. Studies on indigenous rices in Yunnan and their utilization. 2. A revised classification of Asian cultivated rice. *Acta Agronomica Sinica*, 10: 271-280 (in Chinese).

Dally, A.M. and Second, G. 1990. Chloroplast DNA diversity in wild and cultivated species of rice (genus *Oryza*, section *Oryza*): cladistic-mutation and genetic-distance analysis. *Theoretical and Applied Genetics*, 80: 209-222.

Glaszmann, J.C. 1985. A varietal classification of Asian cultivated rice (*Oryza sativa* L.) based on isozyme polymorphism. In: *Rice Genetics* I. Manila, Philippines, IRRI, pp. 83-90.

Glaszmann, J.C. 1987. Isozymes and classification of Asian rice varieties. *Theoretical and Applied Genetics*, 74: 21-30.

Glaszmann, J.C. 1988. Geographic pattern of variation among Asian native cultivars (*Oryza sativa* L.) based on fifteen isozyme loci. *Genome*, 30: 782-792.

Glaszmann, J.C. and Arraudeau, M. 1986. Rice plant type variation: Japonica-Javanica relationships. *Rice Genetics Newsletter*, 3: 41-43.

Glaszmann, J.C., Benoit, H., and Arnaud, M. 1984. Classification des riz cultivés (*Oryza sativa* L.): utilisation de la variabilité isoenzymatique. *L'Agronomie tropicale*, 39: 51-66.

Glaszmann, J.C., Mew, T., Hibino, H., Kim, C.K., Vergel de Dios-Mew, T.I., Vera Cruz, C.H., Notteghem, J.L., and Bonman, J.M. 1995. Molecular variation as a diverse source of disease resistance in cultivated rice. In: *Rice Genetics III*. Manila, Philippines, IRRI, pp. 460-465.

Ishii, T., Brar, D.S., Second, G., Tsunewaki, K., and Khush, G.S. 1995. Nuclear differentiation in Asian cultivated rice as revealed by RFLP analysis. *Japanese Journal of Genetics*, 70: 643-652.

Jackson, M.T. 1997. Conservation of rice genetic resources: the role of the international rice genebank at IRRI. *Plant Molecular Biology*, 35: 61-67.

Kameswaro-Rao, N. and Jackson, M.T. 1997. Variation of seed longevity of rice cultivars belonging to different isozyme groups. *Genetic Research and Crop Evolution*, 44: 159-164.

Lolo, O.M. and Second, G. 1988. Peculiar genetic characteristics of *Oryza rufipogon*. *Rice Genetics Newsletter*, 5: 67-70.

Mackill, D.J. 1995. Classifying japonica rice cultivars with RAPD markers. *Crop Science*, 35: 889-894.

Morishima, H. 1969. Differentiation of pathogenic races of *Pyricularia oryzae* into two groups, indica and japonica. *SABRAO Newsletter*, 1: 81-94.

Nei, M. 1978. Estimation of average heterozygosity and genetic distance from a small number of individuals. *Genetics*, 89: 583-590.

Oka, H.I. 1958. Intravarietal variation and classification of cultivated rice. *Indian Journal of Genetics and Plant Breeding*, 18: 78-89.

Parsons, J.B., Newbury, H.J., Jackson, M.T., and Ford-Lloyd, B.V. 1997. Contrasting genetic diversity relationships are revealed in rice (*Oryza sativa* L.) using different marker types. *Molecular Breeding*, 3: 115-125.

Pressoir, G., Albar, L., Ahmadi, N., Rimbault, I., Lorieux, M., Fargette, D., and Ghesquiere, A. 1998. Genetic basis and mapping of the resistance to rice yellow mottle virus. 2. Evidence of a complementary epistasis between two QTLs. *Theoretical and Applied Genetics*, 97: 1155-1161.

Rabary, E., Noyer, J.L., Benyayer, P., Arnaud, M., and Glaszmann, J.C. 1989. Variabilité génétique du riz (*Oryza sativa* L.) à Madagascar: origine de types nouveaux. *L'Agrononomie tropicale*, 44: 305-312.

Second, G. 1982. Origin of the genetic diversity of cultivated rice (*Oryza sativa* L.): study of the polymorphism scored at 40 isozyme loci. *Japanese Journal of Genetics*, 57: 25-57.

Second, G. 1985. Evolutionary relationships in the *Sativa* group of *Oryza* based on isozyme data. *Génétique, sélection, évolution*, 17: 89-114.

Second, G. and Ghesquiere, A. 1995. Cartographie des introgressions réciproques entre les sous-espèces indica et japonica de riz cultivé (*Oryza sativa* L.) In: Colloque techniques et utilisations des marqueurs moléculaires. Paris, INRA, les Colloques de l'INRA no. 72, pp. 83-93.

Second, G., Parco, A., and Caiole, A. 1995. The hybrid origin hypothesis of cultivated rice (*Oryza sativa*) explains some of the gaps in its RFLP maps and suggests an efficient mapping population for useful genes and QTLs. In: *Plant Genome and Plastome: Their Structure and Evolution*. K. Tsunewaki, ed., Tokyo, Japan, Kodansha Scientific, pp. 129-136.

Second, G. and Wang, Z.Y. 1992. Mitochondrial DNA RFLP in genus *Oryza* and cultivated rice. *Genetic Research and Crop Evolution*, 39: 125-140.

Virk, P.S., Ford-Lloyd, B.V., Jackson, M.T., and Newbury, H.J. 1995. Use of RAPD for the study of diversity within plant germplasm collections. *Heredity*, 74: 170-179.

Virk, P.S., Ford-Lloyd, B.V., Jackson, M.T., Pooni, H., Clemeno, T.P., and Newbury, H.J. 1996. Predicting quantitative traits in rice using molecular markers and diverse germplasm. *Heredity*, 76: 296-304.

Wan, J. and Ikehashi, H. 1997. Identification of two types of differentiation in cultivated rice (*Oryza sativa* L.) detected by polymorphism of isozymes and hybrid sterility. *Euphytica*, 94: 151-161.

Wang, Z.Y. and Tanksley, S.D. 1989. Restriction fragment length polymorphism in *Oryza sativa* L. *Genome*, 32: 1113-1118.

Yang, G.P., Sanghai Maroof, M.A., Xu, C.G. Zhang, Q., and Biyashev, R.M. 1994. Comparative analysis of microsatellite DNA polymorphism in landraces and cultivars of rice. *Molecular and General Genetics*, 245: 187-194.

Zhang, Q., Saghai Maroof, M.A., Lu, T.Y. and Shen, B.Z. 1992. Genetic diversity and differentiation of *indica* and *japonica* rice detected by RFLP analysis. *Theoretical and Applied Genetics*, 83: 495-499.

Zhou, H., Glaszmann, J.C., Cheng, K.S., and Shi, X. 1988. A comparison of methods in classification of cultivated rice. *Chinese Journal of Rice Science*, 2: 1-7 (in Chinese).

Zhu, J., Gale, M.D., Quarrie, S., Jackson, M.T., and Bryan, G.J. 1998. AFLP markers for the study of rice biodiversity. *Theoretical and Applied Genetics*, 96: 602-611.

Zhuang, J.Y., Qian, H.R., and Zheng, K.L. 1998. Screening of highly-polymorphic RFLP probes in *Oryza sativa* L. *Journal of Genetics and Breeding*, 52: 39-48.

Banana

Christophe Jenny, Françoise Carreel, Kodjo
Tomekpe, Xavier Perrier, Cécile Dubois,
Jean-Pierre Horry and Hugues Tézenas du Montcel

The world production of banana was estimated at 84 million tonnes in 1996. It has the largest production in the world among fruits and the largest international trade, greater than apple, orange, grape, and melon. Banana is cultivated in more than 120 tropical and subtropical countries across five continents. Banana production is important in terms of food, society, and economy, as well as the environment.

There are two major channels of production: bananas cultivated for export and bananas reserved for local markets.

The bananas cultivated for export, known as Grande Naine, Poyo and Williams, belong to the subgroup of triploid bananas of the Cavendish type. They differ from each other only in somatic mutations such as plant height and conformation of regimes and fruits. Their production relies on an intensive monoculture of the agro-industrial type, with no rotation, and a large number of inputs.

The cultivation of bananas for local consumption is based on a large number of cultivars adapted to different conditions of production as well as the varied uses and tastes of consumers. The production systems of these bananas generally use little or no input.

Diploid bananas, close to the ancestral wild forms, are still cultivated in Southeast Asia. In other regions, triploid cultivars belonging to different subgroups—Plantain, Silk, Lujugira, Gros Michel, Pisang Awak—are the most widely distributed.

Bananas have many uses. They are consumed as fresh fruits but also cooked, as with Plantains. They are processed in various ways, into chips, fries, fritters, purees, jams, ketchup, and alcohol (banana wine and beer, production of which is particularly high in East Africa). The daily per capita consumption of bananas varies from 30 g to over 500 g in some East African countries. Apart from the fruit, other parts of the plant are used: the pseudostem is used for its fibres and as floaters (*M. textilis* or abacca) in the Philippines, and the leaves are used to make shelters or roofs, or as wraps for cooking (Plantain in Africa). In Thailand, the floral buds of particular varieties (Pisang Awak) are used in various culinary preparations. Finally, some varieties are considered to have medicinal properties.

Cultivated throughout the world, bananas are threatened by several diseases and pests. The greatest constraints are exerted by the nematodes—*Radopholus similis* and several representatives of the genus *Pratylenchus*—and by the black weevil of banana, *Cosmopolites sordidus*. Various fungal diseases are also major constraints in industrial production and, to a lesser degree, in local production. For example, yellow sigatoka due to *Mycosphaerella musicola* and black leaf streak disease caused by *M. fijiensis* result in harvest losses in large industrial plantations and necessitate costly pest treatment. In certain production zones, Panama disease due to the soil fungus *Fusarium oxysporum* f. sp. *cubense* prevents the cultivation of varieties of the Gros Michel type. Finally, viral diseases are spreading, or perhaps are simply better detected. Those of greatest concern are due to CMV (cucumber mosaic virus), BSV (banana streak virus), BBTV (banana bunchy top virus), and BBMV (banana bract mosaic virus).

BOTANY AND TAXONOMY

Botany

Bananas are monocotyledons originating from Southeast Asia and belong to the genus *Musa*, of the family Musaceae, order Zingiberales. The banana is a giant plant with a pseudostem formed by the interlocking of leaf limbs and measuring from 1 to 8 m. The leaves emerge from the terminal meristem of the true stem, incorrectly called the bulb, which is underground and small. The bud located at the leaf axil of each leaf may develop as a shoot. At the end of the vegetative phase, the change in the function of the central meristem induces the growth and elongation of the true stem at the heart of the pseudostem, then the emergence of the inflorescence. The inflorescence, which can be vertical, pendant, or subhorizontal, forms a cluster. It is made up of imbricate spathes arranged in a helix, at the axils of which emerge simple or double rows of flowers.

The flowers of the first rows are female (developed ovary, presence of stamenoids) or in some rare cases hermaphrodite (developed ovary and stamens). The cavity of the inferior ovaries may be filled with seeds and pulp to form the fruit. These rows of flowers, or 'hands', form the bunch. After the female hands, the hands of neutral flowers appear (neither male nor female), then the hands of male flowers (reduced ovary, well-developed stamens). In some cultivars, the growth of the apical meristem is interrupted after the appearance of female hands, but in general the inflorescence grows indefinitely to form what is incorrectly called the male bud.

Wild bananas have fruits filled with seeds and containing little pulp. Cultivated bananas, on the other hand, are of the true parthenocarpic type. The cavity of the inferior ovaries is filled with pulp to form the fruit, without pollination or formation of zygote. Most of these bananas have high, nearly total, female sterility and usually have no seeds in their fruits. For all bananas, the perenniality of the plant is ensured by natural vegetative propagation by new shoot formation; each variety is thus a clone.

Basis of Classification

The genus *Musa* comprises four sections: *Australimusa* (2n = 20), *Callimusa* (2n = 20), *Rhodochlamys* (2n = 22), and *Eumusa* (2n = 22). In this last section are found almost all of the cultivated bananas with the exception of Fe'i, diploid cultivated bananas of the Pacific region belonging to the *Australimusa*. The *Callimusa* and *Rhodochlamys* essentially contain species of floral interest.

Morphotaxonomy has enabled us to characterize the varieties of bananas and draw up a basis for the botanical classification used at present. In 1865, Kurz proposed the hypothesis of a bispecific origin of cultivars—*Musa acuminata* and *M. balbisiana* (Fig. 1). Simmonds and Shepherd (1955), using a method of scores, specified the relative contribution of two species of origin in the constitution of cultivars. Among the numerous morphological characters from which we can characterize a banana, these authors have retained 15, chosen for their stability and capacity to discriminate among the different groups of cultivated bananas. Each character has been quantified on a scale of 1 to 5, in which 1 corresponds to a phenotypic expression of wild bananas of the species *M. acuminata*, called A, and 5 corresponds to that of wild bananas of the species *M. balbisiana*, called B. For each cultivar, the level of ploidy and the score obtained by the addition of notes for each of the 15 characters determine its genomic constitution and consequently its position in a given group (Table 1). The majority of cultivars are categorized in the groups AA, AAA, AAB, and ABB.

Within each genomic group, cultivars that have a high proportion of common morphotaxonomic characters and are derived from one another by mutations are grouped together into subgroups (Table 2).

Information provided by molecular markers has demonstrated the marginal implication of *M. schizocarpa* (genome S, section *Eumusa*) and *M. textilis* (genome T, *Australimusa*, 2n = 2x = 20) in the genomic constitution of some cultivars, which are classified as AS, AT, or AAT.

Sexual Reproduction and Interfertility

Wild bananas reproduce sexually as well as by vegetative propagation. Some species, such as *M. schizocarpa* and *M. acuminata* ssp. *banksii*, are preferentially autogamous, the first flower hands of the inflorescence being hermaphrodites.

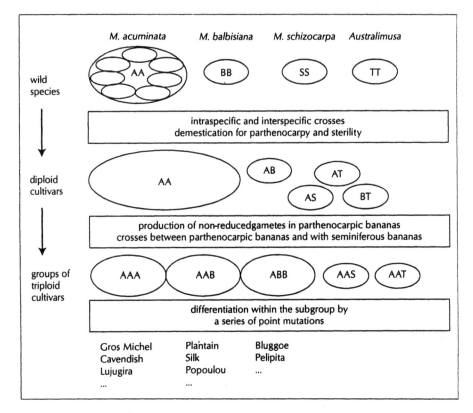

Fig. 1. Domestication of bananas

In the other species, the basal flowers are female and the banana plants allogamous. Nevertheless, these species often cross in bunches, which enables self-pollination by interposed shoots (geitonogamy).

In Asia, where the seminiferous forms are always present, some hybrids between the sections have been identified, between *M. acuminata (Eumusa)* and *M. laterita (Rhodochlamys)* and between *M. balbisiana (Eumusa)* and *M. textilis (Australimusa)*. Several interspecific hybrids, AB and AS, have also been found.

The species *M. acuminata* has a wide diversity of forms. It has been organized into interfertile subspecies on the basis of morphological characteristics and geographic distribution. Moreover, from cytological observations, Shepherd (1987) established a nomenclature of this species in seven groups within each of which the clones are structurally homozygous. The hybrids between these groups are heterozygous for one to four translocations or inversions, which limit the introgressions. These groups for the most part confirm the organization into subspecies.

Table 1. Deduction of the genomic constitution of a variety from information on its level of ploidy and its score. The intervals of scores are indicated in parentheses (Simmonds and Shepherd, 1955)

Theoretical score	Level of ploidy		
	2x	3x	4x
15	AA (16-23)	AAA (15-21)	AAAA (15-20)
30			AAAB (27-35)
35		AAB (26-46)	
45	AB (46-49)		AABB (45-48)
55		ABB (59-63)	
60			ABBB (63-67)
75	BB	BBB	BBBB

Table 2. The different types of cultivated triploid bananas

Genomic group	Subgroup	Type*	Present geographic distribution
AAA	Gros Michel**	dessert	world, regions
	Cavendish**	dessert	industrial production
	Red/Green Red	dessert	Pacific, Antilles, Philippines
	Lujugira-Mutika	cooking, beer	East Africa
	Ibota	dessert	Thailand, Central Africa
AAB	Silk	dessert	Far East, Latin America, Caribbean
	Pome-Prata	dessert	India, Australia, Hawaii, Brazil, Africa
	Mysore	dessert	India, Brazil
	Pisang Kelat	mixed	India, Malaysia
	Pisang Rajah	mixed	Malaysia, Indonesia
	Plantain**	cooking	Philippines, Latin America, Central and West Africa, Caribbean
	Popoulou-Maia Maoli	cooking	Pacific
	Laknao	cooking	Philippines
	Iholena	cooking	Pacific
ABB	Bluggoe**	cooking	Philippines, Pacific, Latin America, Caribbean, East Africa
	Monthan	cooking	India
	Pelipita	cooking	South America
	Pisang Awak	dessert	India
	Peyan	cooking	India
	Saba	cooking	Philippines

*The dessert or cooking quality of a fruit is highly subjective; most of the dessert types can also be consumed cooked, but the reverse is only rarely true.
**The most widespread subgroups.

Cultivated bananas are highly sterile. Several phenomena causing meiotic abnormalities lead to gametic sterility. These phenomena may be genomic (Bakry et al., 1990) (partial homology of two genomes *acuminata* and *balbisiana*), related to chromosomal structure (such clones are structurally heterozygous), or related to chromosome number (triploidy of the majority of cultivars leading to the formation of unbalanced gametes). Other phenomena that can be called genic lead to morphological and physiological abnormalities such as asynchronies or time lag in flower receptivity.

GENETIC RESOURCES

The species complex of bananas has characteristics that lend themselves to a particular method for genetic resource management. Clonal multiplication, associated with frequent sterility or incompatibility, and the coexistence of several levels of ploidy make it essential to have a profound understanding of the potentialities of each clone and of phylogenic relations between the different known types, to exploit the genetic resources for varietal improvement. Several tools have been developed for this purpose.

The major banana improvement programmes across the world have collections *ex situ*. Among the most important and best documented are those of the CARBAP (Centre africain de recherches sur bananiers et plantains) in Cameroon, IITA (International Institute of Tropical Agriculture) in Nigeria, FHIA (Fundación Hondureña de Investigación Agrícola) in Honduras, EMBRAPA (Empresa Brasileira de Pesquisa Agropecuária) in Brazil, MARDI (Malaysian Agricultural Research and Development Institute) in Malaysia, CRIH (Central Research Institute for Horticulture) in Indonesia, and CIRAD in Guadeloupe. These collections hold 300 to 500 clones each. Apart from conservation of the genetic patrimony, they serve as a basis and reference for varietal improvement. The objectives of the varietal creation programme associated with each collection determine the clones that are conserved *in vivo*. The collections of Cameroon, Nigeria, and Brazil are largely directed at local varieties for local consumption: Plantain in Africa, Pome and Silk in Brazil. The FHIA and CIRAD have larger collections, dedicated more to the improvement of export varieties. The CIRAD collection is particularly rich in diploid varieties related to the dessert types, material that is the basis for the procedure to create triploid hybrids developed in Guadeloupe.

On the international scale, INIBAP (International Network for the Improvement of Banana and Plantain) coordinates the different research programmes and mediates the exchange of certified material. INIBAP maintains in the name of the international community an *in vitro* collection of 1100 accessions at the international transit centre on the campus of the Catholic University of Leuven, in Belgium. It has also established two centres for indexing plant material to secure international exchanges, one at QDPI

(Queensland Department of Primary Industry) in Australia and the other at CIRAD in Montpellier, France.

Over the past twenty years, several markers of diversity have successively been perfected and applied to banana. Morphotaxonomic descriptors have been the primary markers developed and perfected in banana. Variations in vegetative organs occur mainly in the colour of the pseudostem, the presence and colour of blotches at the base of petioles, the shape of the petiolar canal section, and the size and shape of the plant. Particular variations are also recognized to be due to true dwarfism (shortening of plants, narrow leaves, shoot inhibition) and to chimeras of colour. The most significant variations are, however, those of the inflorescence and consequently the bunch. The size, shape, and colour of fruits as well as the colour of the pulp are among the criteria by which fruits are differentiated from one another. The Plantains, for example, have a very firm, yellow-orange pulp, which is not found in other cooking bananas such as Laknao, Popoulou, Bluggoe, and Monthan (Tezenas du Montcel, 1979). The bananas of East Africa are highly specific; they are intended for cooking or beer-making, depending on the clones. The fragrances of dessert bananas are varied, as well as their tastes: very sweet in certain diploid cultivars such as Pisang Mas, slightly acidulate in the Silk subgroup widely cultivated in India and Brazil, standard and generally agreeable for the Cavendish type meant for export. The morphological variability of the male bud is expressed partly by differences in the shape and colour of the bracts and the floral pieces.

A set of 119 agromorphotaxonomic descriptors has been defined as a norm of description for bananas (IPGRI-INIBAP and CIRAD, 1996). These descriptors serve as a basis for a system of information exchange between collections, the MGIS (Musa Germplasm Information System), run by INIBAP. Information tools have been developed to help identify plants on the basis of these descriptors (Perrier and Tezenas du Montcel, 1990).

The 119 descriptors were studied for 273 clones of the CIRAD collection and for 223 clones of the CRBP collection, 99 completely described clones being common to the two collections. The statistical analysis was done from an index of dissimilarity weighted for probabilities of observation error (see Appendix).

For ten years, different types of molecular markers have been used for banana (Lagoda et al., 1998): polyphenols (Horry, 1989), isozymes (Lebot et al., 1993), RFLP (Jarret and Litz, 1986; Carreel et al., 1994), and sequence tagged microsatellite sites (Grapin et al., 1998). Within CIRAD, these molecular data are grouped in the Tropgene database (Raboin and Lagoda, 1998). Other tools such as flow cytometry and *in situ* hybridization have also contributed to the characterization of genetic resources and allow a more refined determination of the genomic nature of the accessions studied: ploidy, aneuploidy, and number of chromosomes linked to each of the original species (Dolezel et al., 1994; Jenny et al., 1997; Horry et al., 1998; D'Hont et al., 1999).

The most complete results are related to the study of cytoplasmic and nuclear genomes with the help of RFLP markers.

The two cytoplasmic genomes are characterized by a single-parent heredity. Within the species M. *acuminata*, Faure et al. (1994) showed that the mitochondrial genome was transmitted by the paternal route and the chloroplast genome by the maternal. To date, all the monospecific and interspecific hybrids studied confirm this original transmission of cytoplasmic genome in banana.

The cytoplasmic and nuclear genomes of clones coming from the CIRAD collection at Guadeloupe and the *in vitro* collection of INIBAP were analysed by RFLP for 3 chloroplast probe-enzyme combinations, 14 mitochondrial probe-enzyme combinations, and 51 nuclear probe-enzyme combinations distributed throughout the genome. In total, 115 wild accessions and 243 cultivars were studied. Some 158 accessions, of which 89 are diploids comprising 34 wild species, and 69 are triploid, were compared with the morphological descriptions of the Guadeloupe collection. The list of clones studied having common morphological as well as molecular characteristics is presented in the appendix to this chapter (see Table 5).

From these successive inputs, the genetic resources of bananas were structured and their characterization was refined. This chapter presents a synthesis of those studies.

Diversity of Diploids

The diploid bananas, wild and cultivated, are presently much less widespread than the cultivated triploids. However, they are still found in the endemic state throughout Southeast Asia. The presence of seeds often makes the fruits unsuitable for consumption and sale. The plants are generally feeble and have smaller yields than the triploids. Only cultivars of the Sucrier type, which have small, very sweet fruits, are cultivated outside their zone of origin. These diploid clones are nevertheless indispensable for genetic improvement programmes, especially because of the high sterility of the triploids.

WILD BANANAS

The seminiferous wild bananas of the genus *Musa* are found in the humid but well-drained valleys and glades of forests at low and medium altitude, in the tropical zone, in South and Southeast Asia and in the Pacific, the Indian peninsula, and the Samoan islands. More than 25 species have been described and included within the genus *Musa*. Only those that have contributed to the genome of parthenocarpic bananas are discussed here.

Species belonging to the section *Australimusa* and M. *schizocarpa* of the section *Eumusa* are present east of the range of *Musa*: in the eastern area of Indonesia, Papua New Guinea, and the Pacific. The *Australimusa* are identified by their erect inflorescence. The various species of this section were described by Cheesman (1947) and Argent (1976) as related and morphologically very close. The 16 accessions studied on the molecular scale are found to have little variation.

Musa schizocarpa is characterized by a water-green stem colour, identifiable in the interspecific hybrids, as well as the green colour of the bracts of the male bud. The two accessions studied and present in the collection are very similar from the morphological as well as molecular point of view.

Musa balbisiana, of the section *Eumusa*, is found in India, the Philippines, and occasionally the Indochina peninsula. The tree representation of morphotaxonomic data (Fig. 2), according to the NJtree method applied on the dissimilarity Δ_{PP}, clearly shows the differentiation of *M. balbisiana* in relation to *M. acuminata*, as well as their low variability (see Appendix). Isozyme analysis also reveals reduced polymorphism. Horry (1989) classifies the eight accessions present in the collection into four distinct types, which

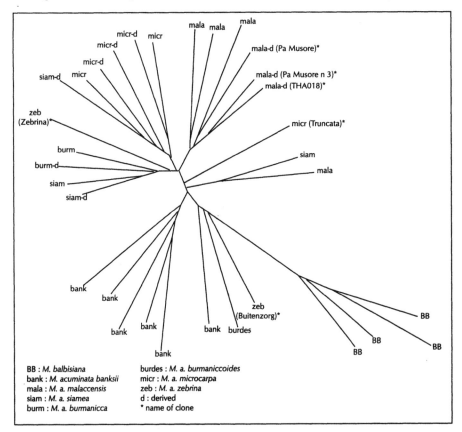

BB : *M. balbisiana*
bank : *M. acuminata banksii*
mala : *M. a. malaccensis*
siam : *M. a. siamea*
burm : *M. a. burmanicca*

burdes : *M. a. burmaniccoides*
micr : *M. a. microcarpa*
zeb : *M. a. zebrina*
d : derived
* name of clone

Fig. 2. Morphological diversity of seedy bananas (Guadeloupe collection): tree representation according to the NJtree method, realized on the Δ_{PP} of dissimilarity between 32 accessions on the basis of 99 morphological descriptors.

confirms the RFLP analysis. It is surprising that *M. balbisiana*, widespread over a vast geographic area, varies so little, but that could be due to lack of information.

The area of extension of *M. acuminata*, section *Eumusa*, covers most of the area of distribution of the genus *Musa*, from east to west, from Burma to New Guinea. Its botanical classification refers to 7 subspecies. The tree representation of the morphotaxonomic data spreads from four pools (Fig. 2), which correspond to *M. a. malaccensis*, *M. a. banksii*, and *M. a. microcarpa*, and to accessions of two subspecies, *M. a. siamea* and *M. a. burmanicca*. Nuclear RFLP analysis confirms these results and allows us to identify the intersubspecific nature of certain representatives of *M. acuminata*. Parallel analysis of two types of markers structures the species into five pools

The *M. a. banksii* pool constitutes a highly homogeneous set, more so in molecular terms than in morphological terms. Its representatives are highly autogamous and have diverged from other subspecies in an isolated fashion in Papua New Guinea. It is thus logical to observe that they are highly homozygous and morphologically original and related at the same time.

The *M. a. burmanicca* and *M. a. siamea* pool also comprises the subspecies *burmaniccoides*. These species originate from a vast continental zone from northern India to Thailand. Some very specific morphotaxonomic characters prevent the subspecies *burmanicca* and *burmaniccoides* from being grouped together, and de Langhe and Devreux (1960) therefore dissociated them. Analysis using molecular markers confirms the work of Shepherd (1990), who grouped them together.

The case of accessions of *M. a. zebrina* is a little peculiar: the Buitenzorg, heretofore considered the holotype, is highly atypical. The RFLP analysis proves its hybrid nature. It is not surprising, therefore, that these accessions are seen to diverge within the tree. The *zebrina* pool in the collection is made up of a single accession, Zebrina.

For the pool corresponding to *M. a. malaccensis*, it is necessary to introduce the notion of accessions derived from a subspecies. This appellation arises from the phenotypic observation of the accessions close to the holotypes, but not identical to them. These accessions (THA018, Pa Musore ..) are genetically identified as very close to the subspecies of origin by nuclear and cytoplasmic RFLP markers. They are more highly homozygous and are thus classified within the same group as *M. a. malaccensis*.

The same notion of derived accessions exists for *M. a. microcarpa*. The grouping is simple in morphological terms. Nevertheless, the accessions are polymorphic from the molecular point of view and RFLP analysis demonstrates their intersubspecific hybrid nature. It is nearly accepted that the accession Truncata, classified previously within *M. a. microcarpa*, constitutes an entire subspecies in morphological as well as molecular terms.

Musa acuminata is thus the most variable and best-structured species. The topography of its area of origin has led to geographic and thus reproductive isolation, which is a source of differentiation. The chromosomal modifications of the translocation type observed by Shepherd (1987) between the subspecies of *M. acuminata* can be associated with this reproductive isolation.

CULTIVATED DIPLOIDS

For the cultivated diploids, and despite the large number of clones known, no further subgroups have been established. Most of the cultivated diploid bananas present in the collection belong only to the species *M. acuminata*. Out of 135 diploid clones of the CIRAD collection, only about 10 accessions have been identified as interspecific and classified in the groups AB, AS, or AT, with the help of specific alleles of the species.

Factorial analysis on morphotaxonomic and molecular nuclear data reveals a nearly continuous cloud, and the structuration does not emerge. On the other hand, analysis of cytoplasmic genomes and rates of heterozygosity suggests elements of structuration.

The AA cultivars are separated into four of the nine chloroplastic profiles identified (designated I to IX) when all the *Musa* are analysed. Similarly, they present 47 of the 111 mitochondrial profiles identified. These last could be grouped into nine sets of clones having comparable mitochondrial profiles (designated α to ι). The AA cultivars are separated into five of these mitochondrial sets. Information on the chloroplast genome, indicating a maternal relationship, and that on the mitochondrial genome, indicating a paternal heredity, are complementary and can be compiled (Table 3). The AA cultivars can thus be divided into nine cytoplasmic types, or cytotypes, but most of them correspond to three cytotypes: II-α, V-α, or V-Φ.

The study of the nuclear genome has also made it possible to estimate the rate of heterozygosity using 43 probe-enzyme combinations that can be assimilated at loci, and thus appreciate the complexity of the genome and the intersubspecific or other origin of the clones. Two groups of clones can be distinguished within the three principal cytotypes. For example, within the V-φ, which comprise mainly clones from studies in Papua New Guinea, several clones have a low rate of heterozygosity (from 10% to 20%) and are closely related to *acuminata banksii*, in cytoplasmic as well as nuclear studies. The other cultivars have much higher rates of heterozygosity (40% to 70%), characteristic of an intersubspecific origin.

From the grouping into cytotypes and rates of heterozygosity, 12 sets of clones can be defined within the AA cultivars. This overall organization is found during nuclear genome analysis or morphological analysis. Nevertheless, the analyses show that the limits are somewhat arbitrary because the variation is continuous. From the classification, however, we can define the base populations for breeding programmes.

Table 3. Structuration of bananas according to their cytotype

Chloroplast profile	Mitochondrial profile								
	α	β	χ	δ	ε	φ	γ	η	τ
I									
II	M. a. errans AAcv (36)* AAA (Orotova, Red, Cavenish, Gros Michel) ABA (Mysore, Nadan, Pome)	AAcv (1)*	M. a. burmanicca M. a. burmanicoides M. a. siamea	M. a. malaccensis AAcv (2)* AB ABA (Silk)	AAcv (1)*	AAcv (3)* AAA (3)**	M. a. microcarpa		ABB (Saba, Bluggoe, Ney, Mannan)
III	AAcv (4)* AAA (1)**				M. a. malaccensis				
IV	AAcv (1)*		M. a. siamea						
V	AAcv (49)* AAA (4)** ABA (Laknao, Maia Maoli) ABB (1)**				AAA (Lujugira-Mutika)	M. a. banksii AAcv (18)* ABA (Plantain, Popoulou, Laknao)		ASs (2)*	ABB (Pelipita, Saba)
VI	AScv (1)*					ASs (1)* AScv (2)*		M. schizo-carpa	
VII								M. balbisiana type 2	
VIII	BAA (P. Rajah Bulu)				BAA (P. Kelat)				
IX								M. balbisiana type 1, 3, 4 BAB (Peyan, Saba, P. Awak)	BAB (P. Awak)

*Number of clones having this cytotype. **Number of clones of indeterminate subgroup. In italics: seedy bananas. s: wild, cv: cultivated.

Triploid Bananas

Analysis on morphological data and molecular data has been done on triploid bananas (Fig. 3). The two types of markers clearly differentiate the triploid cultivars that are purely *acuminata* from interspecific cultivars. Simmonds and Shepherd (1955) based the classification of bananas into genomic groups

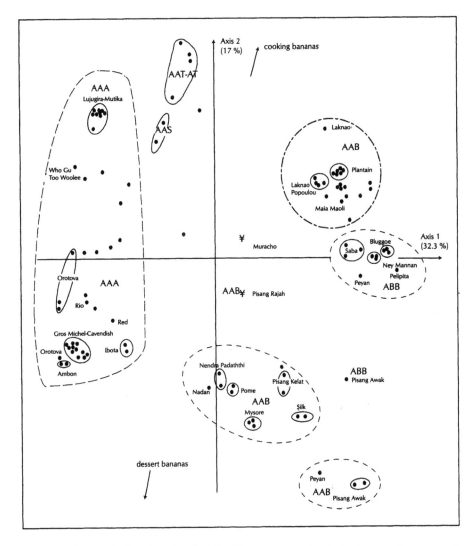

Fig. 3. Nuclear molecular diversity of triploid bananas according to their genomic group and subgroup. First plane of a factorial analysis on a Jaccard dissimilarity between 109 cultivars on the basis of 267 variables.

on the observation of 15 morphotaxonomic descriptors. The use of more complete descriptors and their analysis by multivariate statistical methods allow us to easily recover this organization in genomic groups, from observations from the Cameroon collection as well as the Guadeloupe collection.

On the other hand, the molecular data do not indicate as clear a distinction between AAB and ABB as morphological markers do. *Musa balbisiana* is found to be hardly polymorphic with the molecular markers, and the two genomes B of ABB are rarely differentiated. Several AAB cultivars also have their two A genomes nearly identical. In these two cases, the AAB and ABB clones have molecular profiles of the AB type, and only a reading of the relative intensity of RFLP bands—which is possible only for some probe-enzyme combinations—allows us to clearly differentiate the AAB from the ABB.

Although certain subgroups emerge from the morphological analysis, it is not always possible to identify the subgroups as easily as for Plantain or Cavendish. First of all, environmental variations strongly influence the stability of morphotaxonomic descriptors. Second, the mutations—themselves quite small—are the basis of the differentiation of cultivars within the subgroups causing phenotypic modifications (dwarfism, for example) that are strong enough to eliminate any hope of advanced structuring. In contrast, certain phenomena of convergence render illusory this type of analysis using only morphotaxonomic characters. One of the most glaring examples is probably the Mnalouki cultivar, AAB of the Comoro Islands, the appearance of which causes it to be mistaken for a Plantain of the French type: molecular markers and just a taste of the fruit, however, prove that it has nothing to do with the Plantain.

This botanical classification into subgroups is found more at the molecular level, the cultivars having identical or very similar nuclear and cytoplasmic profiles in most cases. The existence of some subgroups, such as the Orotova AAA, is nevertheless called into question. Certain clones that are not attached to any subgroup in morphological terms are somewhat related to each other, as with AAB Muracho, related to Pisang Rajah. Other clones that were earlier not grouped have been identified as closely related to each other, as with AA, Who Gu, and Too Woolee.

It is also possible to distinguish clones of the dessert type from clones of the cooking type within the AAB on a morphological basis, and within the three groups AAA, AAB, and ABB by means of molecular markers. Thus, among the AAB, bananas of the subgroups Mysore, Pome-Prata, and Pisang Kelat are differentiated from the typical cooking bananas: Plantain, Popoulou, Maia Maoli, and Laknao. It is to be noted that the clone Pisang Raja Bulu of the subgroup Pisang Rajah, of the mixed type, has a profile intermediate between the dessert and cooking types. This classification can be ascribed to the genome A of each of these subgroups.

The classification of triploid bananas is much easier to establish than that of the diploid clones because of the mode of evolution of the triploid clones. At this stage, all these cultivars are highly sterile and propagation is exclusively vegetative. The clones are differentiated among each other only through small mutations, which lead rapidly to the identification of true subgroups. The degree of variability within each subgroup depends on the intensity with which each type of clone will be used and thus multiplied. The greatest morphological variability is found in two subgroups particularly exploited in Africa: Plantains throughout the Central African zone and West Africa, and Lujugira in East Africa. On the other hand, within these subgroups, the nuclear and cytoplasmic RFLP profiles are identical or very similar.

Relationships between Different Levels of Variability

The crosses carried out during breeding involve clones of varying ploidy. The tetraploid hybrids are a result of crosses between the triploid cultivar to be improved and a diploid progenitor that carries resistance. The procedure developed by CIRAD aims to produce triploid cultivars related to cultivated varieties from diploid progenitors (Bakry et al., 1997). In these two situations, a thorough knowledge of the origin of cultivated bananas is essential. An understanding of the processes of domestication—parthenocarpy, sterility, and evolution towards triploidy—is also necessary to best manage and exploit the genetic resources of banana. Through the study of the genetic organization of bananas, we try to identify populations of present diploids most closely related to different types of triploid cultivars.

Several morphological resemblances are known between diploid and triploid clones. The bunches of several AA cultivars—Pongani, for example—that come from collections in Papua New Guinea are similar to those of the Plantains. The taste and type of consumption of other AA cultivars, such as IDN110, relate them to triploids of the Silk type. These relationships have been confirmed by molecular analysis of cytoplasmic and nuclear genomes, and other relationships have been brought to the fore.

The analysis of cytoplasmic genomes, presented for the cultivated diploid clones, has been extended to all the wild and cultivated clones, diploid or triploid. In Table 3, two clones shown on the same line are related from the maternal side and two clones in the same column are related on the paternal side, independent of their level of ploidy. Many triploids are of the same cytoplasmic group as wild or cultivated diploids. Thus, cultivars of the subgroups Cavendish, Gros Michel, Ambon, Rio, and Ibota have the same cytoplasmic profiles as several AA cultivars. AAB clones of the dessert type in the Silk subgroup have the same cytoplasmic profile as the AB of the Silk type. AAB clones of the cooking type in the subgroups Plantain, Popoulou, and Laknao have two cytoplasmic genomes identical to that of *M. a. banksii*. This relationship between the two genomes A of Plantain and *M. a. banksii* has been revealed by morphological observations (Tezenas du Moncel, 1990). It has been confirmed by genome analysis with isozymes (Horry, 1989) as well as by RFLP marking. For three subgroups of triploids, no diploid clone belongs to the same cytoplasmic group or to the same mitochondrial profile. Among them, the AAA clones of East Africa, which belong to the subgroup Lujugira-Mutika, belong on the maternal side to *M. a. banksii* and on the paternal side to *M. a. zebrina*, which demonstrates an intersubspecific origin, confirmed by the analysis of their nuclear genome. This bisubspecific origin has also been found for the AA cultivars of nearby geographic origin, of Madagascar or the Comoro Islands.

The set of clones has also been analysed with the same nuclear RFLP markers. The projection of data on cultivated triploids on the results obtained for *M. acuminata* and the cultivated diploids reveals the relationships between the A genomes of wild clones, cultivated diploids, and triploids (Fig. 4). It is interesting to observe that the diploid and triploid cultivars of the sets related

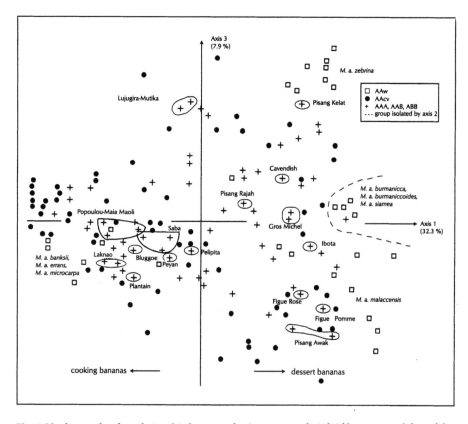

Fig. 4. Nuclear molecular relationship between the A genomes of triploid bananas and that of the wild and cultivated diploids: projection of data relative to triploids on the 1-3 plane of the factorial analysis on a Jaccard dissimilarity of wild and cultivated diploids on the basis of 19 probe-enzyme combinations that reveal significant differences between the cytotypes V-α and V-Φ

to the *banksii, errans,* and *microcarpa* Borneo axes have floury fruits and are thus consumed cooked, while bananas of the groups close to the *malaccensis* and *zebrina* axes have sweeter fruits.

The study of nuclear genome has also been used to estimate the rates of heterozygosity from a profile at two levels of bands, for the set of clones, and a profile at three levels of bands, for the triploids, using 43 probe-enzyme combinations that can be assimilated at loci. These rates allow us to assess the complexity of the genome of triploids to be improved. The different A genomes of a single genotype are more related—their rates of heterozygosity lower—for the cooking types than for the dessert types, which thus have a more complex origin.

The cytoplasmic and nuclear data on diploids and triploids allow us to target the diploid progenitors most closely related to the triploids to be improved. Thus, on the basis of molecular markers, crossing procedures of diploids to be improved, doubled with colchicine or not, have been established in order to obtain hybrids similar to the triploids, such as those that were proposed by Raboin et al. (forthcoming) for Gros Michel and Cavendish.

CONCLUSION

From all the morphological and molecular data, hypotheses have been advanced as to the origin of parthenocarpy and principal subgroups of triploids.

Dispersal and Evolution of Banana

Simmonds (1962) placed the centre of primary domestication of cultivated bananas in the Malaysian region. However, several results suggest that it is the genomes of *M. a. banksii* accessions of Papua New Guinea and the accession Agutay of the Philippines, representing *M. a. errans*, that are the origin of parthenocarpy and thus of cultivated bananas. The nuclear or cytoplasmic genomes of these two subspecies have been identified in almost all the parthenocarpic bananas, diploid or triploid. Moreover, the genetic distances are minimal between *M. a. banksii* and the homozygous AA cultivars. Simmonds (1962) indicated moreover a significant development of the pulp, which is an index of parthenocarpy, in a representative of the subspecies *banksii*. The Philippines and Papua New Guinea area, where the two subspecies *banksii* and *errans* originated, thus seems to be the centre of primary domestication.

The fruits of certain wild bananas are consumed immature before the seeds become hard. The *M. a. banksii*, which presented the beginning of parthenocarpy, were selected by humans. Transported to Southeast Asia, they were crossed with other subspecies of *M. acuminata* and thus acquired the genetic and structural heterozygosity that characterize all the cultivars. They integrated genes of sterility and regulatory genes of parthenocarpy to give the first edible diploids. For these diploids, the sterility of cultivars limited the occurrence of a panmictic unity in each region and vegetative propagation conserved the intermediate forms. Thus, despite geographic isolation, no strong structuration appeared. The partial fertility of these cultivars has nevertheless, gradually and under human action, allowed the establishment of a cline. Vegetative propagation allowed the genotypes resulting from each of these steps to be conserved.

The AAB Plantains, present in the entire forest zone of Africa (Champion, 1967), were introduced in ancient times on that continent. The introduction had an Austronesian origin and dated from at least 3000 years BCE (de Langhe, 1995). Champion notes that it took place at a time when other bananas did not exist, otherwise those also would have been introduced into Africa. These bananas, like the Popoulou, also called Pacific Plantain, have their two A genomes very closely related to that of *M. a. banksii* and do not belong to any other subspecies. The parthenocarpic bananas of the subspecies *banksii* could have been crossed with *M. balbisiana* in areas covered by the two species. The genomic differences between *M. a. banksii* and *M. balbisiana* are sufficient for their hybrids to give diploid gametes resulting from a single restoration. The AAB derived from these few crosses expanded towards the west, into India and then Africa, where they were differentiated and propagated. Central Africa is thus the centre of secondary diversification of Plantains. The crosses between AA of *banksii* origin and *M. balbisiana* have also led to Laknao in the Philippines, and Popoulou-Maia Maoli in the Pacific.

The Lujugira-Mutika AAA were not observed in East Africa. They are related to the two subspecies *banksii* and *zebrina*. These clones or the diploids from which they were derived were introduced directly from Indonesia, the zone of origin of *M. a. zebrina*, perhaps via the Comoro Islands and Tanzania, as Simmonds has hypothesized. Transported to Indonesia by the Austronesians, the partly parthenocarpic bananas of *banksii* coming from Papua New Guinea could have been crossed with *M. a. zebrina*. Since the hybrids were structurally heterozygous, their partial sterility was sufficient to give diploid gametes required for the formation of triploids.

The formation of triploid cultivars of the dessert type relies on the existence of hybrid cultivars between the *banksii, errans, malaccensis*, and *zebrina* genomes. Diploid and then triploid cultivars that relate these genomes appeared progressively in Malaysia. This region is thus, according to our hypothesis, a secondary centre of domestication and not, as Simmonds has proposed, the primary centre.

Conservation of Genetic Resources

The collected cultivated clones—about a thousand, which is a small number compared to that of other plants—are conserved *in vitro* and managed by

INIBAP. Each research centre maintains an *in vivo* collection of clones representative of the diversity and several clones corresponding to the type of fruit the centre wishes to improve. In the natural environment, the variability is mostly of somaclonal origin because of the high sterility of cultivars. The variability resulting from recombination arises essentially from artificial crosses carried out during the creation of a variety. By characterizing the created clones accurately, and comparing that characterization with that of clones of natural origin, we can verify that there is no genetic drift.

The situation is entirely different for the seminiferous bananas. Several species and subspecies of the genus *Musa*, which have been studied very little, are represented in the collection only as holotypes that are maintained by vegetative propagation. These varieties are difficult to study owing to the geographic and political situation of their countries of origin, but it is essential to study them because of the accelerated deforestation of the zones they grow in. A number of wild populations are at risk of extinction in the short or long term. The conservation and knowledge of these varieties are essential for future improvement programmes, which look for new sources of resistance to parasitic threats.

APPENDIX

Choice of a Distance for the Analysis of Morphotaxonomic Descriptors

One of the characteristics of morphotaxonomic descriptors is their subjectivity. The choice of a modality depends on the expression of the character in the plant as a function of the environment and on the researcher's evaluation of this expression. Despite all efforts to reduce this subjectivity, it is impossible to eliminate it completely. The variations of expression, however, constitute one means of access to the plant's responses to its environment. A particular dissimilarity has been defined, taking into account these errors of evaluation.

Definition of the Dissimilarity Δ_{PP}

The descriptors retained are qualitative. In order to refine the responses, a matrix of probability of error has been associated with each descriptor. In the Musaid software (Perrier and Tezenas du Montcel, 1990), the use of these probabilities means we do not have to systematically consider that two accessions are the same or not but rather we can assign a weight to their similarity. In Table 4, for example, in which the colour of the ventral face of the midrib is recorded, the notation 'light green' can easily be confused with 'yellow' or 'green', while the notation 'black' is unequivocal.

Table 4. Matrix of probability for the descriptor 'colour of ventral face of the midrib'

Response	Probability of identity					
1—yellow	43	30	10	05	01	01
2—light green	30	33	30	05	01	01
3—green	20	30	53	05	01	01
4—pink purple	05	05	05	64	10	01
5—red purple	01	01	01	01	86	01
6—black	01	01	01	01	01	95

The index of dissimilarity defined derives from the Sokal and Michener index (or simple matching coefficient, designated Δ_{SM}), which, for two accessions, is equal to the ratio of the number of equal responses to the number of descriptors compared. Here our index, written Δ_{PP}, is weighted by the probabilities of error of observation. The contribution of the descriptor v to the dissimilarity between two accessions i and j is expressed as follows:

$$\Delta_V = \frac{1}{2}\left[\frac{P_v(v_i, v_i) - P_v(v_i + v_j)}{P_v(v_i, v_i) - P_v(v_i + v_j)} + \frac{P_v(v_j, v_j) - P_v(v_j + v_i)}{P_v(v_j, v_j) - P_v(v_j + v_i)}\right]$$

where P_v represents the matrix of probability associated with the descriptor v and where v_i and v_j correspond to the responses of two accessions for this descriptor. The overall dissimilarity Δ_{PP} between i and j is equal to the mean of Δ_v over all the descriptors.

For a descriptor, Δ_v varies between 0 and 1, it is null when the two individuals have the same modality and large to the extent that the probabilities attached to the responses v_i and v_j are different, without, however, reaching 1.

Thus, for the descriptor of the colour of the ventral face of the midrib, we obtain, if $v_i = 2$ (light green) and $v_j = 3$ (green):

$$\Delta_{PP} = \frac{1}{2}\left(\frac{33-30}{33+30} + \frac{53-30}{53+30}\right) = 0.16$$

If $v_i = 2$ (light green) and $v_j = 6$ (black):

$$\Delta_{PP} = \frac{1}{2}\left(\frac{33-1}{33+1} + \frac{95-1}{95+1}\right) = 0.96$$

In these two situations, the index Δ_{SM} would be equal to 1.

The index Δ_{PP} tends to minimize the differences observed between accessions. It offers the advantage of preventing an excessive increase of the index calculated by accumulation of minor differences between morpho-taxonomic criteria that are difficult to evaluate. This gain in precision for small distances is a desired advantage, since small distances have a great importance in the first state of ascending algorithms of classification.

The graphic representation of Δ_{PP} as a function of Δ_{SM} indicates that the linear correlation between the indexes is good ($r = 0.896$). The two indexes have a clear tendency to measure the divergences between accessions in the same direction. The extreme points are the same, but the values in the middle are divided by a factor of about 2 with the index Δ_{PP}.

Comparison of the Index Δ_{PP} in the Various Collections

For this analysis the official descriptors of bananas (IPGRI-INIBAP and CIRAD, 1996) were used. The observations were made on two collections, in Cameroon (CRBP) and Guadeloupe (CIRAD). Only the bananas for which the descriptors were complete in the two collections were used. Thus, 99 accessions were retained: 58 diploids and 41 triploids.

The means of the Δ_{PP} indexes were calculated for each collection and for each genomic group. Observation of the same cultivars on the two sites allows us to quantify and distinguish the variability observed between different cultivars within a single collection from the variability between collections for the same cultivar. The comparison of results allows us to highlight some main points.

There is less divergence between sites than between cultivars for the collections in general and, more precisely, for the AAA and AAB groups. On these scales of diversity, the environmental effects do not disturb the classification of clones in the genomic group. On the other hand, the divergence becomes higher for the subgroups. For a subgroup in which homogeneity between cultivars is large, variations linked to the environment have more influence than the differences between clones. This is the case of Cavendish, which differ among themselves essentially in often minute variations of size. The influence of the environment on these variations is high, and the divergence will be larger between descriptions of the Grande Naine clone at Cameroon and at Guadeloupe than between the description of a Grande Naine and that of a Williams in a single collection. Since the clones respond in the same way to the influence of the environment, the variation within a single site is possibly displaced, but it is not amplified.

The Δ_{PP} index, which takes possible errors into account using tables of probabilities, allows a more reliable determination of the Δ_{SM} index genomic group than even when data from another collection is used. On the other hand, to identify a cultivar, it is not always reliable to use a reference base (Cameroon, for example) to determine an accession described in another geographical area (Guadeloupe, for example), especially when there are accessions resulting from somaclonal mutations. This confirms the recommendations to use a probability model (Musaid software) to identify an accession.

Table 5. List of clones studied having common morphological as well as molecular characteristics

Section	Subgroup	Clone	GLP	
● Group species	Subspecies	Variety	CMR*	RFLP**
Rhodochlamys				
● *Musa velutina*		Velutina	x	x
● *Musa laterita*		Jamaïque		x
Eumusa				
● *Musa acuminata*	*malaccensis*	Pahang, Sélangor	x	x
		Cici (Brazil), Malaccensis		x
	burmanicca	Long Tavoy		x
	burmaniccoides	Calcutta 4	x	x
	banksii	Banksii, Madang, Paliama	x	x
		Hawain 2, Higa, Waigu		x
	siamea	Khae (Phrae), Pa Rayong	x	x
	microcarpa	Bornéo, Microcarpa, Truncata	x	x
	zebrina	Zebrina	x	x
		Burtenzorg		x

(Contd.)

(Contd.)

Section ● Group species	Subgroup Subspecies	Clone Variety	GLP CMR*	RFLP**
	malaccensis derivative	Pa (Musore) n° 3, THA018	x	x
		Pa (Musore)		x
	burmanicca derivative	Pisang Prentel		x
	siamea derivative	Pa (Songhkla)	x	x
		Pa (Abyssinea)		x
	microcarpa derivative	Pisang Cici Alas	x	x
		AAs IDO113, EN13		x
● *Musa balbisiana*		Balbisiana (CMR), Balbisiana (HND), Klue Tani, Pisang Batu	x	x
● Cultivated AA		Akondro Mainty, Ato, Bie Yeng, Chicame, Galeo, Gorop, Guyod, IDN110, Kenar, Khai Nai On, Kirun, M48, M53, Manameg Red, Mapua, Niyarma Yik, Pa (Patthalong), Pisang Bangkahulu, Pisang Berlin, Pisang Sasi, pongani, Sa, Samba, Sepi, SF215, SF265, Sowmuk, Thong Det, Tomolo, To'o, Tuu Gia, Vudu Papua, Wikago	x	x
		Bebeck, Beram, Dibit, Fako Fako, Gu Nin Chiao, Hom, Inori, Khi Maeo, Kumburgh (Kunburg), Manang, N° 110-THA052, Padri, Pallen Berry, Pisang Gigi Buaya, Pisang Lilin, Pisang Sapon, Pitu, Saing Todloh, Ta, Wudi Yali Yalua		x
● Cultivated AB		Figue Pomme (Ekona), Safet Vetchi	x	x
		Kunnan		x
● AAA	Gros Michel	Cocos, Gros Michel	x	x
		Bout Rond, Dougoufoui, Highgate		x
	Cavendish	Americani, Mutant Rouge, Padji, Seredou	x	
		Grande Naine, Petite Naine, Poyo	x	x
		IRFA901 Williams		x
	Red/Green Red	Figue Rose Naine		x
	Orotova	Pisang Sri	x	x
		Hom (Sakhon Nakhon), Orotova, Pisang Kayu, Pisang Umbuk		x
	Ibota	Khom Bao, Yangambi km5	x	x
	Lujugira	Foulah	x	x
		Bui Se Ed, Nakitengwa, Nshika		x
	Ambon	Hom (Thong Mokho), Pisang Ambon, Pisang Bakar		x
	Indeterminate AAA	Ouro Mel, Too Woolee Who Gu	x	x

(Contd.)

(Contd.)

Section	Subgroup	Clone variety	GLP	
● Group species	Subspecies		CMR*	RFLP**
		Lagun Vunalir, Palang		x
● AAB	Plantain	Amou, Kelong Mekintu, Kwa,	x	
		Madre del Platano	x	
		Big Ebanga		x
	Pisang Kelat	Pisang Kelat, Pisang Pulut		x
	Pisang Rajah	Pisang Raja Bulu	x	x
	Popoulou	Poingo, Popoulou (CIV)		x
	Nendra Padaththi	Pisang Rajah	x	x
		Rajapuri India		x
	Mysore	Pisang Ceylan	x	x
	Silk	Kingala 1, Supari	x	
		Figue Pomme Naine		x
	Nadan	Lady Finger		x
	Pome	Guindy, Rois	x	
		Foconah, Prata Ana		x
	Laknao	Laknao	x	x
		Kune, Mugus		x
	Indeterminate AAB	Mnalouki	x	
		Pisang Nangka	x	x
		Kupulik, Muracho, Teeb Kum, Tomnam	x	
● ABB	Bluggoe	Dole, Poteau Nain	x	
	Pisang Awak	Bom, Praha	x	
		Namwa Knom, Pisang Kepok		x
	Monthan	Monthan	x	x
	Ney Mannan	Ice Cream	x	x
		Radjah		x
	Peyan	Brazza IV, Pisang Kepok Bung		x
	Saba	IDN107		x
		Saba	x	x
	Indeterminate ABB	Auko-PNG034, Dwarf Kalapua	x	x
		Auko-PNG125, Klue Tiparot***		x
● AAAA		Champa Nasik	x	
● AAAB		Langka 08	x	x
		Ouro da Mata, Platina	x	
● AABB		Pisang Slendang		x
● ABBT		Yawa 2		x
Total			99	158
		cultivated diploids	37	55
		wild diploids	21	34
		triploids and tetraploids	41	69

*Population common to the Guadeloupe and Cameroon collections, morphotaxonomic descriptors.
**Accessions of Guadeloupe described by RFLP markers as well as by morphotaxonomic descriptors.
***New classification, according to Jenny et al. (1997).

REFERENCES

Argent, G.C.G. 1976. The wild bananas of Papua New Guinea. *Royal Botanic Garden Edinburgh*, 35(1): 77-114.

Bakry, F., Carreel, F., Caruana, M.L., Cote, F.X., Jenny, C., and Tezenas du Montcel, H. 1997. Les bananiers. In: *L'amélioration des Plantes Tropicales*. A. Charrier et al. eds., Montpellier, France, Cirad-Orstom, collection Repères, pp. 109-139.

Bakry, F., Horry, J.P., Teisson, C., and Tezenas du Montcel, H. 1990. L'amélioration génétique des bananiers à l'Irfa-Cirad. *Fruits*, spec. no.: 25-40.

Carreel, F., Noyer, J.L., Gonzalez de Leon, D., Lagoda, P.J.L., Perrier, X., Bakry, F., Tezenas du Montcel, H., Lanaud, C., and Horry, J.P. 1994. Evaluation de la diversité génétique chez les bananiers diploïdes (*Musa* sp.). *Génétique, sélection, évolution*, 26(suppl. 1): 125s-136s.

Champion, J. 1967. Les bananiers et leur culture. 1. Botanique et génétique. Paris, IFAC, 212 p.

Cheesman, E.E. 1947. Classification of the bananas. 2. The genus *Musa* L. *Kew Bulletin*, 2: 106-117.

de Langhe, E. 1995. Is plantain the oldest fruit crop in the world? In: *Celebration of the King Baudouin award to IITA*. Ibadan, Nigeria, IITA, pp. 44-47.

de Langhe, E. and Devreux, M. 1960. Une sous-espèce nouvelle de *Musa acuminata* Colla. *Bulletin du Jardin Botanique de Bruxelles*, 30: 375-388.

D'Hont, A., Paget-Goy, A., Escoute, J., and Carreel, F. 1999. The interspecific genome structure of cultivated banana, *Musa* spp., revealed by genomic DNA *in situ* hybridization. *Plant & Animal Genome Conference*, San Diego, Poster.

Dolezel, J., Dolezelova, M., and Novak, F.J., 1994. Flow cytometric estimation of nuclear DNA amount in diploid bananas (*Musa acuminata* and *M. balbisiana*). *Biologia Plantarum*, 36(3): 351-357.

Faure, S., Noyer, J.L., Carreel, F., Horry, J.P., Bakry, F., and Lanaud, C. 1994. Maternal inheritance of chloroplast genome and paternal inheritance of mitochondrial genome in bananas (*Musa acuminata*). *Current Genetics*, 25: 265-269.

Grapin, A., Noyer, J.L., Carreel, F., Dambier, D., Baurens, F.C., and Lagoda, P.J.L. 1998. Diploid *Musa acuminata* genetic diversity with STMS. *Electrophoresis*, 19: 1374-1380.

Horry, J.P. 1989. Chimiotaxonomie et organisation génétique dans le genre *Musa*. Doct. thesis, Université Paris XI, Orsay, 105 p.

Horry, J.P., Dolezel, J., Dolezelova, M. and Lysak, M.A. 1998. Les bananiers naturels tétraploïdes A × B existent-ils? *InfoMusa*, 7(1): 5-6.

IPGRI-INIBAP, and CIRAD, 1996. *Descripteurs pour le Bananier (Musa* spp.). Rome, IPGRI, 55 p.

Jarret, R.L. and Litz, R.E. 1986. Enzyme polymorphism in *Musa acuminata* Colla. *Journal of Heredity*, 77: 183-188.

Jenny, C., Carreel, F., and Bakry, F. 1997. Revision on banana taxonomy: Klue Tiparot *(Musa* sp.) reclassified as a triploid. *Fruits*, 52(2): 83-91.

Lagoda, P.J.L., Amson, C., Aubert, G., Auboiron, E., Bakry, F., Baurens, F.C., Carreel, F., Costet, L., Dambier, D., Faure, S., Foure, E., Ganry, J., Gonzales de Leon, D., Grapin, A., Horry, J.P., Jenny, C., Noyer, J.L., Ollitrault, P., Raboin, L.M., Reschœur, F., Tezenas du Montcel, H., Tomekpe, K., and Lanaud, C. 1998. Acronyms to study the Musaceae, reviewed paper. *Acta Horticulturae*, 461: 113-121.

Lebot, V., Aradhya, M.K., Manshardt, R.M., and Meilleur, B.A. 1993. Genetic relationships among cultivated bananas and plantains from Asia and the Pacific. *Euphytica*, 67: 163-175.

Perrier, X. and Tezenas du Montcel, H. 1990. Musaid, a computerized determination system. In: *Identification of Genetic Diversity in the Genus Musa*. R.L. Jarret, eds., Montpellier, France, INIBAP, pp. 76-91.

Raboin, L.M., Carreel, F., Noyer, J.L., Baurens, F.C., Horry, J.P., Bakry, F., Tezenas du Montcel, H., Ganry, J., Lanaud, C., and Lagoda, P.J.L. The putative common diploid ancestor of Cavendish and Gros Michel banana cultivars (forthcoming).

Raboin, L.M. and Lagoda, P.J.L. 1998. *Eléments moléculaires d'Aide à la Décision pour la Sélection de la Banane et du Plantain*. Montpellier, France, CIRAD-Agetrop.

Shepherd, K. 1987. Translocations in *Musa acuminata*. Photocopied document, 3 p.

Shepherd, K., 1990. Observations on Musa taxonomy. In: *Identification of Genetic Diversity in the Genus Musa*. R.L. Jarret, eds., Montpellier, France, INIBAP, pp. 158-165.

Simmonds, N.W. 1962. *The Evolution of the Bananas*. London, Longmans, 170 p.

Simmonds, N.W. and Shepherd, K. 1955. The taxonomy and origins of the cultivated bananas. *Journal of the Linnean Society of London, Botany*, 55: 302-312.

Tezenas du Montcel, H. 1979. Les plantains du Cameroun: propositions pour leur classification et dénominations vernaculaires. *Fruits*, 34(2): 83-97.

Tezenas du Montcel, H. 1990. *Musa acuminata* subspecies *banksii*: status and diversity. In: *Identification of Genetic Diversity in the Genus Musa*. R.L. Jarret, ed., Montpellier, France, INIBAP, pp. 211-218.

Cacao

Claire Lanaud, Juan-Carlos Motamayor and
Olivier Sounigo

Cacao has been domesticated from ancient times. It was cultivated about 2000 to 3000 years ago by the pre-Colombian civilizations, particularly the Mayas and the Aztecs (Paradis, 1979). Cacao beans were used in a beverage called 'chocolatl' for which cacao, maize, pimento, and other aromatic plants were mixed and ground. Cacao beans were so highly valued that they constituted a currency of exchange. Even before the arrival of the Spanish, cacao travelled through the trade routes of the Mayas and Aztecs, but also with the help of the Pipil-Nicaraos (Young, 1994; Coe and Coe, 1996). In 1585, the Spanish exported cacao to Spain, where the secret of chocolate—a sugared version of the beverage—was guarded for about 40 years. The popularity of this chocolate drink then spread throughout Europe. To respond to the growing demand, cacao plantations were extended into Central America and new plantations were established in several of the Caribbean islands, such as Trinidad, in 1525, and Jamaica. Cacao trees from Central America, particularly Costa Rica, were introduced in Venezuela by the Spanish, but it is possible that cacao trees were already cultivated in the southwestern part of the country before Spanish colonization (Pittier, 1933; Bergman, 1969; Motamayor, 2001). Around 1750, the French planted cacao in Martinique and in Haiti, and the Portuguese planted it in Belem and Bahia.

Cacao trees have grown in Asia and the Pacific islands since 1560 (Wood, 1991; Young, 1994). During that time, Criollo of Venezuela were introduced in the island of Celebes and then in Java, by the Dutch. In 1614, Criollo from Mexico were transplanted in the Philippines by the Spanish. In 1798, cacao cultivation was introduced in Madras, India, from the island of Amboine and was brought from Trinidad to Ceylon by the British. From Ceylon, the cacao tree was introduced to Singapore and Fiji in 1880, to Samoa in 1883, to Queensland in 1886, and to Bombay and Zanzibar in 1887. Cacao was cultivated in Malaysia from 1778 and in Hawaii in 1831.

The introduction of cacao to Africa is more recent. Spanish or Portuguese navigators imported it to the island of Sao Tome in 1822, then to the island of Fernando Poo in 1855 (Burle, 1952). Other introductions were then made by

Swiss missionaries from Surinam. The first cacao seeds were sown on the African continent in 1857. The first cacao that spread through West Africa from Ghana originated from the Amazon basin (Forastero Amelonado) and then, in 1920, hybrid forms Trinitario and Criollo were imported and were hybridized with the local Amelonado (Toxopeus, 1985). Each of these transfers was made with a small number of genotypes. As a consequence, the genetic basis of the cacao plants initially cultivated in West Africa was very narrow and their origin uncertain.

The technique of chocolate manufacture was perfected with the invention of presses to extract cacao butter in 1828 (Enriquez, 1985). The technique of fermenting and roasting beans made it possible to develop the chocolate aroma from cultivars of cacao other than the varieties of Criollo traditionally cultivated, which require little fermentation.

At the beginning of the 20th century, nearly 80% of the world production of cacao, about 115,000 tonnes, was produced in Central and South America (Braudeau, 1969). In 1997, the production reached nearly 2.7 million tonnes. Côte d'Ivoire is presently the primary world producer with 1.12 million tonnes per year, followed by Ghana and Indonesia, with 330,000 tonnes each.

TAXONOMY AND GENETIC RESOURCES

Taxonomy

Cacao belongs to the family Sterculiaceae and the genus *Theobroma*. *Theobroma cacao* L. ($2n = 2x = 20$) is a tree originating from humid tropical regions of the northern part of South America and Central America. Even though the first attempts at domestication and cultivation were made in Central America, Cheesman (1944) considers the centre of origin of cacao to be the upper course of the Amazon, near the equatorial Andes. It is in this region that Pound (1938) observed the greatest morphological variation. Cacao has a small genome, estimated at 0.4 pg per haploid genome (Figueira et al., 1992; Lanaud et al., 1992).

The taxonomy of the genus *Theobroma* has been studied by botanists since the end of the 19th century (Bernoulli, 1869; Schumann, 1886; Pittier, 1930; Chevalier, 1946). A later and more complete study is that of Cuatrecasas (1964), which divided the genus *Theobroma* into 6 sections and 22 species, which are spread across the American continent between 18°N and 15°S. One of these sections is made up of the single species *T. cacao*. The classification proposed by this author is based on the mode of germination, the architecture of the trees, and the characters of the fruits and flowers. Among all these species, only *T. cacao* is economically important. However, *T. grandiflora*, the cupuassu, is also exploited in Brazil, the pulp of its fruits being used in the manufacture of drinks and sorbets.

The species *T. cacao* is composed of a large number of interfertile populations, highly variable morphologically. The plants are autogamous or allogamous depending on their genetic origin. A system of gametosporophytic self-incompatibility enforces the allogamy in certain populations (Knight and Rogers, 1955; Bouharmont, 1960; Cope, 1962; Glendinning, 1966).

Classifications proposed for the species *T. cacao* are never entirely satisfactory. They are very often schematic, given the intense genetic exchange between populations of varying genetic origin. Morris (1882) was the first to propose a classification of cacao trees into two groups: Criollo and Forastero[1]. Pittier (1930) placed each species into a group: *T. leiocarpum* for the Forastero and *T. cacao* for the Criollo. Cuatrecasas (1964) considered them two subspecies: *T. cacao* ssp. *cacao* for the Criollo and *T. cacao* ssp. *sphaerocarpum* for the Forastero. At present, a classification into two morphogeographic groups, the Criollo and the Forastero, is generally used. A third group, the Trinitario, combines the hybrid forms of the two former groups.

THE FORASTERO

The Forastero group combines a large number of wild populations and cultivated varieties originating from South America. The cacao trees of this group are found from Ecuador to the Guyanas. It is the vigorous trees that have several types of disease resistance. Their beans, generally purple and flat, but sometimes white and rounded in certain populations, give a medium to good quality cacao. The varieties of the Nacional type, which originated from Ecuador, and the classification of which in this group is now under question (Enriquez, 1992), produce a fine, aromatic cocoa much in demand by chocolate manufacturers.

The Forastero of Brazil and Venezuela
The Forastero of the Amazon basin and the Orinoco valley are widely cultivated throughout the Amazon basin. The most typical form is the Amelonado. It is this type of cacao tree, self-compatible, that was first introduced in Africa. Many collections of wild material in the Amazon region in Brazil have demonstrated high morphological diversity (Barriga et al., 1984).

The Forastero of Peru, Ecuador, and Colombia
During the course of several research expeditions on genotypes resistant to witches' broom disease, a disease caused by the fungus *Crinipellis perniciosa*, Pound (1938, 1943) described the populations of Peru and Ecuador for the

---

[1] The terms Criollo and Forastero come from Venezuela and distinguish the traditionally cultivated local cacao trees, the Criollo, from foreign cacao trees, the Forastero, also called Trinitario, introduced later from Trinidad. Thus, the term Forastero originally related to all the different varieties of Criollo. By extension, the term Criollo, associated with cacao of high quality resulting from thick, light-coloured, aromatic beans, has been used to designate cacaos of the same quality coming from other countries such as Nicaragua and Mexico.

first time. These have high morphological diversity, with pods of variable shape and colour, green or pale green or slightly pigmented in the zones close to Colombia. Later studies in the upper Amazon and, in particular, eastern Ecuador (Allen and Lass, 1983) indicated indigenous and wild populations of cacao. Contrary to the observations of Pound at Peru, all the populations of eastern Ecuador seem morphologically quite similar. They are characterized by large trees, young leaves of a pale green colour, and green and rough pods containing a high portion of white beans. These cacaos have characters in common with the Criollo varieties of Central or South America, but the pods of Criollo may be more elongated and completely red, contrary to those of Ecuador cacao, which are only slightly pigmented.

Hybrid forms are also found in Ecuador, particularly the Refractario. These are descendants of an introduction of Trinitario made in Ecuador around 1890 from some pods from Trinidad. These Trinitario were then naturally crossed with the Nacional variety originally cultivated in Ecuador. This particularly vigorous and early-producing material, even on poor soils, has been used by most planters. There has thus been considerable genetic exchange between cacao types of different origins. During the occurrence of witches' broom disease, which spread throughout almost all the districts and caused serious losses in production in the 1920s to 1930s, some trees, called Refractario, appeared more tolerant. A small percentage of seeds from these surviving trees were then planted.

The Forastero of Guyana

The first observations of wild cacao in Surinam revealed populations highly homogeneous in terms of pod shape, of the Amelonado type, and distinct from the cultivated types in the region (Myers, 1930). Spontaneous cacao were observed in southwestern French Guyana from 1729 (Lecomte and Challot, 1897). Studies organized in French Guyana revealed certain populations, particularly those of the upper Camopi, that have a significant phenotypic variability (Sallee, 1987). These populations are distinguished from Forastero of the lower Amazon by their pods, the shape of which varies from that of the Amelonado to that of the Angoleta, with more or less marked verrucosity and medium to large size (Lachenaud and Sallee, 1993). The origin of the types cultivated in Guyana is uncertain. According to Guisan (1825), they come from the natural cacao forests located above the Camopi.

THE CRIOLLO

The Criollo were the first cacao domesticated by the Mayan and Aztec civilizations. These are the only varieties that were cultivated in all of Central America and northern South America up to the 17th century.

The Criollo, self-compatible, are found from Mexico to Colombia and in Venezuela in the cultivated or subspontaneous form (Soria, 1973; Reyes, 1992). They are slow-growing, are more sensitive to disease and insects than the

Forastero, and have a high morphological diversity (Soria, 1970a, 1970b). The fruits are elongated, with an acuminate point, a smooth or rough surface, red or green colour before maturity, and a poorly lignified cortex. The beans are of variable size but, most often, large and thick and white or pink in colour. Criollo beans are generally rounded, unlike the flat Forastero beans of the Amazon basin and Peru. These bean characteristics, associated with poor lignification of the cortex, are normally used to classify the cacao clones in the Criollo group. The cultivated forms of Criollo vary from a smooth type of pod, such as Porcelana of Venezuela, to Cundeamor type, with highly verrucose pods, found in Mexico, Colombia, and Venezuela. A peculiar form, the Pentagona, is found in the old plantations of Mexico, Guatemala, Nicaragua, and Venezuela.

Apart from some traditional varieties of Criollo present in the collections, more vigorous and productive Criollo types have been collected in the plantations of the last decades. It is these clones of Criollo, generally representing this genetic group in the collections, that are called 'modern Criollo' in the rest of this study.

THE TRINITARIO

The Trinitario combine all the hybrid forms between Criollo and Forastero of the lower Amazon or the Orinoco valley, which are the source of these varieties. The Trinitario were cultivated first of all in Trinidad, then in Venezuela and Central America, where they were mixed with the Criollo in the traditional plantations.

Genetic Resources

About 40 collections of wild or cultivated material have been put together since 1930, mainly on the Forastero of Peru, Brazil, Ecuador, Colombia, Venezuela, and Guyana and, more recently, on the Criollo of Central America, Venezuela, and Colombia.

The most important collections from these studies are conserved at CRU, Cocoa Research Unit, at Trinidad (2500 genotypes), at CEPLAC (Comissão Executiva do Plano de Lavoura Cacaueira, in Brazil (2000 genotypes), and at CATIE, the Centro Agronómico Tropical de Investigación y Enseñanza, in Costa Rica (700 genotypes). Each of these collections has its speciality: The one in Trinidad is rich in Forastero of the upper Amazon. That of Brazil includes material from collection expeditions of the Amazonian region in Brazil and the varieties cultivated at Bahia. The CATIE collection has mainly Trinitario and Criollo varieties. The CIRAD collection in French Guyana has descendants of about 200 spontaneous mother plants collected in that country.

International exchanges of plant material are subjected to a quarantine of two years in order to prevent viral and fungal diseases. Three quarantine stations are in operation: that of the University of Reading in the United Kingdom, that of CIRAD in France, and that of CRU in Barbados.

An international database was formed during the 1990s. It includes information on more than 27,000 genotypes conserved in 43 collections and is updated regularly (Wadsworth and Harwood, 2000). Another database, Tropgene-db, presently established in CIRAD, contains molecular data of more than 400 genotypes.

The genetic improvement of cacao, targeted on different objectives depending on the country—quality, production, resistance—relies always on the creation of hybrids between progenitors belonging to different populations. Since it is difficult to establish a classification solely on the basis of morphological characters, biochemical and molecular markers are used to refine the genetic organization of species and to study the processes of domestication.

ORGANIZATION OF GENETIC DIVERSITY

Morphological Diversity

To describe the morphological diversity, several authors have attempted to define the most efficient morphological descriptors that take into account the intra- and interclonal variance (Enriquez and Soria, 1968; Engels et al., 1980; Soria and Enriquez, 1981; Engels, 1983a, b; Bekele, 1992; Bekele et al., 1994; Raboin and Paulin, 1993). The IBPGR (1981) published a list of 65 morphological descriptors to characterize the genetic resources of cacao. However, in order to reduce the time required for characterization in studies on a large number of genotypes, this list has often been shortened.

Engels (1986) studied the diversity of 294 clones of the CATIE collection using 39 qualitative and quantitative descriptors pertaining to the morphology of flowers, leaves, and fruits. This analysis indicates a structuration between the Criollo and Forastero types as well as a significant diversity in each of these groups. In this analysis, the Trinitario are grouped with the Criollo.

N'goran (194) analysed the diversity of 52 clones belonging to Forastero, Criollo, and Trinitario groups for 9 characters of beans and pods. This analysis confirmed the structuration observed by Engels, with a differentiation between Forastero on the one hand, and Criollo and Trinitario on the other.

A later study (Bekele and Bekele, 1998) on 100 clones resulting from 24 populations indicated a structuration of populations depending on geographic origin.

These studies together show that morphological markers allow an overall structuring of the diversity of different populations of cacao in collections. They are accessible to all, but they are time-consuming, and they are difficult to use because they vary according to the environment.

Enzymatic Diversity

Enzymatic diversity of clones or of cacao populations has been evaluated by several authors (Lanaud and Berthaud, 1984; Amefia, 1986; Atkinson et al., 1986; Lanaud, 1986a, b, 1987; Yidana et al., 1987; Ronning and Schnell, 1994; Warren, 1994; Sounigo et al., 1996).

Lanaud (1987) analysed the diversity of 296 genotypes using 6 polymorphic enzymatic systems representing 9 loci. On the set of the samples analysed, a total of 30 alleles were identified. With the exception of *PAC1*, all the alleles are found in the populations originating in the upper Amazon. *PAC1* is on the other hand frequent among the Criollo and Trinitario. Another allele, *MDHA1*, is also frequent among the Crillo and the Trinitario and rare among the Forastero.

The populations that have the largest number of alleles per locus are those of the upper Amazon, with 1.8 to 2.2 alleles per locus. The populations of Venezuela and Guyana are the least variable with 1 to 1.5 alleles per locus. The percentage of polymorphic loci is the highest in populations of the upper Amazon, where it is higher than 50%. The mean heterozygosity is highest among the American Trinitario (0.36), the modern Criollo (0.29), and the Refractario EQX of Ecuador (0.35). The heterozygosity is lowest for the African Amelonado (0.04), the Forastero of Guyana (0.06), and the two populations of Venezuela that were analysed (0.1 and 0).

This variability in rates of heterozygosity within the populations of Forastero could result from the system of reproduction of the trees and their system of self-incompatibility. The cacaos of the upper Amazon are highly self-incompatible and preferentially allogamous, while the Forastero of the lower Amazonian basin are self-compatible and thus preferentially autogamous.

A correspondence analysis (CA) was done using 31 variables observed on 286 individuals. No clear structuration between Forastero and Criollo or Trinitario appeared on the first CA axis, while a slight structuration was visible on the 3-4 plane. The Forastero from the upper Amazon present the largest genetic diversity. Those of Ecuador are as variable as those of Peru, contrary to the morphological observations of Allen and Lass (1983). Comparatively, the other populations of Forastero analysed are much less diverse. The Forastero cultivated in Guyana are clearly differentiated from wild Forastero collected in the south of Guyana on the first CA plane (Fig. 1). The direct use of spontaneous local cacao seems thus excluded in the explanation for the origin of cultivated cacao, the provenance of which remains unknown. A Venezuelan origin is, however, probable (Lachenaud and Sallee, 1993). The diversity of the wild population of Guyana is relatively low, taking into account the large number of individuals analysed (92). The two populations of Guyana, cultivated and wild, are clearly differentiated from the lower Amazonian Forastero (Amelonado) cultivated in West Africa

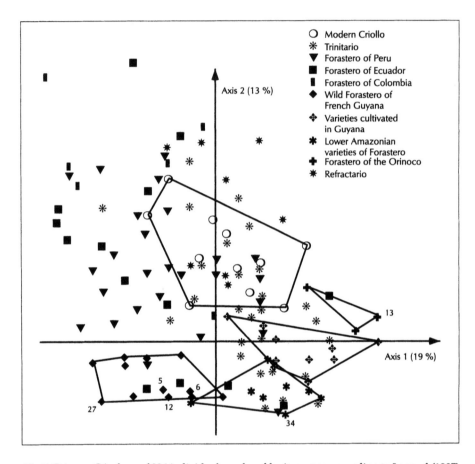

Fig. 1. Primary CA plane of 296 individuals analysed by isozymes, according to Lanaud (1987). The numbers shown on the figure indicate the number of individuals superimposed on a given point.

and from populations of the Orinoco valley in Venezuela. The cultivated Guyanese cacao, however, are more closely related to the Forastero of Venezuela. On the first axis of the CA, the wild Guyanese cacao are closer to certain Forastero of Peru and Ecuador than to Forastero of Venezuela and the cultivated forms of Guyana.

There is close genetic proximity between the modern cultivated varieties of Criollo and the Trinitario cultivated in America and Africa, which is easily explained by the significant genetic exchange that occurs on plantations. The variability of Trinitario cultivated in Africa is very high and extends from the Forastero Amelonado types to the Criollo. This situation is linked to the history of the introduction of cacao in Africa, with a first wave of Forastero Amelonado and a second wave of Trinitario. The genetic exchange that followed resulted in certain hybrid types closer to the Forastero Amelonado type.

Using four enzymatic systems, Warren (1994) analysed the diversity of seven populations of Forastero of Colombia, Ecuador, and Peru and two populations of Trinitario, at the rate of 10 individuals per population. He observed a large number of alleles for the PGI system (10 alleles against 5 in the study of Lanaud, 1987). The Shannon diversity indexes indicate a greater diversity in the populations of Forastero of Ecuador than in those of Peru. The hybrid Trinitario forms also appear variable. The author concludes that, if there really is a centre of diversity of wild cacao, it must be located not in Peru, as suggested by Cheesman (1944), but further north, in Ecuador and Colombia.

Ronning and Schnell (1994) studied the enzymatic diversity of 86 clones of cacao resulting from Forastero, modern Criollo, Trinitario, and an undefined hybrid group using six enzymatic systems corresponding to eight loci. The allelic frequencies and the genetic distances obtained confirm the differentiation between Forastero and Criollo. The values of parameters of genetic diversity are overall similar to those of other perennial ligneous species: $H_T = 0.295$, $H_S = 0.266$, $G_{ST} = 0.096$ (Hamrick and Godt, 1990).

Sounigo et al. (1996) studied the enzymatic diversity of 487 clones of cacao belonging to 28 populations or groups of accessions present in the CRU collection at Trinidad, using five polymorphic enzyme systems. The analysis of parameters of genetic diversity showed highly variable allelic richness and heterozygosity. Certain populations such as the Forastero of Guyana and the Trinitario of the Dominican Republic or Martinique are poorly diversified, with an average of 1.3 to 1.8 alleles per locus and 6% to 16% of mean heterozygosity per locus and per individual. The reverse is observed for certain hybrid populations of Trinitario and Refractario, which have 2.2 to 2.5 alleles per locus and 42% to 50% of mean heterozygosity per locus and per individual. Other populations of Trinitario show high heterozygosity but a low Shannon index of diversity or allelic richness. On the other hand, certain populations of Forastero and of Refractario of Ecuador have low heterozygosity but great allelic richness.

The low diversity of certain populations, such as those of Martinique and the Dominican Republic, reflect the homogeneity of the material resulting probably from a small number of introductions into these Caribbean islands. On the other hand, the high diversity values of Refractario clearly indicate their multiple origins.

Estimated from the Nei diversity index, the mean intrapopulation diversity, H_{pop}, is 0.77 and the interpopulation diversity is 0.23. Thus, three quarters of the total diversity is explained by the intrapopulation diversity. These results agree with those of Ronning and Schnell (1994) and Russel et al. (1993). They are characteristic of allogamous perennial species in which significant genetic exchange has occurred (Hamrick et al., 1992). The hierarchical clustering (HC) based on the genetic distances of Nei is represented in Fig. 2. Two major groups are observed at $d = 0.2$. One comprises the populations of Refractario and Trinitario as well as the population of IMC Forastero of Peru. This group is in turn structured into two subgroups: the first subgroup comprises the Trinitario differentiated according to their geographic origin, and the second subgroup includes all the Refractario as well as the populations of IMC (Forastero), ICS, and CC (Trinitario). The other major group combines all the other populations of Forastero, including the wild clones of Guyana, which seem closely related to the MO and PA populations of Peru. In this analysis, the population of Scavina is close to that of Forastero of Ecuador LCTEEN.

This analysis, despite the small number of loci analysed, thus allows us to reveal an overall structuration corresponding to the different genetic groups, and a geographic structuration of populations within certain groups.

Molecular Diversity

THE LEVEL OF HETEROZYGOSITY OF CLONES

The percentage of heterozygous loci in 300 clones has been evaluated using RFLP after hybridization of 33 genomic probes and cDNA. These probes have been mapped on a reference map (Lanaud et al., 1995), which can be used to find the genetic determinism of the markers used.

The clones that have the highest rate of heterozygosity are the Trinitario (86% for the UIT clones). High heterozygosity is also observed for certain modern varieties of Criollo (73% for CHO42), as well as in certain Forastero of the upper Amazonian region, such as IMC105 (42%). The lowest levels of heterozygosity are found in the old varieties of Criollo (Porcelana, 3%) or lower-Amazonian Forastero (Pará, 4%). Some Forastero of Guyana (GU154), Venezuela (VENC20), Colombia (EBC5), Peru (P2), and the lower Amazon (Catongo) are totally homozygous.

MOLECULAR DIVERSITY REVEALED BY RFLP

The molecular diversity of cacao populations, revealed by RFLP at the nuclear or cytoplasmic level, was evaluated by several authors (Laurent et al., 1993a, b, 1994; Figueira et al., 1994; Lerceteau et al., 1997; Motamayor et al., 1997).

Nuclear diversity
Laurent et al. (1994) analysed 201 genotypes belonging to different morphogeographic groups. The diversity of total nuclear DNA was analysed using 31 cDNA probes that enabled the identification of 87 polymorphic bands. Despite continuous variation between the groups, due particularly to the large number of Trinitario hybrids, a fairly clear structuration appeared on axis 1 of a CA (Fig. 3) between Forastero on the one hand, and modern

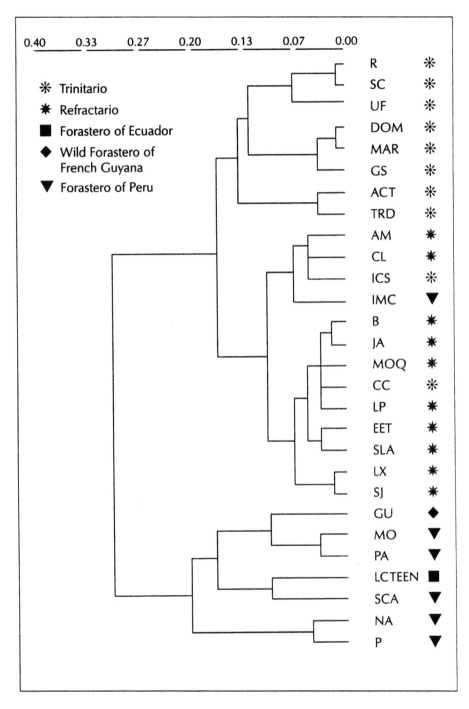

Fig. 2. Tree constructed by the neighbour joining method from non-biased Nei distances of various populations analysed with isozymes (Sounigo et al., 1996)

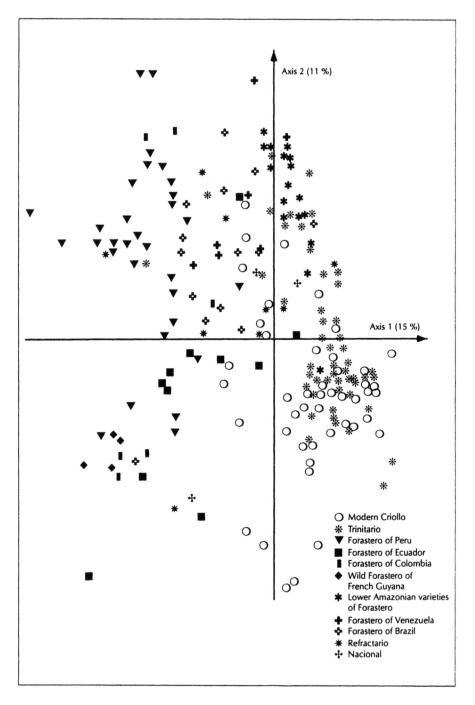

Fig. 3. Primary CA plane of 201 individuals studied for their nuclear diversity using 31 cDNA probes (Laurent et al., 1994).

Criollo and Trinitario on the other. The upper-Amazonian populations of Forastero and the modern varieties of Criollo have high variability. The cultivated varieties of Criollo are superposed on the pool of Trinitario hybrids.

In this analysis, a marked differentiation appears between the wild Forastero of Guyana on the one hand and the lower-Amazonian or Venezuelan Forastero on the other. The Refractario are found on the same area of distribution as the Forastero. The two clones of Nacional are close to certain Refractario.

A second RFLP study was focused on the diversity of the forms of Criollo cultivated in Central America and Venezuela (Motamayor et al., 2001). It covered 208 genotypes, which came from collections from the oldest plantations of Venezuela, without taking into account agronomic criteria of production, from the Lacandona forest of Mexico, close to Mayan archaeological sites where subspontaneous cacao are found that are probably the descendants of cacao cultivated in ancient times by the Mayas, and from the Yucatan. These collections comprise representations of pure varieties of Criollo cultivated in ancient times, which present varied forms of pod: oval and smooth, like those of Porcelana, or highly verrucose, like those of Pentagona. The analysis of this material was complemented by that of Trinitario, Forastero, and modern varieties of Criollo in the collections in Venezuela, Mexico, and Costa Rica.

About 50 clones and 30 probes common to the study of Laurent et al. (1994) were then used to compare the results of the two types of study. The allelic frequencies of 26 mapped loci and the mean heterozygosity per locus for each group show a high level of polymorphism among the Forastero of the upper Amazon and the Trinitario. The largest number of alleles per locus (2.2) is observed in the groups of Trinitario and upper-Amazonian Forastero, and the highest Nei genetic diversity value (0.41) is found in the Trinitario group. Similar values of mean genetic diversity (0.37) were found among the upper-Amazonian Forastero studied and the group of clones of modern Criollo. This diversity of the modern Criollo has also been indicated by other authors (Engels, 1986; Figueira et al., 1994; Laurent et al., 1994; Lerceteau et al., 1997).

On the other hand, the Criollo corresponding to the ancient cultivated types are poorly polymorphic and show the lowest values of genetic diversity (0.02). Similarly, they have a low level of mean heterozygosity per locus (0.02), contrary to the modern varieties of Criollo (0.43), which, with the Trinitario (0.40), present the highest values. Moreover, within these ancient varieties, almost no genotypic difference was found between morphological types that are quite different, such as Porcelana, Pentagona, and Guasare of Venezuela or the Criollo of the Lacandona forest in Mexico.

A CA from 64 polymorphic RFLP bands revealed a clear differentiation between the Forastero and the old varieties of Criollo (Fig. 4). The Trinitario are superposed on the modern varieties of Criollo as in the preceding analyses.

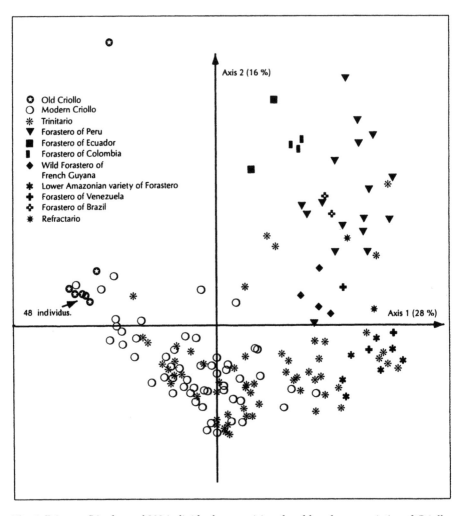

Fig. 4. Primary CA plane of 208 individuals comprising the old and new varieties of Criollo, studied for their nuclear diversity using 26 nuclear RFLP probes (Motamayor et al., 1997).

The modern Criollo clones in the collections appear generally more heterozygous than the ancient varieties, which can be explained by a selection of clones based not only on characters of bean quality, but also on certain agronomic characters of vigour, production, or disease resistance. The more vigorous types could correspond to the more or less hybridized forms resulting from introgressions of Forastero genes in the pure Criollo types cultivated in ancient times. The diversity observed among the modern varieties of Criollo could be explained by introgressions of Forastero genes— more or less significant and involving different portions of the genome—and by the diversity of Forastero at the origin of these introgressions.

In a study by Laurent et al. (1994), the diversity revealed on the first CA plane showed a differentiation between Forastero, on the one hand, and Criollo and Trinitario on the other hand, along axis 1. However, the variability appeared identical for these two groups along axis 2, which could suggest that the variability of Forastero for the characters of axis 2 may be the source of the diversity found within the modern Criollo and Trinitario. It appears that this second hypothesis can be ruled out. Indeed, in a study by Motamayor et al. (1997), for which supplementary probes were used, the diversity revealed by the first CA plane was structured differently. The diversity of Forastero could not explain that of the modern Criollo and the Trinitario, and the majority of clones of these two latter groups range from the ancient Criollo to the lower-Amazonian Forastero varieties. Moreover, the highly homozygous nature of the ancient varieties of Criollo, indicated by RFLP and microsatellites over the entire genome (Lanaud et al., 1995; Risterucci et al., 2000), allows us to make use of a reference genotype for all the chromosomes characterizing the pure Criollo. The observation of about 137 Trinitario, 10 modern varieties of Criollo, and 8 Forastero from Lower Amazonia using 16 highly polymorphic microsatellite markers seems to indicate that the same genotype or the same homogeneous population of lower-Amazonian Forastero is the source of the majority of Trinitario. For all the loci, it is always the same Forastero allele, different from the allele present in the ancient Criollo, that is found in most of the Trinitario, while the microsatellites reveal great allelic diversity in the Forastero group (Motamayor, 2001). Thus, according to these results, the different combinations of parts of the Forastero genome introgressed in the pure Criollo seem to be the source of the variability observed within Trinitario and the modern varieties of Criollo. The results of Laurent et al. (1994), which show an identical variability between the groups revealed by axis 2 of the CA, are undoubtedly linked to the low polymorphism revealed by the RFLP markers, which for the most part have only two possible alleles, common to all the groups. The use of a larger number of RFLP markers and more polymorphic markers, such as microsatellites that have revealed numerous alleles, enables us to identify the probable genetic origins of the pool of modern Criollo and Trinitario. These results thus indicate a very similar genetic structure between what is called the Criollo, which includes a majority of modern hybrid Criollo, and the Trinitario.

Lerceteau et al. (1997) analysed the genetic diversity of 59 Nacional clones from Ecuador, as well as 29 Forastero, 29 Trinitario, and 9 modern varieties of Criollo. Forty-three genomic probes, coded in terms of locus, were used. The intragroup genetic diversity was identical for Forastero (0.33), Trinitario (0.31), and the modern Criollo (0.31). It was lowest in Nacional (0.19). The heterozygosity was highest in Trinitario and modern Criollo, and lowest for Nacional. Among the latter, certain genotypes sampled in the very old plantations of Ecuador seem almost totally homozygous.

Hierarchical clustering was done using modified Rogers distances, calculated either between morphogeographic groups or between the geographic origins of clones alone. The discrimination between groups was best when the clones were classified as a function of their geographic origin (G_{ST} = 0.23 against 0.16). Considering the morphogeographic groups, the Criollo and Trinitario are the closest groups, while the Nacional are the most distant from all the other groups. These results confirm the genetic specificity of Nacional and their marked differentiation in relation to the Criollo.

The percentage of heterozygous loci of clones was estimated on the basis of 31 RFLP probes. Among the Nacional clones, two groups of individuals can be identified, BCH and SA, which have a very high rate of homozygosity varying from 90% to 100% and contain the same alleles. These trees come from two plantations that date from about 100 years ago, located about 450 km apart in the northern and southern parts of Ecuador. The present Nacional variety thus constitutes a highly hybrid population and, just like the genesis of modern Criollo and Trinitario, the BCH and SA clones, which are highly homozygous, could represent the ancestors of this population.

The diversity of cytoplasmic DNA
Yeoh et al. (1990), in a study on the structure of the chloroplast genome of cacao, pointed out the small size of this genome, which is of the order of 100×10^3 bases. The cytoplasmic diversity of 177 genotypes was analysed at the mitochondrial and chloroplast level by Laurent et al. (1993b) using mitochondrial heterologous probes (ATP-synthetase of sunflower, cytochrome oxidase of wheat) and a chloroplast heterologous probe (Rubisco of spinach).

Two chloroplast profiles were observed, A and B. Seventy per cent of clones have the profile A and belong to all the morphogeographic groups. Clones having a profile B are mostly modern Criollo, Trinitario, and some Forastero of the lower Amazon, Colombia, and Peru (Scavina).

The mitochondrial probes enabled the detection of 44 mitochondrial profiles, of which 35 contain 5 clones or fewer each. These highly variable and minor types are essentially made up of Criollo and Trinitario. Among the 9 remaining types, type 1 includes two thirds of Forastero types, including genotypes of Guyana, Venezuela, Brazil, Peru, Colombia, and Ecuador. The other major type, type 2, combines a majority of Criollo and Trinitario (26 clones) and some lower-Amazonian genotypes. By means of a CA, the diversity of mitochondrial DNA of clones studied can be visualized (Fig. 5). It is interesting to observe that, unlike with the nuclear DNA, the diversity of mitochondrial DNA is much greater in modern varieties of Criollo than in Forastero, which seem poorly variable. Recently, this mitochondrial polymorphism was indicated also among the pool of old Criollo of Mexico and Venezuela, which are also nearly totally homozygous at the nuclear level.

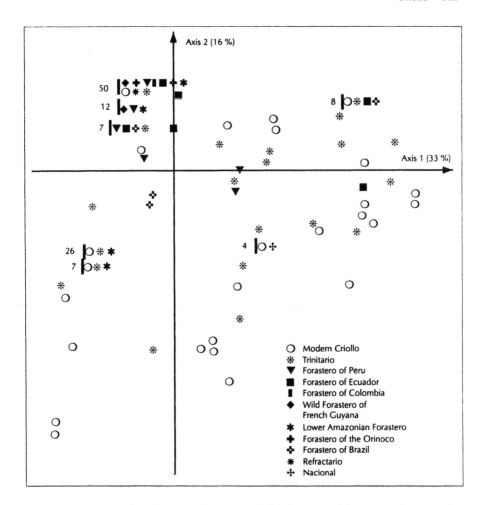

Fig. 5. Diversity of mitochondrial DNA of 177 individuals belonging to different morphogeographic groups and represented on the first plane of a CA (Laurent et al., 1993b).

MOLECULAR DIVERSITY REVEALED BY RAPD

Easy access to RAPD markers has allowed several researchers to use them to identify or study cacao diversity (Wilde et al., 1992; Russel et al., 1993; Figueira et al., 1994; N'goran et al., 1994; Ronning and Schnell, 1994; de la Cruz et al., 1995; Ronning et al., 1995; Sounigo et al., 1996; Lerceteau et al., 1997; Whitkus et al., 1998).

Figueira et al. (1994) analysed the diversity of genotypes belonging to three groups—Criollo, Trinitario, and Forastero—using 128 RAPD markers corresponding to amplified fragments taken from 23 primers. A continuous variation was observed between these groups, and they could not be differentiated from each other. On the other hand, a fairly clear structuration was observed between wild and cultivated clones. That structuration is confirmed by analysis of ribosomal DNA. Considering these results, the authors proposed a new classification of cacao based not only on the three traditional groups, Criollo, Trinitario, and Forastero, but on the groups of wild cacao and cultivated cacao.

At the same time, Russel et al. (1993) studied the diversity of 25 clones resulting from three Forastero populations of Peru and Ecuador using 9 RAPD primers generating 75 bands. Despite the reduced sample, the Shannon indexes of diversity reveal greater diversity within populations than between populations. Multivariant analyses, CA and HC, reveal a genetic differentiation among the three populations of upper-Amazonian Forastero in relation with their geographic origin.

N'goran et al. (1994) analysed the genetic diversity of 106 genotypes that belong to different morphogeographic groups including the modern varieties of Criollo, using 19 primers. After thoroughly screening the bands, 49 polymorphic and reproducible bands were retained for the analyses. Out of 36 hybridized bands of DNA, 12 corresponded to highly repeated and dispersed sequences, 12 to unique sequences, and 12 to sequences that were repeated a few times. The HC, established from factorial coordinates of a CA (Fig. 6), shows a clear structuration on the first plane between Forastero and modern Criollo, as well as a clear differentiation between upper-Amazonian Forastero and lower-Amazonian Forastero (N'goran et al., 1994). If only those bands corresponding to unique sequences are considered, no clear diversity structuration appears. On the other hand, a structuration appears between modern Criollo and Forastero when only those bands corresponding to highly repeated sequences are considered, while they do not differentiate the lower-Amazonian Forastero from the upper-Amazonian Forastero.

Lerceteau et al. (1997) studied the diversity of 155 clones belonging to the groups Nacional, Forastero, Trinitario, and modern Criollo, using 40 RAPD bands resulting from 18 primers. The diversity was structured using a principal components analysis (PCA). Axis 2 indicates essentially the specificity of wild Forastero of Guyana, while the 1-3 plane reveals the structuration of other populations. Despite the continuous variations from one group to another due to intense genetic exchange, a clear differentiation is observed between Nacional and Criollo. The indication of this structuration as well as the information supplied by the RFLP on these same clones suggest that the highly homozygous Nacional clones such as BCH and SA, from very old plantations, could represent part of the original pool of Nacional. These clones were subsequently widely hybridized with other introduced clones. The high homozygosity could be a common character of pure varieties of Nacional.

Sounigo et al. (1996) used RAPD to analyse the diversity of 149 clones belonging to the groups Forastero, Trinitario, and Refractario. The calculation of the Shannon index of diversity shows that the most variable populations

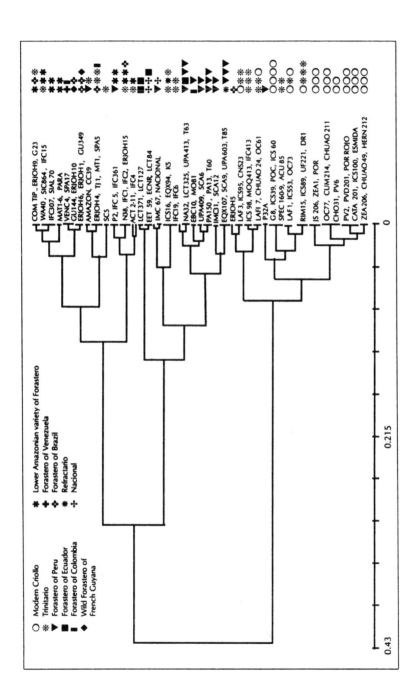

Fig. 6. Hierarchical clustering constructed by the neighbour joining method from coordinates of individuals of the seven first axes of the CA (N'goran et al., 1994). The diversity of individuals has been evaluated using 49 polymorphic RFLP bands.

are LCTEEN (Ecuador Forastero, 0.85), IMC (Peru Forastero, 0.72), and CL (Refractario, 0.71). The least variable are the populations B, JA (Refractario, 0.29 and 0.47), NA (Peru Forastero, 0.42), and GU (Guyana Forastero, 0.50). An additive tree constructed from Rogers distances shows a structuration of the populations into three groups. The first includes the three populations of Peru, NA, IMC, and SCA. The second combines the populations of Forastero, PA and MO of Peru, LCTEEN of Ecuador, and GU of Guyana. The third group contains the populations of Refractario and Trinitario.

In this RAPD analysis, the populations of Forastero are grouped differently from what is observed with the enzymatic markers (Sounigo et al., 1996). In particular, the IMC population is close to other populations of Forastero, while it is associated with Trinitario in enzymatic analysis. The SCA population appears closer to the IMC and the NA of Peru, while with isozymes it is closer to the population of the Ecuador LCTEEN. The population of wild Guyana cacao seems most distant from all the other populations (Fig. 7), while with the isozymes it is quite close to the MO and NA populations.

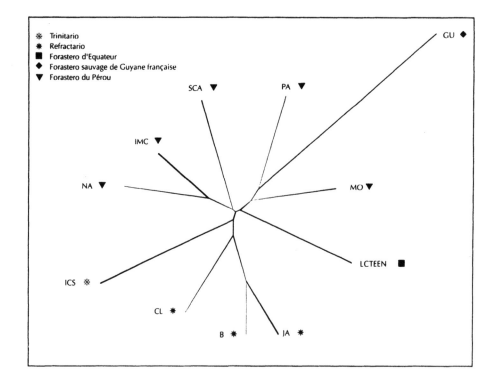

Fig. 7. Tree constructed by the neighbour joining method from Rogers distances obtained on RAPD data of Sounigo et al. (1996).

De la Cruz et al. (1995) used RAPD to analyse 42 genotypes corresponding to wild plants from either Mexico (forested islands of the Yucatan and Chiapas) or the upper Amazon, to cultivars of Criollo, Forastero, and Trinitario, and to a representative of a related genus, *Herrania*. From the calculations of dissimilarity indexes, a phylogenic tree was constructed according to the method of closest neighbours. The tree shows a greater similarity between the cultivated Criollo and the wild cacao of South America than between the cultivated Criollo and the wild plants of Mexico. The authors suggest that the wild trees found in the ancient sacred forests of the Mayans do not exist in the present collections and that they could be the closest representatives of the ancient Mayan cultivars.

In a later study, Whitkus et al. (1998) used RAPD to analyse a sample of 86 individuals including, in particular, 26 wild genotypes collected in the Lacandona forest in the state of Chiapas and 5 individuals from the forested islands of Yucatan, in Mexico, as well as the wild and cultivated clones of Central America. Unlike the populations and cultivars of Central America, where the intrapopulation diversity is always higher than the interpopulation diversity, the diversity found between the two Mexican populations is higher than that within each region. This situation reflects the low level of polymorphism found in the two populations. A structuration of the diversity similar to that of de la Cruz et al. (1995) has been observed. The two populations of Mexico seem well differentiated and original. These results conform to the hypothesis of Cuatrecasas (1964) about the natural distribution of cacao in Central America. The authors emphasize nevertheless the lack of affinity between the cultivated Criollo and the wild plants observed in Mexico. According to them, the cacao collected in the Lacandona forest could represent the wild cacao. The plants sampled in the island forests cultivated in ancient times by the Mayans could on the other hand have been introduced from wild populations. In this case, they represent a subsample of populations present in the Lacandona forest, which would explain the difference between the two Mexican populations. They also would correspond to the material closest to the cacao cultivated by the Mayans.

Our own analyses, done simultaneously on the population of the Lacandona forest and on that of the Yucatan, indicate a perfect genetic similarity between these two populations, unlike the analyses of Whitkus et al. (1998). This contradiction perhaps originates in the fact that the analyses of Whitkus et al. (1998) possibly included young wild plants, morphologically very similar to those of *T. cacao*, but not belonging to that species. These plants, observed in the Yucatan, could explain the genetic distances obtained by Whitkus et al. (1998) between cacao from the two places.

CONCLUSION

The intense genetic exchange that has occurred during the past three centuries between the wild and cultivated populations of cacao often makes them difficult to classify.

Morphological descriptors can be used to differentiate the populations of different geographic origin and to structure their diversity. However, they sometimes give a biased picture of the diversity of the genetic resources. For example, our analyses suggested that the ancient varieties of Criollo, highly varied as to their morphology, ranging from the Porcelana type with smooth pods to the Pentagona type with particularly verrucose pods, are genetically highly homogeneous and highly homozygous. Human selection could in this case have contributed to the fixation and conservation of very different morphological types, resulting in occasional mutations.

Genetic markers, which reflect the structure of the entire genome, are the tools of choice to reveal the diversity and relationships between wild and cultivated cacao. Different types of markers, nuclear (RAPD, RFLP from genomic or cDNA probes, microsatellites) or cytoplasmic, contribute varying and complementary types of information, indicating the variation of sequences that evolve at different rates. Codominant markers such as RFLP and isozymes provide more precise information on the genetic structure of populations and individuals.

From identification of alleles present in the genotypes as well as the gradients observed in the levels of heterozygosity of clones within hybrid cultivated populations such as Nacional of Ecuador or Criollo and Trinitario, we can understand the genesis of these populations and discover their most probable ancestors (Lerceteau et al., 1997; Motamayor et al., 1997). In these two cases, the ancestors probably constitute a very restricted genetic base.

A wide genetic diversity has been indicated between the wild populations of Forastero and within some of these populations originating in the upper Amazon, particularly Ecuador. A continuous variation can be observed among them. However, in several analyses, the Forastero of Guyana seem differentiated from other types of Forastero, including those of the lower Amazon or of Orinoco, which they are closest to geographically. A differentiation between Forastero and Criollo has been observed with all the markers, with a more marked differentiation between Forastero and ancient Criollo revealed by nuclear RFLP. Analysis of Nacional has also shown their originality with respect to Forastero and Criollo simultaneously. Nacional is clearly distinct from these two groups.

Thus, the diversity of the species could be structured diagrammatically around four unequal groups: ancient Criollo, pure varieties of Nacional, wild Forastero of Guyana, and a wider pool comprising the Amazonian Forastero (of the upper and lower Amazon) and those of the Orinoco, even though this pool comprises populations that are differentiated from one another according to their geographic origin. This structuration differs from the usual classification of cacao into three groups: Forastero, Criollo, and Trinitario. On the one hand, the genetic structure of Trinitario is similar to that of the Criollo group, which in this classification includes a majority of modern Criollo. On the other hand, within the group called Forastero are included cacao that are well-differentiated, such as the wild populations of Guyana or those of the original variety Nacional.

The classification that we propose can be supported by the paleoclimatic history of the South American continent during the Quaternary era, during which considerable ecological upheavals occurred. The succession of periods of glaciation interrupted by phases of heating of the climate resulted in an alternation of dry and humid periods in the tropical regions. During the dry periods, the tropical forest shrank so as to be limited to the forest islands, favouring the differentiation of some populations and the disappearance of others. During the humid phases, the forests extended anew from these reserve zones. Such reserve zones were identified in Guyana, Brazil, Colombia, and Venezuela, as well as the upper Amazon, where a vast reserve zone is indicated. This phenomenon, recognized for numerous animal and plant species (Simpson and Haffer, 1978), could explain the diversity and differentiation of certain populations of cacao. Through molecular study on the sequence of certain genes, the analysis of relationships and differentiation between populations could be refined, and the time of their divergence could be established. It would thus be possible to situate the differentiation of populations in relation to major ecological and geographic events.

The importance of the sampling to the conclusions drawn from various analyses must be emphasized. Genetic resources have often been collected with criteria of agronomic interest, which could give a biased picture of the diversity actually existing in a given region: this is the case of Forastero collected in Peru with a specific objective, resistance to witches' broom disease. For the Criollo, the primary studies pertained only to the clones present in the collections and collected for their agronomic characters. The great diversity revealed within this group reflected in fact the diversity of introgressions from a unique ancestral type identified subsequently using markers and studies conducted directly on the oldest plantations or close to remains of Mayan civilizations, which were the first to have domesticated cacao. From a wider sample and better representation of certain populations the structuration between different origins of cacao that has appeared in these analyses could be corroborated.

These analyses confirmed that only a small part of the genetic diversity is exploited in selection, as several authors have already mentioned (Bartley, 1979; Lockwood and End, 1992). The hybrid forms between Criollo and Forastero, which correspond most often to traditional forms of Trinitario, implied a very narrow genetic base in each of the two groups of origin. Other types of hybrids between Criollo and Forastero, exploiting all the genetic richness of Forastero, could thus be used in selection. Moreover, only a few meioses separate the ancient Criollo from the formation of hybrid types, and the linkage disequilibrium between characters and markers could have been maintained. This situation would facilitate the exploitation of genetic resources of the Criollo and Trinitario type, using markers that frame the useful genes as markers of early selection. Similarly, for the Forastero, only a very small part of the genetic diversity has actually been included in the selection programmes, and a small number of Peru Forastero, often having strong parental links, have been widely diffused and integrated in selection programmes of all the producer countries.

Even though the existing diversity in the collections is far from having been fully exploited, the extent of diversity revealed on a sometimes small sample of certain populations testifies in favour of the pursuit of collection expeditions in regions rich in diversity to preserve the genetic resources of cacao.

APPENDIX

Plant Material

The enzymatic analyses of Lanaud (1987) were done on 12 modern Criollo, 17 Trinitario selected from America, 22 Trinitario selected from Africa, 64 lower-Amazonian Forastero selected from Africa, 19 Forastero collected by Lanaud (1986) along the Orinoco in Venezuela (VENC), 92 wild Forastero collected by Sallee (1987) in French Guyana, 19 cultivated cacao collected by Clément (1986) in French Guyana, 74 Forastero collected by Pound (1938, 1943) in Peru (39 individuals of GO, 8 IMC, 11 P, 7 NA, 8 PA, and 1 SCA), 10 Refractario collected in Ecuador by Chalmers (1968) (6 individuals of EQX) and Pound (1938) (4 individuals of MOQ), 7 Forastero collected during the Colombian expedition of Ocampo (1985) in Colombia (EBC), and 18 Forastero collected by J.B. Allen between 1980 and 1985 in Ecuador (LCTEEN).

The enzymatic analyses of Sounigo et al. (1996) were done on 482 clones of cacao belonging to 28 populations or groups of accessions present in the Trinidad collection: 6 populations of Forastero collected in Peru (IMC, MO, PA, NA, P, SCA: 112 individuals), 1 population of Forastero from Ecuador (LCTEEN: 16 individuals), 1 population of Forastero collected in French Guyana (16 individuals), 9 populations corresponding to Refractario (AM, B, CL, JA, LP, LX, MOQ, SJ, SLA: 245 individuals), and 10 populations of Trinitario (ACT, CC, DOM, ICS, GS, MAR, R, SC, TRD, UF: 114 individuals).

The RFLP analyses (Laurent et al., 1993b, 1994) were done on 201 genotypes belonging to different morphogeographic groups: 45 modern Criollo (ICS, CATA, BOC, CHUAO, OC, POR, JS, MT, PV, ZEA, LAF), 20 Forastero of Peru (P, PA, NA, MO, IMC, SCA), 12 Forastero selected from Africa from clones of Peru (T, UPA), 11 Forastero of Ecuador (LCTEEN), 6 Forastero of Colombia (EBC, SPA), 7 Forastero of the Orinoco valley in Venezuela (VENC), 4 wild Forastero of French Guyana (GU), 22 Forastero of Brazil (ERJOH, Comun, Para, SIAL, SIC, MAT, Catongo), 9 lower-Amazonian Forastero selected from Africa (Amelonado) (IFC, SF), 37 Trinitario selected from Central and South America and from the Caribbean (ACT, ICS, GS, CHUAO, CNS, WA, RIM, MT, TJ, SC, SGU, CC, UF), 4 Trinitario selected from Asia (WA, LAFI, DR, G), 12 Trinitario selected from Africa (SNK, IFC, N, W, ACU, K), 10 Refractario (EQX, MOQ), and 2 Nacional.

Enzymatic Analyses

The methods used as well as the genetic determination and protocols of extraction and indication are given in Lanaud (1986a). In the analyses of Lanaud (1987), 6 polymorphic enzymatic systems (PGI, PGM, ADH, MDH, PAC, ICD), corresponding to 9 loci, were used. The analyses of Sounigo et al. (1996) used 5 polymorphic enzymatic systems (PGI, ADH, MDH, PAC, ICD), corresponding to 6 loci.

RFLP Analyses

A genomic library bank and a cDNA library were constructed. Between 31 and 50 of these probes, for the most part mapped (Lanaud et al., 1995), were used to study the nuclear diversity. The heterozygosity level of clones was found using mapped markers with known genetic determinism. Moreover, heterologous probes were employed to study the cytoplasmic diversity (mitochondrial and chloroplast). The protocols are described in Laurent et al. (1993a, b, 1994).

RAPD Analyses

After a thorough selection of primers from Operon kits that generate reproducible bands, 19 primers producing 49 polymorphic bands were used to analyse the diversity (N'goran et al., 1994). The nature of the RAPD bands observed—unique or repeated sequence—was tested by Southern hybridization on the total DNA of cacao restricted by restriction enzymes.

Data Analysis

Several genetic parameters of diversity were calculated on the enzymatic and molecular data: percentage of polymorphic loci, percentage of heterozygosity, mean number of alleles per locus, Shannon index of diversity based on genotype frequencies, and Nei diversity index (Nei, 1978) based on allelic frequencies.

From the Nei and Shannon indexes, the total diversity was broken down into intrapopulation and interpopulation diversity according to the following formulas:

$$H_{intrapop} = H_{pop}/H_{total}$$

$$H_{interpop} = (H_{total} + H_{pop})/H_{total}$$

For all the data, multivariate analyses were generally used—correspondence analysis (CA) (Benzecri, 1973) or principal components analyses (PCA) for quantitative characters.

For the enzymatic data, the genetic distances of Nei on the allelic frequencies were used to measure distances between populations. A hierarchical clustering (HC) was constructed according to the UPGMA method.

For the RAPD data (N'goran et al., 1994), a HC was constructed from the coordinates of clones on the seven primary axes of the CA.

For the RAPD data of Sounigo et al. (1996), the Rogers and Tanimoto distances were calculated, and a tree was constructed according to the neighbour joining method.

REFERENCES

Allen, J.B. and Lass, R.A. 1983. London cocoa trade Amazon project: final report, phase 1. Cocoa Grower's Bulletin, 34: 1-71.

Amefia, Y.K. 1986. Sur le polymorphisme enzymatique des cacaoyers introduits au Togo. Doct. thesis, INA, Paris- Grignon, 146 p.

Atkinson, M.D., Withers, L.A., and Simpson, M.J.A. 1986. Characterisation of cacao germplasm using isoenzyme markers. 1. A preliminary survey of diversity using starch gel electrophoresis and standardisation of the procedure. *Euphytica*, 35: 741-750.

Barriga, J.P., Machado, P.R.F., de Almeida, C.M.V.C. and de Almeida, C.F.G. 1984. A preservação e utilização dos recursos genéticos de cacau na Amazônia brasileira. In: *IXe Conférence Internationale sur la Recherche Cacaoyère*. Lagos, Nigeria, Cocoa Producers' Alliance, pp. 73-79.

Bartley, B.G. 1979. Global concepts for genetic resources and breeding in cacao. In: *VIIe Conférence Internationale sur la Recherche Cacaoyère*. Lagos, Nigeria, Cocoa Producers' Alliance, pp. 519-525.

Bekele, F.L. 1992. Use of botanical descriptors for cocoa characterization: CRU experiences. In: *International Workshop on Conservation, Characterisation and Utilisation of Cocoa Genetic Resources in the 21st Century*. Port of Spain, Trinidad and Tobago, CRU, pp. 77-91.

Bekele, F. and Bekele I. 1998. A sampling of the phenetic diversity of cacao in the international cocoa gene bank of Trinidad. *Crop Science*, 36: 57-64.

Bekele, F.L., Kennedy, A.J., McDavid, C., Lauckner, B., and Bekele, I. 1994. Numerical taxonomic studies on cacao (*Theobroma cacao* L.) in Trinidad. *Euphytica*, 75: 231-240.

Benzecri, J.P. 1973. L'Analyse des Données. 2. L'Analyse des Correspondances. Paris, Dunod, 616 p.

Bergman, J.F. 1969. The distribution of cacao cultivation in pre-Columbian America. *Annals of the Association of American Geographers*, 59: 85-96.

Bernoulli, G. 1869. Ubersicht der bis jetzt bekannten Arten von *Theobroma*. *Neue Denkschriften der allgemeinen schweizerischen Gesellschaft für die gesammten Naturwissenschaften*, 24: 1-15.

Bouharmont, J. 1960. Recherches cytologiques sur la frutification et l'incompatibilitié chez *Theobroma cacao* L. INEAC, Série scientifique no. 89, 117 p.

Braudeau, J. 1969. *Le Cacaoyer*. Paris, Maisonneuve et Larose, 304 p.

Burle, L. 1952. La production du cacao en Afrique occidentale française. Centre de recherches agronomiques de Bingerville, Bulletin no. 5, pp. 3-21.

Chalmers, W.S. 1968. Cacao collecting trip to the Oriente of Ecuador (November 1968). Port of Spain, Trinidad and Tobago, CRU, 4 p.

Cheesman, E.E. 1944. Notes on the nomenclature, classification and possible relationships of cacao populations. *Tropical Agriculture*, 21: 144-159.

Chevalier, A. 1946. Révision du genre *Theobroma* d'après l'herbier du Muséum national d'histoire naturelle de Paris. *Revue de Botanique Appliquée*, 26: 265-285.

Clement, D. 1986. Cacaoyers de Guyane: prospections. Café, Cacao, Thé, 30: 11-35.

Coe, S.D. and Coe., M.D. 1996. The True History of Chocolate. London, Thames and Hudson, 280 p.

Cope, F.W. 1962. The mechanism of pollen incompatibility in *Theobroma cacao*. *Heredity*, 17: 157-182.

Cuatrecasas, J. 1964. Cacao and its allies: a taxonomic revision of the genus *Theobroma*. *Contributions from the United States Herbarium*, 35: 379-614.

de la Cruz, M., Whitkus, R., Gomez-Pompa, A., and Mota-Bravo, L. 1995. Origins of cacao cultivation. *Nature*, 375: 542-543.

Engels, J.M.M. 1983a. A systematic description of cacao clones. 1. The discriminative value of quantitative characteristics. *Euphytica*, 32: 377-385.

Engels, J.M.M. 1983b. A systematic description of cacao clones. 2. The discriminative value of qualitative characteristics and the practical compatibility of the discriminative value of the quantitative and qualitative descriptors. *Euphytica*, 32: 719-733.

Engels, J.M.M. 1986. The systematic description of cacao clones and its significance for taxonomy and plant breeding. PhD thesis, Agricultural University, Wageningen, Netherlands, 125 p.

Engels, J.M.M., Bartley, B.G.D., and Enriquez, G.A. 1980. Cacao descriptors, their states and modus operandi. *Turrialba*, 30: 209-218.

Enriquez, G.A. 1985. Cursos sobre el cultivo del cacao. Turrialba, Costa Rica, CATIE, Serie Materiales de Enseñanza, no. 22, 239 p.

Enriquez, G.A. 1992. Characterisitcs of cacao Nacional of Ecuador. In: International workshop on conservation, characterisation and utilisation of cocoa genetic resources in the 21st century. Port of Spain, Trinidad and Tobago, CRU, pp. 269-278.

Enriquez, G.A. and Soria, J. 1968. The variability of certain bean characteristics of cacao (*Theobroma cacao* L.). *Euphytica*, 17: 114-120.

Figueira, A., Janick, J., and Goldsbrough, P. 1992. Genome size and DNA polymorphism in *Theobroma cacao*. *Journal of the American Society for Horticultural Science*, 117: 673-677.

Figueira, A., Janick, J., Levy, M., and Goldsbrough, P. 1994. Re-examining the classification of Theobroma cacao L. using molecular markers. *Journal of the American Society for Horticultural Science*, 119: 1073-1082.

Glendinning, D.R. 1966. The incompatibility alleles of cocoa: report for the period 1st October 1963–31st March 1965. Ghana, Cocoa Research Institute.

Guisan, 1825. De la culture du cacaoyer. In: *Traité sur les Terres Noyées de Guyane*, 2nd ed. Imprimerie du Roy, pp. 170-189.

Hamrick, J.L. and Godt, M.J.W. 1990. Allozyme variation in plant species. In: *Plant Population Genetics, Breeding, and Genetic Resources*. A.H.D. Brown et al. eds., Sunderland, USA, Sinauer, pp. 43-63.

Hamrick, J.L., Godt, M.J.W., and Sherman-Broyles, S.L. 1992. Factors influencing levels of genetic diversity in woody plant species. *New Forests*, 6: 95-124.

IBPGR. 1981. Report of the IBPGR working group on genetic resources of cocoa. Rome, IBPGR, ACP-IBPGR/80/56, 28 p.

Knight, R. and Rogers, H. 1955. Incompatibility in *Theobroma cacao*. *Heredity*, 9: 67-69.

Lachenaud, P. and Sallee, S. 1993. Les cacaoyers spontanés de Guyane: localisation, écologie et morphologie. *Café, Cacao, Thé*, 37: 101-114.

Lanaud, C. 1986a. Utilisation des marqueurs enzymatiques pour l'étude génétique du cacaoyer *Theobroma cacao* L. 1. Contrôle génétique et linkage de neuf marqueurs enzymatiques. *Café, Cacao, Thé*, 30: 259-270.

Lanaud, C. 1986b. Utilisation des marqueurs enzymatiques pour l'étude génétique du cacaoyer *Theobroma cacao* L. 2. Etude du polymorphisme de six systèmes enzymatiques. *Café, Cacao, Thé*, 30: 271-280.

Lanaud, C. 1987. Nouvelles données sur la biologie du cacaoyer (*Theobroma cacao* L.): diversité des populations, systèmes d'incompatibilité, haploïdes spontanés; leurs conséquences pour l'amélioration génétique de cette espèce. Doct. thesis, Université Paris XI, Orsay, 106 p.

Lanaud, C. and Bertaud, J. 1984. Mise en évidence de nouveaux marqueurs génétiques chez *Theobroma cacao* L. par les techniques d'électrophorèse. In: IXe conférence internationale sur la recherche cacaoyère. Lagos, Nigeria, Cocoa Producers' Alliance, pp. 249-253.

Lanaud, C., Hamon, P., and Duperray, C. 1992. Estimation of nuclear DNA content of *Theobroma cacao* L. by flow cytometry. *Café, Cacao, Thé*, 36: 3-8.

Lanaud, C., Risterucci, A.M., N'goran, A.K.J., Clement, D., Flament, M.H., Laurent, V., and Falque, M. 1995. A genetic linkage map of *Theobroma cacao* L. *Theoretical and Applied Genetics*, 91: 987-993.

Laurent, V., Risterucci, A.M., and Lanaud, C. 1993a. Variability for nuclear ribosomal genes within *Theobroma cacao*. *Heredity*, 71: 96-103.

Laurent, V., Risterucci, A.M., and Lanaud, C. 1993b. Chloroplast and mitochondrial DNA diversity in *Theobroma cacao*. *Theoretical and Applied Genetics*, 87: 81-88.

Laurent, V., Risterucci, A.M., and Lanaud, C. 1994. Genetic diversity in cacao revealed by cDNA probes. *Theoretical and Applied Genetics*, 88: 193-198.

Lecomte, H. and Challot, C. 1897. Le Cacaoyer et Sa Culture. Paris, Carré et Naud, 121 p.

Lerceteau, E., Robert, T., Petiard, V., and Crouzillat, D. 1997. Evaluation of the extent of genetic variability among *Theobroma cacao* accessions using RAPD and RFLP markers. *Theoretical and Applied Genetics*, 95: 10-19.

Lockwood, G. and End, M. 1992. History, technique and future needs for cacao collection. In: International workshop on conservation, characterisation and utilisation of cocoa genetic resources in the 21st century. Port of Spain, Trinidad and Tobago, CRU, pp. 1-14.

Morris, D. 1882. *Cacao: How to Grow and How to Cure It*. Jamaica, pp. 1-45.

Motamayor, J.C. 2001. Etude de la diversite génétique et de la domestication des cacaoyers du groupe Criollo (*T. cacao* L.) a l'aide de marqueurs moléculaires. Doct. thesis, University of Paris XI, Orsay, 18 July 2001.

Motamayor, J.C., Risterucci, A.M., Laurent, V., Moreno, A., and Lanaud, C. 1997. The genetic diversity of Criollo cacao and the consequences for quality breeding. In: *I° Congreso Venezolano del Cacao y Su Industria*. Aragua, Venezuela.

Motamayor, J.C., Risterucci, A.M., and Lanaud, C. 2001. Cacao domestication I: genetic evidence of a South American origin of cacao cultivated by the Mayans. *Heredity*.

Myers, J.G. 1930. Notes on wild cacao in Surinam and in British Guiana. *Kew Bulletin*, 1: 1-10.

Nei, M. 1978. Estimation of average heterozygosity and genetic distance from a small number of individuals. *Genetics*, 89: 583-690.

N'goran, J.A.K. 1994. Contribution à l'étude génétique du cacaoyer par les marqueurs moléculaires: diversité génétique et recherche de QTLs. Doct. thesis, Université Montpellier II, Montpellier, France, 131 p.

N'goran, J.A.K., Laurent, V., Risterucci, A.M., and Lanaud, C. 1994. Comparative genetic diversity of *Theobroma cacao* L. using RFLP and RAPD markers. *Heredity*, 73: 589-597.

Ocampo, R.F. 1985. Informe sobre recolección de germoplasma de *Theobroma cacao* L. *El Cacaotero Colombiano*, 31: 24-29.

Paradis, L. 1979. Le cacao précolombien: monnaie d'échange et breuvage des dieux. *Journal d'agriculture traditionnelle et de botanique appliquée*, 26: 3-4.

Pittier, H. 1930. A propos des cacaoyers spontanés. *Revue de botanique appliquée*, 10: 777.

Pittier, H. 1933. Degeneration of cacao through natural hybridization. *Journal of Heredity*, 36: 385-390.

Pound, F.J. 1938. Cocoa and witches' broom disease (*Marasmius perniciosus*) of South America with notes on other species of *Theobroma*. Port of Spain, Trinidad and Tobago, Yuille's Printery.

Pound, F.J. 1943. Cocoa and witches' broom disease: report on a recent visit to the Amazon territory of Peru. In: *The Archives of Cocoa Research*, vol. 1. H. Toxopeus, eds., London, ACRI-IOCC.

Rabouin, L.M. and Paulin, D. 1993. Etude de l'héritabilité de descripteurs floraux. In: XIe Conférence Internationale sur la Recherche Cacaoyère. Lagos, Nigeria, Cocoa Producers' Alliance, pp. 449-450.

Reyes, H. 1992. Criollo cacao germplasm in Venezuela. In: International workshop on conservation, characterisation and utilisation of cocoa genetic resources in the 21st century. Port of Spain, Trinidad and Tobago, CRU.

Risterucci, A.M., Lanaud, C., N'goran, J.A.K., and Pieretti, I. 2000. A saturated linkage map of *Theobroma cacao*. In: XIIe Conférence Internationale sur la Recherche Cacaoyère. Lagos, Nigeria, Cocoa Producers' Alliance.

Ronning, C.M. and Schnell, R.J. 1994. Allozyme diversity in a germplasm collection of *Theobroma cacao* L. *Journal of Heredity*, 85: 291-295.

Ronning, C.M., Schnell, R.J., and Kuhn, D.N. 1995. Inheritance of random amplified polymorphic DNA (RAPD) markers in *Theobroma cacao* L. *Journal of the American Society for Horticultural Science*, 120: 681-686.

Russel, J.R., Hosein, F., Waugh, R., and Powell, W. 1993. Genetic differentiation of cocoa (*Theobroma cacao* L.) populations revealed by RAPD analysis. *Molecular Ecology*, 2: 89-97.

Sallee, B. 1987. Compte rendu de prospection dans les populations naturelles de *Theobroma cacao* L. du Haut Camopi (Guyane française). Kourou, Guyana, CIRAD, 45 p.

Schumann, K. 1886. Fam. Sterculiaceae g. *Theobroma*. In: *Flora Brasiliensis*, vol. XII, part III, pp. 70-77.

Simpson, B.B. and Haffer, J. 1978. Speciation patterns in the Amazonian forest biota. *Annual Review of Ecology and Systematics*, 9: 497-518.

Soria, J. 1970a. Principal varieties of cocoa cultivated in tropical America. *Cocoa Growers' Bulletin*, 19: 12-21.

Soria, J. 1970b. The present status and perspectives for cocoa cultivars in Latin America. Proceedings of the Tropical Region American Society for Horticultural Science, 14.

Soria, J. 1973. Survey of crop genetics resources in their centers of diversity: first report. Rome, FAO, pp. 119-125.

Soria, J. and Enriquez, G.A. 1981. International cacao cultivar catalogue. Turrialba, Costa Rica, CATIE, Technical Series, Technical Bulletin no. 6, 156 p.

Sounigo, O., Christopher, Y., and Umaharan, R. 1996. Genetic diversity assessment of *Theobroma cacao* L. using isoenzyme and RAPD analyses. In: Annual Report 1996, CRU. Port of Spain, Trinidad and Tobago, CRU, pp. 35-51.

Toxopeus, H. 1985. Botany, types and populations in cocoa. In: *Cocoa*, 4th ed. G.A.R. Wood and R.A. Lass, eds., London, Longman, pp. 11-37.

Wadsworth, R.M. and Harwood, T. 2000. International Cocoa Germplasm Database ICDG version 4.1. London International Financial Futures and Options Exchange and the University of Reading, UK.

Warren, J.M. 1994. Isozyme variation in a number of populations of *Theobroma cacao* L. obtained through various sampling regimes. *Euphytica*, 72: 121-126.

Whitkus, R., de la Cruz, M., and Mota-Bravo, L. 1998. Genetic diversity and relationships of cacao (*Theobroma cacao* L.) in southern Mexico. *Theoretical and Applied Genetics*, 96: 621-627.

Wilde, J., Waugh, R., and Powell, W. 1992. Genetic fingerprinting of *Theobroma* clones using randomly amplified polymorphic DNA markers. *Theoretical and Applied Genetics*, 83: 871-877.

Wood, G.A.R. 1991. A history of early cocoa introductions. *Cocoa Growers' Bulletin*, 44: 7-12.

Yeoh, H.H., Chung, D.K., and Fritz, P.J. 1990. *Theobroma cacao* chloroplast DNA: isolation, molecular cloning and characterization. *Café, Cacao, Thé*, 34: 173-178.

Yidana, J.A., Kennedy, A.J., and Withers, L.A. 1987. Variation in peroxidase isozymes of cocoa (*Theobroma cacao* L.). In: Xe Conférence Internationale sur la Recherche Cacaoyère. Lagos, Nigeria, Cocoa Producers' Alliance, pp. 719-723.

Young, A.M. 1994. *The Chocolate Tree: A Natural History of Cacao*. Washington, D.C., Smithsonian Institution Press, 200 p.

Cassava

Gérard Second, Jean-Pierre Raffaillac and
Carlos Colombo

Cassava is cultivated for its roots, which develop into tubers. It is part of the diet of the poorest countries of the tropics, particularly in Africa, which accounts for more than half of the world production (Fig. 1). Young cassava leaves are also consumed: they are a significant source of supplementary protein, vitamins, and minerals for the populations of Central Africa and northeastern Brazil that essentially survive on cassava tubers. The leaves are sometimes used as forage.

Cassava is cultivated most often in traditional agricultural systems, which rarely use improved cultivation practices. It is in the Asian countries, where cassava is a cash crop grown for export and industry, that improved varieties are adopted most readily by farmers. A typical case is that of Thailand, which in two decades has become among the major world producers along with Brazil, Nigeria, Zaire, and Indonesia.

Cassava is usually planted from a stem cutting of around 20 cm, but domestication has involved sexually reproduced plants using techniques that are still found in Africa and the Amazon region (Emperaire et al., 1998). Seeds are used exclusively for varietal improvement, but improved techniques are being researched for their direct use in cultivation (Iglesias et al., 1994).

Grafting wild cassava with robust vegetative growth and resistance to leaf diseases, such as *Manihot glaziovii*, on a cultivated cassava stock at the beginning of the cycle may triple the individual yield. This is the *mukibat* system described by de Bruijn and Dhamaputra (1974) and it is easy to implement on small areas.

Cassava prefers light and well-drained soils that are predominantly sandy, but it can tolerate heavier, more clayey soils, if they are broken down. Mineral fertilization is still rarely applied, but the response to various manures is well known. Nitrogen favours the development of above-ground parts, sometimes to the detriment of tuberization, while potassium fertilization considerably increases yields (Howeler, 1990). The presence of endomycorrhizae on roots improves phosphorus nutrition.

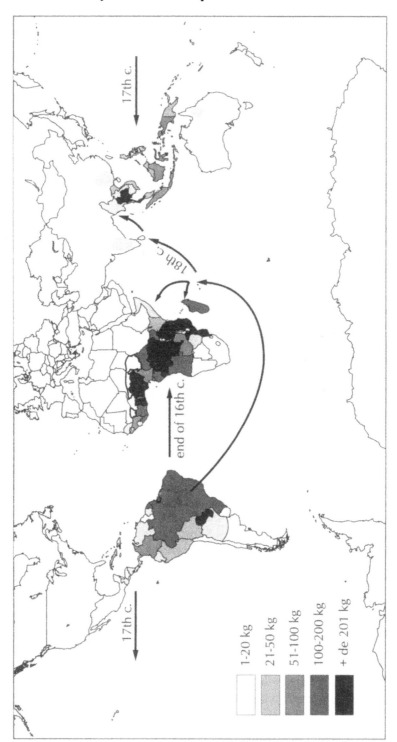

Fig. 1. Production of cassava according to FAO data (in kg per year per inhabitant, average of three years: 1995, 1996, 1997). The arrows indicate the diffusion of cassava from its centre of origin, Brazil, and also probably from Mexico (Purseglove, 1992).

The major phytosanitary constraints are: African mosaic—a geminiviral disease transmitted by the whitefly *Bemisia*; vascular bacteriosis, caused by *Xanthomonas campestris* pv. *manihotis*); and attacks of mealy bugs (*Phenacoccus manihoti*) and acarids (*Mononychellus tanajoa*). These diseases and pests are significant in Africa.

The length of the culture cycle depends on the varieties and environmental factors—temperature, length of dry season, insolation; it varies from six months for the highly precocious varieties to three years and more under unfavourable climatic conditions. In some systems of production, one or two tubers of the plant are harvested for domestic consumption without uprooting the plant.

Cassava is a source of good quality starch, which lends itself to a wide variety of uses: basic human nutrition, modern baking, traditional beer making, industrial uses, and animal nutrition. It ranks fifth worldwide among food plants produced, at 166 million t (FAO, 1998). However, genetic improvement of cassava has begun only recently (McKey and Beckerman, 1996; Raffaillac and Second, 1997).

The root of cassava produces hydrocyanic acid in variable quantities depending on the variety and the environment. Some varieties are dangerous from the nutritional point of view if techniques of transformation are not employed to eliminate most of the hydrocyanic acid. Cyanogenesis is known to be a system of chemical defence in many plants (Jones, 1998).

BOTANY AND TAXONOMY OF THE GENUS *MANIHOT*

Morphology

Cassava plants grown from cuttings simultaneously develop one or several principal stems from buds, while plants grown from seed have a single stem. The vegetative apparatus of cassava is characterized by two types of orthotropic ramification (Medard et al., 1992). The proleptic ramifications, or secondary branches, result from the development of lateral buds after the breaking of apical dominance. The sylleptic ramifications, or floral branches, come from the transformation of the vegetative meristem into floral meristem: two to four branches develop simultaneously. The aptitude for flowering, which determines the architecture of the plant at the end of the cycle, is one of the criteria of identification of varieties. In Fig. 5 (in the Appendix), which presents the major morphological characters by which varieties are identified, the final shape of the stem is illustrated by examples drawn from the works of Cours (1951).

Cassava leaves are highly heteroblastic: during the life of the plant, the number of lobes, an odd number, is greater at the beginning of the cycle, then diminishes to end in a single lobe at the end of the cycle in certain varieties. Leaf morphology is one of the criteria of identification (Table 1, in the

Appendix). Cours (1951) emphasizes that a comparison is valid only if made between varieties planted in the same place and under the same conditions.

The root system is made up of nodal roots and the more numerous basal roots. The latter give the largest tubers. The weight of tubers from a single plant often varies, but their dry matter content is the same. Any root can develop into a tuber. The percentage of roots that tuberize varies greatly depending on the varieties and on environmental factors and cultivation practices. A plant grown from seed puts forth a seminal taproot, which tuberizes first but is too fibrous to be easily consumed. It then produces secondary roots, attached to the taproot, that develop into tubers.

Physiologically, cassava is a C_3 plant, but it has some characteristics of a C_4 plant (El-Sharkawy and Cock, 1990). The tuberization of roots is triggered during the first weeks of growth. The number of tubers is fixed between the second and fourth months of the cycle, barring some damage to the plant. It does not depend on the photoperiod (Keating et al., 1985).

The shape of wild species ranges from a herbaceous rosette form to creepers, to shrubby or bushy forms, to that of a tree more than 15 m high.

Reproduction

All species of *Manihot* are perennial, monoecious, and allogamous. Their pollination is entomo-anemophilous. Flowering may be highly precocious and frequent during plant growth, or late and occurring only once, or nonexistent. Generally, the first floral axes abort. The inflorescences are bunches of 20 to 60 unisexual and monoperianthate flowers. The female flowers, located at the base of the inflorescence, open before the male flowers, which are 10 to 20 times as numerous (Cours, 1951). For each inflorescence, one to six fruits or capsules mature in three to five months. They burst during the dry period and release three seeds having a mottled tegument, brown to grey, with a caruncule.

Taxonomy and Distribution of *Manihot*

The genus *Manihot*, distributed naturally throughout the tropics of the New World, belongs to the family Euphorbiaceae, which is divided into five subfamilies. It is placed in the subfamily of Crotonoidae, with hevea (Webster, 1975).

The limits of the genus are well defined, but the limits between the species are very difficult to establish. The last monograph on the subject (Rogers and Appan, 1973; Rogers and Fleming, 1973) reduced the number of species from 171 to 98 on the basis of multivariate analysis of 44 vegetative characters observed in the herbarium, in which the leaf characteristics were significant. The distribution of species is very clear. Seventeen sections are described, one of which comprises only cultivated cassava. This last section is close to two others: the Central America section, Parvibracteatae, which includes the species *M. aesculifolia*, thus considered the species most closely related to cassava, and the South America section, Heterophyllae, which comprises most of the taxa now included in the species *M. esculenta* by Allem (1994), such as cassava, ssp. *esculenta*, and the wild forms, ssp. *flabellifolia* and ssp. *peruviana*, considered by that author to be the direct ancestors of cassava.

From 77 known species present in Brazil and described by Rogers and Appan (1973), only 38 were retained as non-synonyms by A.C. Allem (personal communication). These are classified into 16 groups depending on affinity. In comparing these two classifications, we observe great differences in the grouping by affinity, which suggests enormous difficulties in defining clear discontinuities within the genus, at least in its principal area of diversity (Second et al., 1997).

In its zone of origin, the genus *Manihot* extends from southern Arizona, in the United States, to northern Argentina. The species are never dominant in the vegetation but rather sporadic. Most of them are found in regions with a long dry season, but some of them are also found in the wet forests or, most often, along the edge of or in clearings that are natural or a result of deforestation. Some species also have adventitious distribution, in disturbed zones along highways, for example.

The natural distribution of the genus does not go beyond 2000 m altitude, but some species are found in subtropical zones. One species, *M. brachyloba*, has a distribution spanning Central and South America and is also found in the Dominican Republic. This exceptional distribution could be linked to its ability to be transported in moderately saline water; some viable seeds were recovered on the coast of French Guyana, where a strong inflow from the Amazon and its local rivers was observed (D. Loubry, personal communication).

The introduction of cassava in the Old World dates from the 16th century and that of *M. glaziovii* from the end of the 19th century. *Manihot glaziovii*, the manicoba, was the object of a development trial in Africa in the 1930s for the production of latex and it has persisted for other reasons (Lefevre, 1989; Serier, 1989).

Since the works of Rogers and Appan (1973), the distinction between *M. utilissima*, bitter varieties, and *M. dulcis* or *M. aipi*, sweet varieties, has been rejected, and it is acknowledged that several environmental factors have a significant impact on the presence of two glucosides responsible for the production of hydrocyanic acid and thus bitterness.

Genome Structure

The 20 species of *Manihot* observed have the same number of chromosomes as cassava, or 18 pairs (Lefevre, 1989; Bai et al., 1993), and this is probably so for all the species in the genus. From the examination of chromosomes with pachytene, Magoon et al. (1969) suggested that cultivated cassava was a segmentary allopolyploid. The observation of proportions of duplicated loci, for RFLP as well as for microsatellites, did not corroborate this hypothesis (Chavarriaga-Aguire et al., 1998). This presumed allopolyploidy could, however, be ancient and correspond to the entire genus, not to the cultivated species in particular. The genome size of cassava is small; it is estimated at 1.68 pg per diploid genome (Marie and Brown, 1993).

GENETIC RESOURCES

The entire *Manihot* genus could be considered a reserve of genetic resources useful for improvement of cassava by sexual means. No strong barrier to hybridization is known to exist. Moreover, interspecific hybridization has frequently been obtained or observed in nature. Nevertheless, most of the *ex situ* collections are made up of cultivars conserved generally in the collections maintained in field and regenerated vegetatively every year. Part of these cultivar collections is maintained *in vitro*. Seeds are used for conservation and exchange mainly of hybrids and wild species that have been collected recently.

It is estimated that nearly 25,000 accessions are conserved in collections throughout the world, all species included. But this number is probably an overestimation because of the loss of several collections and the duplication of others. At present, the two major collections are those of Brazil and of the CIAT (Centro Internacional de Agricultura Tropical) in Colombia. A combination of networks, formal and informal, work to conserve and use the genetic resources of cassava, on each continent and on the global scale (Second and Iglesias, 2001). A very small proportion of this genetic material has been used in crossing for improvement of varieties.

A bibliography of research prior to 1973 (Cab, 1974) pointed out the existence of old collections of the Belgian Congo (INEAC, Institut National pour l'Etude Agronomique du Congo Belge), in West and Central Africa (IRAT, Institut de Recherches Agronomiques Tropicales et des Cultures Vivrières), Madagascar (IRAM, Institut de Recherches Agronomiques à Madagascar), Kenya (EAAFRO, East African Agricultural and Forestry Research Organization), Tanzania, Indonesia, India (CTCRI, Central Tuber Crops Research Institute), Malaysia (MARDI, Malaysian Agricultural Research and Development Institute), Costa Rica (IICA, Instituto Interamericano de Ciencias Agrícolas), Venezuela, Mexico, Brazil, Colombia, Argentina, and elsewhere. Material was sometimes exchanged without rigorous sanitary control. Each research station used different techniques to manage the collections—mode of planting, quality of cuttings, duration of cycle, fertilization, planting density—and followed its own classification of varieties.

From the early 1980s onward, the IITA (International Institute of Tropical Agriculture) in Nigeria attempted to supervise the activities of several countries in Africa and to collaborate with the CIAT on cassava collection and exchange. In this way, a large part of the cassava collection of the CIAT was transferred to Nigeria. On its part, Thailand did the same to take advantage of the great diversity of clones in the CIAT collection.

Wild genetic resources are useful in the improvement of cassava in various ways: as a source of resistance to diseases and parasites—*M. glaziovii* is a good example because it has been widely used in the search for resistance to bacteriosis and African viral mosaic (Hahn, 1984)—and to stresses such as low temperatures; an architecture of cultivars in a given environment and leaf morphology adapted to cultural constraints such as concurrence with weeds and associated cultivation; starch content and physicochemical properties, which are of great importance but have seldom been studied in the wild species.

In this chapter, we attempt to show how the use of molecular markers complements the studies done earlier on a morphological and physiological basis. Through molecular markers, the genetic origin of cassava can be more precisely defined: the most direct wild parent, other closely related parental species, and other species that are involved in the genetic constitution of cassava by introgression. The technique can help us understand the processes of domestication and management of genetic diversity observed in a traditional environment. Finally, it allows us to comprehend the scope and structure of the genetic diversity of cultivars and the entire genus *Manihot*, the principal centre of which, Brazil, will be particularly explored.

ORGANIZATION OF THE DIVERSITY

Agromorphological Variability

Traditional farmers, particularly the Native Americans, distinguish a very large number of cultivars (Emperaire et al., 1998). They use several criteria to classify clones: taste, linked to concentration of cyanogenic glucosides in the root cortex, for which there is a continuum of variation from very low to extremely toxic concentrations (McKey and Beckerman, 1996); precocity—varieties called short-cycle are harvested in 6 to 8 months while others have a cycle of more than 18 months; and appearance of the plant—e.g., the shape of the aerial part, the coloration of leaf, stems, and roots, and their shape.

A large number of morphological and biochemical markers are available for cassava. The most remarkable description, one that has inspired subsequent works in this field, is that of Cours (1951). Cours proposed a classification of varieties in Madagascar into eight botanical sections. The characters used by Cours and in later works are indicated in Table 1 (in the Appendix). Gulick et al. (1983) compiled them in a document for the IBPGR

(International Board for Plant Genetic Resources) and proposed a universal list. Thus, there is an exhaustive catalogue, which comprises classic descriptors that can be used for the identification of cultivated varieties of cassava.

Following the classification of cassava varieties established by Rogers and Appan (1973) from multivariate analysis of 44 characters, all the specific and subspecific denominations of cassava were made synonymous: *M. utilissima* became *M. esculenta*. Ultimately, 19 'similarity groups' were constituted, which were intended to facilitate the work of the breeder but proved to be difficult to use. In effect, apart from genetic factors, cultivated cassava was subjected to two types of significant factors of variation: one, the stage of the cutting and the way in which it is cut and planted; two, soil fertility and cultivation practices. These factors complicate agromorphological evaluation and the comparison of cultivars that have grown in different environments or during different years.

AGROMORPHOLOGICAL PLASTICITY IN RELATION TO THE CUTTING

The quality of the cutting affects the structuration of the plant population. It determines the rate of renewal, and therefore the density of the population and the number of stems per plant. Moreover, the primary rooting is also conditioned by the stage of the cutting, at the nodal level (number of nodes in contact with the wet soil) as well as at the basal level (orientation and number of primary roots linked to the nature of the cut and the position of the cutting in the soil).

In comparison to a cutting taken from the upper part of the stem, a base cutting has a better rate of regrowth, faster growth initiation, more stems per plant, and better nodal and basal rooting (Raffaillac, 1992). Similarly, for a base cutting, the age of the stem affects the quality of the cutting. In Côte d'Ivoire, more than 13,900 stems per hectare were achieved with stem cuttings of 6 months and 27,400 with stem cuttings of 12 months. A cutting of 6 months yields an average of 17.2 roots whereas a cutting of 12 months yields 32.7 (Raffaillac, 1998).

The stems cut at harvest can be stored for several weeks before being planted in a new field. This storage modifies the behaviour of the young plants. A cutting taken from a stem stored for four to eight weeks roots more quickly. Moreover, the number of stems per plant increases (Raffaillac and Nedelec, 1988). These few examples show that it is advisable to make a rigorous selection of cuttings if one wishes to study varietal behaviour on the basis of criteria such as rate of renewal, initial vigour of the plantlet, structuration of the plant population, and root potential.

AGROMORPHOLOGICAL PLASTICITY IN RELATION TO MANAGEMENT AND ENVIRONMENT

The number of stems per plant depends on the way the cutting is planted and the climatic conditions that prevail in the weeks that follow (rain, overall radiation). Certain varieties develop more stems per plant if the plantation is horizontal rather than oblique. When the overall radiation received at the beginning of planting is limited, the number of stems per cutting declines (Raffaillac, 1998). Nevertheless, for a single variety, a field of cassava is always planted with a combination of single stem plants and plants with two or more stems, most often with one type being dominant. An oblique planting combined with a slanted cut allows location of the primary root axes in one soil sector; on the contrary, a vertical planting with a straight cut favours a radial dispersal of roots around the plant (Raffaillac, 1992). The grouping of roots, and thus of future tubers, in a single sector makes it easier to harvest the tubers. It is pointless to study ways to improve varieties for this character, since it can easily be controlled by planting techniques.

The varieties of cassava present differences in their aptitude to develop secondary branches, the presence of which remains linked to environmental factors. These branches are more numerous when the soil is rich and the radiation high. On the contrary, small differences between plants or the simultaneous presence of other plants (adventitious or associated) reduce the number of branches.

Flowering may be nonexistent during the cultivation cycle or may occur several times depending on the environmental factors: on potassium-deficient soils the number of flowerings is higher than that observed on potassium-rich soils (Howeler, 1990; Raffaillac, 1998). This effect of soil fertility is found for the morphology and quality of tuberized roots. At equal weight, a tuber obtained from a potassium-deficient soil is longer and thinner than one from a fertilized soil (Raffaillac, 1997). Potassium fertilization considerably increases the yield but reduces the dry matter content of the tuberized roots. The rainfall received during the last weeks before harvest also modifies the quality of the tubers: the starch content is low when the quantity of water received is high (Raffaillac, 1985).

The plasticity of wild species is also known. We can recall two examples of this. The first pertains to the wild forms of *M. esculenta*, which naturally has two biological types according to whether it grows in the forest or in the savannah (or on the edge of forests). In the first case, it grows as a false creeper, which climbs on tree branches. In the second, it grows in a bush shape. Seeds taken from the creeper form and cultivated in full light or seeds that have germinated on the edge of the forest give bushes without any traces of creeper-like growth. The second example pertains to *M. quinquepartita*, which forms whole leaves or strongly indented leaves depending on whether it grows in full light or in shade, as has been observed in Saul, French Guyana. The morphological differences are such that, in a herbarium, the two forms may be thought to be different species, even by an expert.

Molecular Diversity

RAPD ANALYSIS OF CULTIVATED CASSAVA

RAPD analysis has been done on 126 accessions (Colombo, 1997). These consist of 71 accessions from the collections of IAC (Instituto Agronômico de Campinas) and EMBRAPA (Empresa Brasileira de Pesquisa Agropecuária) and originating from different geographic zones of Brazil, 33 accessions from a core collection of world diversity at CIAT (Hershey et al., 1994), and 42 plants representing 18 ethnovarieties from a single clearing along the middle of the river Negro in the Amazonian basin. This sample can be considered representative: Brazil probably covers a good part of the zone of origin of cassava and is the source of a majority of cultivars, but the collections there are deficient in Amazonian origins. However, the sample remains limited, with only 7 accessions of Central America and the Caribbean and 6 non-American accessions.

From 21 primers, 193 RAPD bands were observed, of which 88 are polymorphic in presence-absence, which indicates the great diversity of the species. The Jaccard coefficient of similarity varies from 0.99 to 0.45 with a mean of 0.67 (the polymorphic bands alone are taken into account). The distribution of 126 genotypes on the two primary factorial axes of a principal coordinates analysis (PCoA) is given in Fig. 2a. The Lebart test (Lebart et al., 1977) on a correspondence analysis (CA) was used to characterize the bands that contribute significantly to the primary axes: 46 bands, or only 52%, contribute to the five primary axes of the CA.

The variation is continuous and no strong structure appears. However, the ethnovarieties from the Amazonian clearing are separated from the rest of the collection by the primary axis. The other varieties that are closest to it originate from the Amazonian region of Brazil, from countries that share the Amazonian basin (Venezuela, Colombia), or countries that have regions with a similar climate, such as Malaysia and Thailand.

The PCoA done on a 'non-Amazonian' group (Fig. 2b), constituted by eliminating varieties of the traditional field and those closest to them, indicates, from three primary axes, a set of 13 varieties, which have some remarkable characteristics. Their origins are varied: 6 originate from Brazil (4 of those from northeastern Brazil), 2 are artificial Brazilian crosses included in the sample, and 5 come from various countries. Among the varieties originating from northeastern Brazil, variety SRT1316 has a remarkable affinity with interspecific hybrids between cassava and *M. glaziovii*. Among the varieties originating from various countries, NGA2 was artificially introgressed by *M. glaziovii* to acquire a resistance to African viral mosaic and bacteriosis; it is one of the parents of the cross used to construct the first genetic map for cassava (Fregene et al., 1997). Another, Col 22, is a traditional variety, which is involved in the genealogy of several recently selected varieties (see table 2 in Second et al., 1998b). The hypothesis that an introgression of

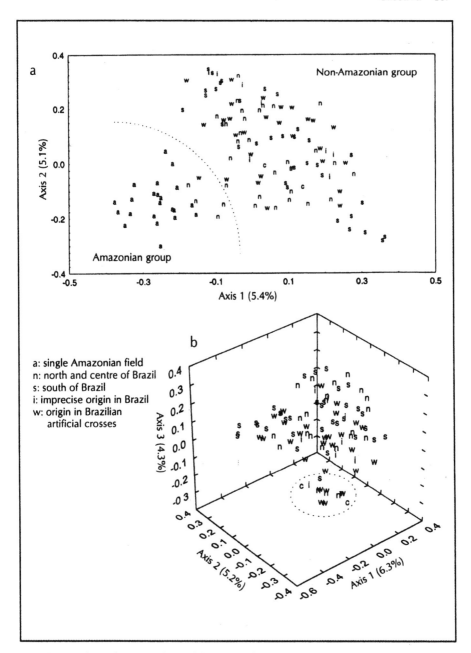

Fig. 2. Principal coordinates analysis of the matrix of similarity (Jaccard) based on the presence or absence of 88 polymorphic RAPD fragments in comparisons of genotypes two by two: plane or volume defined by the 2 or 3 primary axes. a: distribution of 126 genotypes representing a world collection and a single Amazonian field (the dots indicate the separation of an Amazonian group from the rest of the collection). b: analysis of the non-Amazonian group alone (the dotted circle indicates a remarkable group).

M. glaziovii would be at the origin of this group of varieties is worth confirming.

Other peculiarities can be detected by examining certain axes that prove significant despite the low values of total variance extracted. Low values of axis 4 (3.7%) isolate the four varieties from Thailand and Malaysia, the New World varieties closest to them originating from Colombia.

ANALYSIS OF AMAZONIAN ETHNOVARIETIES AND OF A COLLECTION OF CULTIVARS BY AFLP AND RAPD MARKERS

The collection of ethnovarieties from the clearing may clarify the question of the nature of an ethnovariety denoted by a name. Does the multiplicity of names reported by anthropologists cover a real genetic diversity and does a name identify a clone or a family of clones? More generally, it could direct us in verifying the hypothesis of a dynamic traditional management of cassava diversity, which could reveal to us the process of its domestication (Emperaire et al., 1998; Second et al., 1998).

AFLP markers have been used to analyse the 42 plants representing the 18 varieties sampled in the same clearing, at a ratio of 1, 2, or 10 plants per variety, and 40 accessions of distinct geographic origin, coming from the world collection and representing the variability described earlier. The two combinations of primers used enable us to observe 60 polymorphic bands out of a total of 132. The analysis reveals a structuration of diversity similar to that observed on the same samples using RAPD markers. A global analysis has thus been done by combining the polymorphic bands observed with the two techniques, or 143 bands.

The UPGMA dendrogram of 82 plants studied, established by means of the Jaccard coefficient of similarity (Fig. 3), reveals the presence of two groups. One corresponds to the world collection, the other to the clearing. A single variety of the world collection is found in the group of plants from the clearing. It originates from the Brazilian state of Para in the Amazon basin. The two groups, with a widely differing geographic origin, have clearly equivalent total diversity. Only 10 bands (7% of the total), the frequency of which is less than 0.45 in the world collection, are not observed among the varieties from the clearing. A significant coefficient of correlation of 0.7 is found between the frequencies of bands observed in the clearing and in the world collection. An analysis of molecular variance (Amova) reveals that 80% of the total genetic diversity is found within the groups and 20% between the two groups.

All the names correspond to different genotypes and, in general, plants having the same name cannot be differentiated: they probably represent clones. There are, however, four exceptions for which a single name—F, Mn, Bu, and Sn in Fig. 3—corresponds to several genotypes. The varieties Fino (F) and Manipeba (Mn) are each represented by two plants that correspond to two genotypes, with respectively 16 and 26 different bands out of the 143

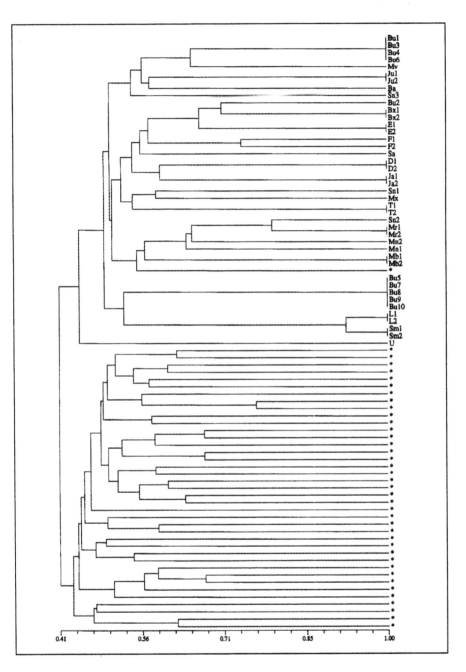

Fig. 3. Comparative molecular diversity of 42 plants from a single Amazonian field (the abbreviations represent the names of varieties, except Sn, which represents nameless varieties; the values denote the order of plants of the same name) and of 40 plants representing the world collections (symbolized by an asterisk). UPGMA dendrogram from the matrix of similarity calculated for 143 polymorphic bands (60 AFLP and 83 RAPD).

studied. The variety Buia (Bu) is represented by three genotypes, which differ by 31, 38, and 40 bands; despite the dispersal of Buia genotypes in the dendrogram of Fig. 3, a PCoA, not presented, confirms the tendency to group genotypes that have the same name in relation to the total diversity of the field. Each of the three 'nameless' varieties (Sn) corresponds to one genotype, which agrees with their presumed origin. They come from three plants grown from seed, in the course of evaluation by the cultivator (Emperaire et al., 1998); their relative grouping agrees with that of genotypes observed respectively in the three varieties F, Mn, and Bu.

Thus, these preliminary results confirm the existence of a dynamic management of cassava diversity: grouping of a considerable diversity in a single field, recombination linked to sexual reproduction, and selection of high-performing genotypes collected subsequently under a single name according to their affinities.

From this perspective, a traditional variety can be considered a family of clones that share certain characteristics of direct interest for cultivators, such as productivity, quality, and resistance to diseases, parasites, and various stresses, as well as characters of less immediate interest by which varieties can be recognized and given a name, for example, colour, shape, or general appearance of the plant. These characteristics do not suffice in themselves to characterize a variety, which must be evaluated up to the point of consumption.

For the sample representative of the world collection, AFLP analyses can only confirm—with a much smaller number of accessions—the results obtained with the RAPD analyses, i.e., the absence of strong structuration in a considerable global diversity. However, taking into account the grouping of certain genotypes on a geographical basis, it seems to us that with a study of a large number of genotypes and more loci, one could determine the loose structuration that can be predicted for this diversity.

ANALYSIS OF DIVERSITY OF *MANIHOT*

The first molecular characterization of the genus *Manihot* was that of Bertram (1993). Based on an RFLP (on Southern blots) of chloroplast and ribosomal DNA, it was limited to six species of Central America, cassava, and a South American species, *M. carthaginensis*. The results show relatively low divergence (0.1% at most for chloroplast DNA) and maximal divergence between the species of Central America, on the one hand, and *M. carthaginensis* and cassava, on the other. These results contradict the hypothesis of a maximal relationship of cassava with *M. aesculifolia*, proposed by Rogers and Appan (1973). On the other hand, they suggest that South America could be the zone of origin of cassava. The study of some species of South and Central America by RAPD (Colombo, 1997) confirms the conclusions of Bertram (1993).

AFLP studies (Roa et al., 1997) and microsatellite studies (Roa et al., 1998) of cassava, of four wild species of South America—wild forms of *M. esculenta*, *M. tristis* (a species related to *M. esculenta* ssp. *flabellifolia*), *M. carthaginensis*, and *M. brachyloba*—and of *M. aesculifolia* confirm the hypothesis of Allem (1994) that the wild forms of *M. esculenta* were the direct ancestors of cassava. Moreover, they show that the diversity of wild forms is greater than that of cassava. In the case of microsatellites, the number of alleles per locus varies from 4 to 22 with a total of 124 alleles for the 10 loci analysed (one or two loci were not detected in *M. aesculifolia*, *M. brachyloba*, and *M. carthaginensis*). The group of species formed by *M. aesculifolia*, *M. brachyloba*, and *M. carthaginensis* (species related to *M. glaziovii*) shows less than 20% similarity with cassava. *Manihot esculenta* ssp. *flabellifolia* and ssp. *peruviana* and *M. tristis* form a group closer to cassava, with 35% to 50% similarity. It is interesting to note, in relation with the exchange of genetic diversity that we have observed in the traditional management of cassava, that the wild species have a deficit of heterozygotes while in cultivated cassava the rates of heterozygosity observed are equivalent to the values expected.

Our study, based mainly on AFLP markers and conducted at the same time as the one just mentioned—independently but with certain common samples—is, to date, the most exhaustive on the genus *Manihot*. It includes all the Brazilian species and four species of Central America; it involves totally 282 wild forms and 82 varieties of cassava. The analyses were done on the presence or absence of 93 polymorphic bands revealed from a single pair of primers.

A UPGMA analysis of similarities (simple matching) shows that the genotype most divergent from the group is that which represents the Central American species *M. aesculifolia*. The distribution of 364 accessions in the plane defined by the two primary axes of a PCoA is represented in Fig. 4a. The first axis isolates most of the cultivars from all the wild species. However, there is a continuum between these two groups, where one finds the wild forms of *M. esculenta* and *M. pruinosa*. The other wild species are distributed along axis 1, while axis 2 allows us to separate these wild species from a group corresponding to *M. glaziovii* in the wide sense, as A.C. Allem has defined it, considered synonymous with *M. carthaginensis* (table 1 in Second et al., 1997). Two groups of forms in an intermediate position between *M. carthaginensis* and *M. esculenta* (isolated by the circles in Fig. 4a) are indicated. They comprise forms classified as *M. glaziovii*, forms of 'tree cassava' frequent in northeastern and central Brazil, and, in the group closest to the cassava group, four cultivars, including SRT1316, originating from northeastern Brazil. We consider these forms hybrids between *M. glaziovii* and *M. esculenta*.

In northeastern Brazil, a similar situation is found to that described in West Africa (Lefevre, 1989). Cultivars of cassava have undoubtedly been introgressed by *M. glaziovii*, probably by the intermediary of forms of tree cassava, even though these are often largely sterile and propagated vegetatively.

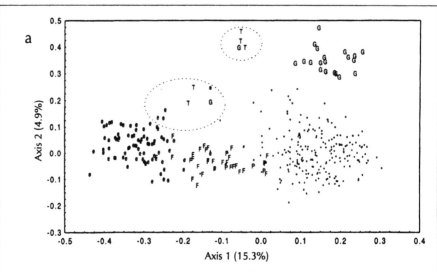

e: cassava, the same genotypes as in Fig. 2. F: the wild forms of *M. esculenta*. P: *M. pruinosa*.
G: *M. carthaginensis* in the wider sense (*M. carthaginensis, M. glaziovii, M. epruinosa*). C: *M. compositifolia*. *: all the other species. T: cassava tree of northeastern and central Brazil.

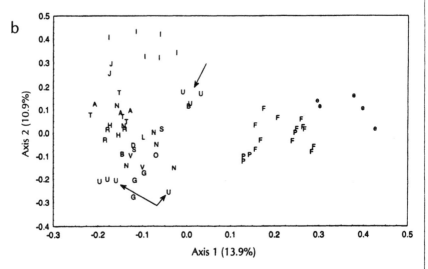

e, F, P as in Fig. 4a, A: *M. hassleriana*. B: *M. brachyloba*. D: *M. pseudo-pruinosa*. E: *M. tenerrima*.
G: *M. gracilis*. H: *M. hilariana*. I: *M. pilosa*. J: *M. sp. nov*. L: *M. longepetiolata*. N: *M. nana*. R: *M. noguerae*. O: *M. procumbens*. S: *M. stipularis*. T: *M. triphylla*. U: *M. fruticulosa* (the arrows indicate the two subspecies distinguished). V: *M. paviaefolia*.

Fig. 4. Diversity of the genus *Manihot*: principal coordinates analyses based on similarities (simple matching) of genotypes two by two in the analysis of 93 polymorphic AFLP fragments. a: distribution of 364 plants representative of the genus. b: distribution of genotypes of *M. esculenta* (wild and cultivated forms), of *M. pruinosa*, and Brazilian species morphologically most closely related that are found on a single AFLP gel. Data of Fig. 4a reduced to this single gel.

Our analysis thus confirms that the wild forms of *M. esculenta* are the closest to cassava among all the species of South America. Moreover, it shows that *M. pruinosa* is not distinguished at the molecular level from the wild forms of *M. esculenta*. This proximity is also found in the morphology (A.C. Allem, personal communication). Our study further suggests that *M. carthaginensis*, and particularly the form recognized as *M. glaziovii*, is part of the genetic constitution of certain cultivars. This contribution needs to be placed in the context of successful introduction of *M. glaziovii* in the Old World and its use as a genetic resource in modern efforts to improve cassava varieties. Finally, certain species, particularly *M. fruticulosa*, could be genetically close to cassava, even though their direct contribution to its domestication is not proved.

The PCoA limited to the species *M. esculenta* and to Brazilian species considered *a priori* the closest to cassava (Fig. 4b) shows a clear separation between *M. esculenta* and *M. pruinosa* on the one hand and the other Brazilian species on the other. *Manihot fruticulosa* splits up into two subspecies in agreement with the morphological analysis that recognizes a subspecies *caiaponia* (A.C. Allem, personal communication). The form *M. fruticulosa* in the strict sense seems closest to *M. esculenta*. *Manihot fruticulosa*, being a herbaceous species, could be used in varietal improvement to modify the architecture of cassava.

The structure illustrated in Fig. 4a is accurate despite the dispersal of samples across six gels and the difficulty of transcribing the gels in terms of presence or absence of specific bands associated with them: AFLP analysis of a subsample of DNA used in Fig. 4a, for three combinations of primers, all placed on the same gel for each primer, confirms this structuration. The three groups suggested by Fig. 4a are found: *M. esculenta*, with its wild forms ssp. *flabellifolia* and *peruviana*, and *M. pruinosa; M. carthaginensis* in the wider sense; a group including all the other South American species represented, among which only *M. compositifolia* appears as an intermediate between this group and *M. carthaginensis*. Also it is noted that grouping according to molecular similarity does not always correspond to the taxonomic classification, which illustrates again the difficulty of classifying this genus.

Manihot carthaginensis (including *M. glaziovii*) appears in this study to be the most divergent species of cassava in South America. The fact that it has indisputably contributed to the domestication of cassava indicates that all the species of *Manihot* could *a priori* be considered direct genetic resources for cassava—of the primary or secondary pool according to the terminology of Harlan.

RELATIONS BETWEEN MORPHOLOGICAL AND MOLECULAR DIVERSITY

The relations between morphological and molecular diversity have not been the subject of rigorous experiments on a significant scale. We have emphasized the agromorphological plasticity of cassava, which complicates this comparison. The plasticity of wild species is also considerable. It is nevertheless possible, on the basis of what has been reported earlier, to make the following parallels between the observations made for these two levels of diversity.

For 31 of the accessions studied, all Brazilian, one description has been drawn from nine characters—colour of three root tissues and roughness of the root, colour of apical bud, colour of young stem, colour of petiole, width and sinuosity of the central lobe—or in total 27 states of characters. This description has enabled a comparison with the diversity observed using RAPD (74 polymorphic bands). The PCoA done independently for morphological descriptors and molecular markers shows similar structures with, notably, a grouping of varieties that present industrially useful characteristics (Colombo, 1997).

More generally, for the cultivated forms, the absence of subspecific division revealed by botanical or agromorphological analysis is found at the molecular level, where the variability seems continuous, without a large linkage disequilibrium, the major part of the diversity of the species being found in a single Amazonian clearing. The Native Americans recognize in fact a large number of varieties, which is consistent with the wide molecular diversity observed, but these varieties are dynamic, in perpetual evolution, notably because of genetic recombination. Even though there is agreement between the two scales of observation, the molecular analysis seems better designed to describe reliably the genetic diversity of cassava, to the extent that it is insensitive to environmental variations.

Classification of the genus *Manihot* into well-defined species is difficult both by morphological and by molecular markers. However, for the genus *Astrocaryum* (palm), molecular AFLP markers unambiguously reveal the large subdivisions and suggest a new one that is not in contradiction with the morphological analysis (Kahn and Second, 1998). Used in the oil palm, RFLP and AFLP markers clearly indicate the cryptic subspecies (Barcelos, 1998; Barcelos et al., 1998). The difficulty of classifying the genus *Manihot* thus appears to be the result of its biological characteristics.

The species *M. aesculifolia*—classified in morphological terms as the closest to cassava (Rogers and Appan, 1973) but highly divergent in molecular terms—represents nevertheless a notable case of disagreement between the two levels of observation. It illustrates the utility of molecular markers from the phylogenetic point of view: the Central American species are geographically isolated, which could explain their phylogenetic distance despite the morphological resemblances that are more subject to natural selection than to molecular diversity.

Moreover, the forms classified as intermediate between the closely related species *M. caerulescens* and *M. quinquepartita* according to the morphological criteria are also found in intermediate position in the molecular analysis.

Besides, several cases of spontaneous hybridization, suspected during the morphological study, were confirmed by molecular analysis (Second et al., 1997). These hybridizations partly explain the difficulties of delimiting the species in the genus *Manihot*.

As with cassava, there is generally no contradiction between the results of the two levels of observation for the entire genus *Manihot*. It is by a combination of morphological observation and molecular characterization that one comes to a better understanding of this difficult genus. It already seems confirmed that entities considered species according to classical botany have generally conserved the possibility of fertile hybridizations. Under these conditions, it is probably inappropriate to represent the relations between these species by a phylogenetic tree. A 'network' of relations gives a more realistic picture. It would be impossible to chart this network exhaustively; it could be illustrated at best by multivariate analysis on several axes of variation. Once we recognize the reality of this situation, we can see that a model of dynamic conservation of the genus is particularly appropriate.

APPLICATIONS IN GENETIC RESOURCE MANAGEMENT

The biological characteristics of the genus *Manihot*, as well as considerations of complementarity with approaches of *in situ* and *ex situ* conservation, have led to a proposed procedure for dynamic conservation for the wild species of the genus (Second et al., 1998; Second and Iglesias, 2001). Artificial populations can be collected in their biotopes of origin by grouping individuals that belong to a single species considered in the strict sense. In biotopes or continents new to these species, the grouping could be done by species in the wider sense. Thus, we end up with a manageable number of species that, by exchange of genetic diversity, have a greater chance of adapting to new conditions. A periodic exchange of seeds between the populations of the same species—excluding populations of the zone of origin, which must not be genetically 'contaminated'—would ensure the maintenance of a high genetic diversity at low cost. This diversity is directly subjected to natural selection and available for integration in breeding programmes. Such a mechanism of dynamic conservation would come to complement *in situ* conservation, when that is applicable, and *ex situ* conservation by seed, which remains to be optimized.

As for cultivated cassava, its status as a plant generally propagated by vegetative means has favoured a procedure of conservation by clones *in vitro*, which has at least two major disadvantages: high cost and the associated risks of transmission of parasites, particularly viral parasites, during exchanges of genetic material.

The confirmation that diversity is dynamically managed in traditional practices—which probably represent the process of domestication of cassava—has led to revived interest in conservation of cassava diversity by seed: the genotypes are modified but diversity is conserved. For dynamic conservation of cultivated cassava in the field (Second and Iglesias, 2001), we must note a peculiarity linked to the consumption of roots. Unlike with cereals, for which the seed-harvest cycle represents a selection pressure in favour of a productive cultivated variety in field collections, selection pressure in a plant cultivated for its roots is imposed directly by the cultivator. Here lies, probably, the explanation of the traditional field model of cassava, where the clones of a single variety are often cultivated in clumps in the field, so that they can be evaluated up to the time they are consumed. Thus, perpetuate this system in an effective and justifiable manner, the association between conservation and improvement must be maintained. When a large diversity is organized in groups of clones in a single field, the clones can be evaluated in the context of domestic agriculture. It is thus possible, as a function of scientific knowledge acquired, to assist this process to the end and to render it more effective, notably through production of hybrids, backcrosses, and genetic transformation. The use of biotechnology such as microchips or DNA chips may make it relatively easy in such a procedure to follow the evolution of the frequency of certain loci in DNA extracts from combinations of individuals in the population. It would thus be possible to associate conservation and improvement of genetic diversity on the farm with the advantages that would result from technologies of genetic manipulation.

CONCLUSION

Use of molecular markers to analyse the scope and organization of genetic diversity of cassava and of the genus *Manihot* yields a set of results that are generally consistent with the morphological data, with just one notable exception. These results have enabled us to make rapid progress in our understanding of the genus.

Molecular analysis unambiguously confirms the relationship between cultivated cassava and its presumably ancestral forms. It also indicates proximity between *M. pruinosa* and *M. esculenta*. *Manihot pruinosa* would also be a wild form of *M. esculenta*, adapted to the Brazilian savannah.

This analysis indicates the singularity of the species *M. carthaginensis* in the wide sense, the validity of its classification with *M. glaziovii* being verified. It suggests the introgression of *M. glaziovii* in certain cultivated varieties, while this species is, among the Brazilian species, the most divergent from *M. esculenta*. *Manihot glaziovii* is thus, since its diffusion in the Old World at the end of the 19th century, an example of dynamic conservation of the genetic diversity associated with varietal evolution on the farm (Second, 1998).

The complex genetic nature of the genus *Manihot* is confirmed by the analysis of molecular diversity; it seems to imply several spontaneous hybridizations.

From these results, new concepts can be advanced. The varieties distinguished traditionally correspond to families of clones resulting from a dynamic management of the diversity, which must be encouraged. This process of dynamic management could include varietal improvement on the farm, which would benefit from inputs of modern genetics. To the extent that the structure of the genus *Manihot* is better described by a network of relations between the species than by a phylogenetic tree, it also seems appropriate to consider a dynamic management for the conservation of this genus.

Cassava is an important food plant for the tropical regions, and the studies we have undertaken are therefore essential. A more representative sample of cassava varieties needs to be examined that includes, apart from the collections already studied, the entire Amazonian basin and the African and Asian continents. The traditional process of management of the diversity must be thoroughly analysed, and the nature of the traditional varieties must be examined from a larger sample, structured according to the field, the village, the region, and so on. At the same time, the perfection of microsatellite and cytoplasmic markers must be pursued. For the genus *Manihot*, it will be necessary to assemble all the Central and South American species in a single analysis. Case studies must be undertaken on the organization of the diversity and gene flows in the species complexes close to the genus *Manihot* as a function of their distribution and their ecology. From the analysis of forms of *M. glaziovii* diffused throughout the world for more than a century and varietal selections, cultivated and spontaneous, that have resulted from them, we can begin to evaluate the possibilities of dynamic conservation of the genus.

APPENDIX

Plant Material

The analyses were carried out on 130 varieties of cassava from collections of Brazil and CIAT and on 278 plants representing all the South American species and four species from the Central American species of the genus *Manihot*. A living collection of all the Brazilian species, such as has been considered by A.C. Allem at CENARGEN (Centro Nacional de Pesquisa de Recursos Genéticos e Biotecnologia, Brazil; table 1 in Second et al., 1997), was constituted from cuttings, transplants, and seeds collected directly in the natural populations. This collection of wild species was supplemented by samples from the herbarium of the CENARGEN collection. Most of the forms distinguished morphologically for which there presently exist known populations have been represented. Six presumed hybrids were included as well as their presumed parents from the same populations. About a third of the samples come from herbarium, the rest, with only a few exceptions, are from direct collections from nature or plants obtained from seeds collected in nature. Seven plants, of which four represent species of Central America, come from *in vitro* collections of CIAT, in Colombia, and from the laboratory of the University of Washington, in the United States. For the analysis, samples of healthy young leaves were dried in a ventilated oven at 50°C for 20 h, then conserved dry in the presence of silica gel.

DNA Extraction

A modified protocol of the CTAB technique (cetyl-trimethyl ammonium bromide) was used (Dellaporta et al., 1983; Colombo, 1997).

RAPD Analysis

For the RAPD analysis, 208 decamer primers were tested and 22 were retained for the quality of profiles of bands obtained (see Colombo, 1997, for details on the primers and PCR amplification).

AFLP Analyses

All the AFLP analyses were entrusted to the commercial laboratory of Linkage Genetics at Salt Lake City, in the United States (presently PE AgGen Inc.). The laboratory used the AFLP technique as published and the Keygene program for computer-assisted reading of the gels. For the arbitrary part of the primers, the AGA/CAG combinations and, eventually, AGT/CTC and AGA/CAA combinations were used, when two or three combinations were used among the 12 combinations tested.

Statistical Analyses

The matrixes of presence or absence of bands were used to calculate the coefficients of similarity (Jaccard coefficients or simple matching) in the comparisons of genotypes two by two or for the purpose of a CA. The matrixes are compared using the Mantel test. The matrixes of similarities are used to construct dendrograms (UPGMA) or to conduct principal coordinates analyses (PCA, PCoA). To construct a UPGMA dendrogram from the value of the coordinates on the primary axes, a Euclidean distance is used.

For all the calculations above, the Ntsys program of Exeter Software (Rohlf, 1998) was used. The analysis of molecular variance is calculated with the Amova program (Excoffier et al., 1992). Ntsys and Statistica were used for the graphics (Statsoft).

Table 1. Characters of vegetation, inflorescence, and roots of cassava used to differentiate varieties (Cours, 1951). The criteria in italics are from later bibliographical courses. The arrows refer to illustrations in Fig. 5

Organs	Type
Wood (stems)	
● dominant colour of leaves that are not fully developed on the terminal shoot (equilibrium between chlorophyll and anthocyanin pigmentations)	1. green 2. dark green with highlights 3. light purple 4. dark purple
● time of impregnation (duration of coloration expressed in number of leaves coloured)	0 to 12 (generally 5 or 6)
● colour of the young part of the stem that is not harvested	1. green 2. yellow-green 3. green and beginning of petiole red 4. green and beginning of petiole red with red ribs 5. green and red in equal area 6. some traces of green 7. entirely light red 8. dark red to purple-brown
● colour of stem (old part) at 1 and eventually at 2 years	1. ash grey to dark green 2. olive green 3. mahogany 4. dark brown
● shape of stem at the end of the cycle in relation with ramification linked to flowering of the terminal vegetative meristem	1. rampant (more than 6 flowerings) 2. spread out (4 to 6 flowerings) → A1 3. tall and spread out (2 or 3 flowerings) → A2 4. erect (1 late flowering) → A3 5. cylindrical (no flowering)
● lateral ramifications (development of lateral buds by lifting of apical dominance)	1. absence 2. presence with:

(Contd.)

(Contd.)

Organs	Type
	2a. erect shape (parallel to stem)
	2b. drooping (stem spread out and drooping)
● *average number of branches developed at each flowering*	1. *2 branches*
	2. *3 branches*
	3. *4 branches*
	4. *more than 4 branches*
● *angle of gap between branches developed by flowering*	1. *none*
	2. *15° to 30°*
	3. *45° to 60°*
	4. *75° to 90°*
● *total height of the main stem*	*in cm*
● *time taken for the first inflorescence to appear*	*in days after planting*
● *height of first branch at flowering*	*in cm*
● *number of stems per plant obtained by cutting (noting the mode of planting)*	*number*

Nodes and internodes on stem

● colour of eye (bud) at latent stage	1. green
	2. coloured base and green scales
	3. entirely coloured
● emergence of eye (bud) at latent stage	1. deep → B1
	2. projecting → B2
● small cushion (bulge that bears leaf scar at the spot where petiole was inserted) with dentate stipules on both sides	1. ephemeral teeth → B3-B4
	2. persistent teeth → B3-B4
● dimension of stipular roll (swelling of stipules and cushion)	in cm → B3-B4
● alignment of internodes (young part)	1. zigzag
	2. straight
● striations on the young part of stems and ribs	1. ephemeral ribs
	2. caducal ribs
	3. persistent ribs
● vigour of stems (measured by the base diameter between two nodes)	in cm

Leaf

● general shape	1. palmipartite (normal shape)
	2. palmisequate (the lobes totally separate)
● dimension of petiole	1. sessile type (absent or less than 1 cm)
	2. intermediate type (associated with the palmisequate form)
	3. long and cylindrical type

(Contd.)

(Contd.)

Organs	Type
• angle at which petiole is attached to stem	1. less than 20° 2. 20° to 55° 3. 55° to 70° 4. 70° to 90° 5. 90° to 120° 6. more than 120°
• area in which deviation of petiole appears	1. submedian 2. median 3. subterminal 4. terminal
• coloration of petiole	1. entirely green 2. entirely yellow-green or light green 3. cross red at the base and the rest green 4. submedian part green and the rest red 5. submedian part green (representing half the length) and the rest coloured 6. red except part of the green submedian zone 7. vivid red, darker than the cross 8. dark red or purple-red
• length and diameter of petiole	5 to 60 cm and 1 to 5 mm
• pubescence of leaf hilum, part at the end of the petiole where principal lobes of the nerve converge	1. absent 2. slight 3. moderate 4. great
• dimension of leaf hilum	1. shrunken → C1 2. spread out → C2
• shape of leaf hilum	1. concave → C3 2. convex → C4
• number of lobes constituting limb, calculated on an average over 12 months of plant growth	1. less than 3 2. 3 to 5.5 3. 5.5 to 7 4. more than 7
• shape of lobes (L/W, ratio between length and width, measured on median lobe)	1. very narrow (L/W > 20) → C5 2. with parallel edges (L/W 6 to 20) 3. normal (L/W 4.5 to 6) 4. wide (L/W 3 to 4.5) → C6 5. rounded (L/W less than 3) with two widening points
• form of lobes defined by position of maximal widening point along median lobe (for types 3 and 4)	1. proximal (in first basal half of lobe) → C7 2. median (in middle of lobe) 3. submedian (between medium and three quarters)

(Contd.)

(Contd.)

Organs	Type
	4. terminal (beyond the last quarter towards the tip) → C8
● ornamentation of lobes	1. presence or absence of median swelling
	2. presence or absence of spurs
	3. incurved lamina
	4. gondolate margins
● colour of upper surface of lamina	1. white without chlorophyll (partial or total albinos) or with motley colours
	2. light green
	3. yellow-green
	4. dark green
	5. purplish
● colour of underside of lamina	1. green
	2. whitish-green
	3. yellow-green
	4. purple or red
● coloration of lamina nerves	1. always green
	2. red on underside of young leaves
	3. red on upper surface of young leaves
	4. red on upper surface of adult leaves
	5. red on underside of adult leaves
	6. red on both sides of young leaves
	7. red on upper surface of young or adult leaves
	8. red on underside of leaves of any age
	9. on both sides of young leaves and subsisting on underside of adult leaves
	10. always on both faces
● lobe sinus (empty space separating two lobes)	closed to the extent that the number of lobes is high → C9
● basiliar sinus (angle formed between the central nerves of the two extreme or lower lobes taken on both sides of petiole)	1. closed (angle less than 180°) → C10
	2. open (lobes at extreme rising, angle greater than 180°) → C11
● dimension of velum (surface of lamina at the confluence of extreme lobes at the level of basiliar sinus	1. band of 0.5 mm
	2. less than diameter of petiole tip → C12
	3. greater than diameter of petiole tip
	4. wider than base of petiole
	5. greater than twice the diameter of base of petiole → C13
● shape of velum	1. straight → C14
	2. spread out

(Contd.)

(Contd.)

Organs	Type
	3. retracted
	4. wrinkled → C15
● ornamentation of velum	1. none
	2. dentate
	3. stipulate → C16
	4. fringed → C17

Inflorescence → D1

● fructification of inflorescence	1. continuous (the first flowers are functional)
	2. late (abortion of first inflorescences, fructification of later ones)
● form of sepals in calix of female flower	1. wide
	2. medium
	3. narrow
● colour of sepals of flower → D2	1. entirely green
	2. green and coloured nerves
	3. red and green
	4. red to purple
● torus, roll with nectaries between calyx and pistil	1. yellow or yellowish
	2. reddish
	3. brown-red
● shape of six wings on the ovary with three carpels (corresponding to points of suture of carpels)	1. straight at any stage of fruit development
	2. straight then sinuous → D3
	3. sinuous then straight at maturity → D4
	4. always sinuous
● colour of wings on the day the flower opens	1. entirely green
	2. red
● colour of ovary body	1. green
	2. red
● colour of stigmata on the day the flower opens	1. absent (white or light pink)
	2. red
● presence of pollen in the male flower with calyx and torus (not used for classification) and androcate with 10 stamens	1. male flower sterile
	2. male flower fertile
● colour of fruit (or capsule) with swelling of peduncle	1. green
	2. light to red purple
	3. predominantly red
	4. entirely red
● *dimension of fruit (length and diameter)*	*in mm*
● colour of seed tegument → D5	1. grey
	2. brown

(Contd.)

Organs	Type
● mottles on seed tegument → D5	1. few 2. dense
● colour of caruncle of seed	1. white or cream 2. pink or red 3. purple
● *dimension of seed (length and diameter)*	*in mm*

Root

Organs	Type
● point of attachment on parent cutting	1. sessile 2. pedunculate 3. long pedunculate (greater than 10 cm)
● length of root	1. short (less than 40 cm) 2. normal (40 to 80 cm) 3. long (greater than 80 cm)
● diameter of root at its maximum	1. thin 2. medium 3. thick
● shape of root	conical → E1 fusiform → E2 cylindro-conical → E3 cylindrical → E4
● *constriction in roots*	1. *absent* 2. *present*
● direction of root	1. horizontal 2. vertical
● number of roots	number
● *homogeneity of weight of roots (numbers per weight class)*	*in %*
● *texture of root surfaces*	1. *smooth* 2. *medium* 3. *rough*
● appearance of external bark (cork)	1. grey and thin 2. brown and thick
● colour of phelloderm	1. white 2. pink 3. purple
● *detachment of phelloderm from central cylinder*	1. *easy* 2. *difficult*
● colour of pulp (or flesh, central cylinder)	1. white 2. yellow
● *index of precocity (dry matter yield of roots at 6-8 months in relation to dry matter yield of roots at 12-14 months)*	*in %*

(Contd.)

(Contd.)

Organs	Type
● *index of harvest (dry weight of tuberized roots in relation to total biomass of plant)*	*in %*
● presence of fibres	1. absent 2. some fibres visible 3. numerous
● taste (significance of release of hydrocyanic acid)	1. sweet 2. bitter, *or three classes defined by the intensity of coloration with picric acid: low, medium, high*

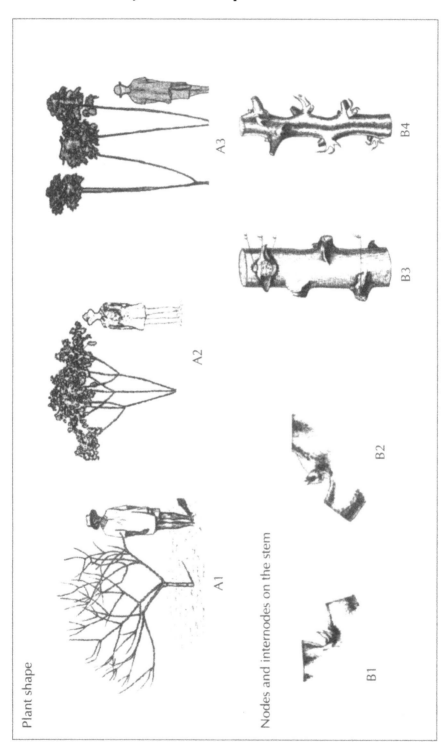

Fig. 5. Characters of vegetation, inflorescence, and roots of cassava used in differentiation of varieties (Cours, 1951)

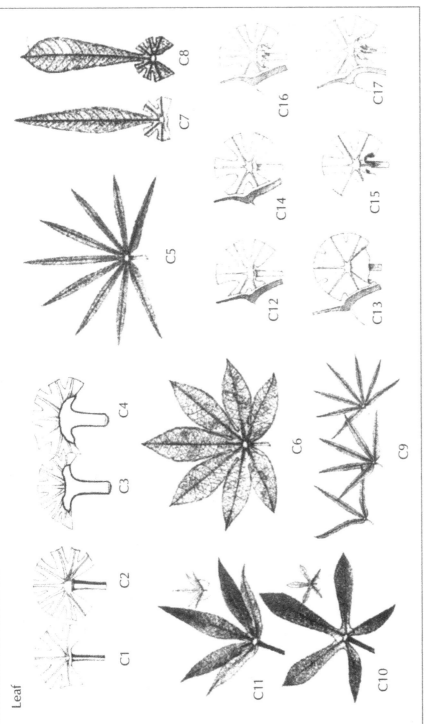

Leaf

C1 C2 C3 C4 C5 C6 C7 C8 C9 C10 C11 C12 C13 C14 C15 C16 C17

Fig. 5 (contd.)

Fig. 5 (contd.)

REFERENCES

Allem, A.C. 1994.The origin of *Manihot esculenta* Crantz (Euphorbiaceae). *Genetic Resources and Crop Evolution*, 41: 133-150.

Bai, K.V., Asiedu, R., and Dixon, A.G.O. 1993. Cytogenetics of *Manihot* species and interspecific hybrids. In: Ist International Scientific Meeting of the Cassava Biotechnology Network, W.M. Roca and A.M. Thro, eds., Cali, Colombia, CIAT, pp. 51-55.

Barcelos, E. 1998. Etude de la diversité du genre *Elaeis* (*E. oleifera* Cortès et *E. guineensis* Jacq.) par marqueurs moléculaires (RFLP et AFLP). Thése de doctorat, ENSAM, Montpellier, France, 137 p.

Barcelos, E., Second, G., Kahn, F., Amblard, P., Lebrun, P., and Seguin, M. 1998. Molecular markers applied to the analysis of the genetic diversity and to the biogeography of *Elaeis*. Memoirs of the New York Botanical Garden.

Bertram, R.B. 1993. Application of molecular techniques to genetic resources of cassava (*Manihot esculenta* Crantz, Euphorbiaceae): interspecific evolutionary relationships and intraspecific characterization. PhD thesis, University of Maryland, USA, 465 p.

CAB, 1974. *Cassava* (*Manihot esculenta*). Maidenhead, UK, CAB, Annotated Bibilography no. G-405, 27 p.

Chavarriaga-Aguire, P., Maya, M.M., Bonierbale, M.W., Kresovich, S., Fregene, M.A., Tohme, J., and Kockert, G., 1998. Microsatellites in cassava (*Manihot esculenta* Crantz): discovery, inheritance and variability. *Theoretical and Applied Genetics*, 97(3): 493-501.

Colombo, C. 1997. Etude de la diversité génétique de maniocs américains (*Manihot esculenta* Crantz) par les marqueurs moléculaires (RAPD et AFLP). Doct. thesis, ENSAM, Montpellier, France, 144 p.

Cours, G. 1951. Le manioc à Madagascar. *Mémoire de l'Institut scientifique de Madagascar, série B*, 3(2): 203-400.

de Bruijn, G.H. and Dhamaputra, T.S. 1974. The Mukibat system, a high-yielding method of cassava production in Indonesia. *Netherlands Journal of Agricultural Science*, 22: 89-100.

Dellaporta, S.L., Wood, J., and Hicks, J.B. 1983. A plant DNA preparation: version II. *Plant Molecular Biology Report*, 4: 19-21.

El-Sharkawy, M.A., and Cock, J.H. 1990. Photosynthesis of cassava (*Manihot esculenta*). *Experimental Agriculture*, 26: 325-340.

Emperaire, E., Pinton, F., and Second, G. 1998. Gestion dynamique de la diversité variétale du manioc en Amazonie du nord-ouest. *Nature, science et société*, 6(2): 27-42.

Excoffier, L., Smouse, P.E., and Quattro, J.M. 1992. Analysis of molecular variance inferred from metric distances among DNA haplotypes: application to human mitochondrial DNA data. *Genetics*, 131: 479-491.

FAO, 1998. Le manioc. In: *Perspectives de l'alimentation* no. 2. Rome, FAO-SMIAR.

Fregene, M., Angel, F., Gomez, R., Rodriguez, F., Chavarriaga, P., Roca, W.M., Tohme, J., and Bonierbale, M.W. 1997. A molecular genetic map for cassava (*Manihot esculenta* Crantz). *Theoretical and Applied Genetics*, 95(3): 431-441.

Gulick, P., Hershey, C., and Esquinar-Alcazar, J. 1983. *Genetic Resources of Cassava and Wild Relatives*. Rome, IBPGR, 56 p.

Hahn, S.K. 1984. Les plantes à racines et tubercules tropicales: amélioration et utilisation. Ibadan, Nigeria, IITA, Conf. rep. no. 2, 32 p.

Hershey, C., Iglesias, C., Iwanaga, M., and Tohme, J. 1994. Definition of a core collection for cassava. In: Ist meeting of the International Network for Cassava Genetic Resources. Rome, IPGRI, International Crop Network Series no. 10, pp. 145-156.

Howeler, R.H. 1990. Long-term effect of cassava cultivation on soil productivity. *Field Crops Research*, 26: 1-18.

Iglesias, C., Hershey, C., Calle, F., and Bolanos, A. 1994. Propagating cassava (*Manihot esculenta*) by sexual seed. *Experimental Agriculture*, 30: 283-290.

Jones, D.A. 1998. Why are so many food plants cyanogenic? *Phytochemistry*, 47(2): 155-162.

Kahn, F. and Second, G. 1998. The genus *Astrocaryum* in Amazonia: classical taxonomy and DNA analysis (AFLP). *Memoirs of the New York Botanical Garden*.

Keating, B.A., Wilson, G.L., and Evenson, J.P. 1985. Effect of photoperiod on growth and development of cassava (*Manihot esculenta* Crantz). *Australian Journal of Plant Physiology*, 12: 621-630.

Lebart, L., Morineau, A., and Tabard, N. 1977. *Techniques de la Description Statistique: Méthodes et Logiciels pour l'Analyse des Grands Tableaux*. Paris, Bordas, pp. 217-244.

Lefevre, F. 1989. Ressources génétiques et amélioration du manioc, *Manihot esculenta* Crantz, en Afrique. Paris, Orstom, Travaux et documents microédités no. 57, 175 p.

Magoon, M.L., Krishnan, R., and Bai, K.V. 1969. Morphology of the pachytene chromosomes and meiosis in *Manihot esculenta* Crantz. *Cytologia*, 34: 612-626.

Marie, D. and Brown, C. 1993. A cytometric exercise in plant DNA histograms, with 2C values for 70 species. *Biology of the Cell*, 78: 41-51.

McKey, D. and Beckerman, S. 1996. Ecologie et évolution des produits secondaires du manioc et relations avec les systèmes traditionnels de culture. In: L'Alimentation en forêt Tropicale. 1. Les ressources Alimentaires: Production et Consommation. C.M. Hladik et al., eds., Paris, UNESCO, pp. 165-202.

Medard, R., Sell, Y., and Barnola, P. 1992. Le développement du bourgeon axillaire du Manihot esculenta. Canadian Journal of Botany, 70: 2041-2052.

Purseglove, J.W. 1992. Tropical Crops. Londres, Royaume-Uni, Longman, 2 volumes.

Raffaillac, J.P. 1985. Pluviométrie et qualité de la production chez le manioc dans le sud de la Côte d'Ivoire. In: Eau et Développement Agricole. Adiopodoumé, Côte d'Ivoire, Orstom, pp. 78-81.

Raffaillac, J.P. 1992. Enracinement de la bouture de manioc (Manihot esculenta Crantz) au cours des premières semaines de croissance. L'Agronomie Tropicale, 46(4): 273-281.

Raffaillac, J.P. 1997. Le manioc: quelles priorités de recherche pour améliorer la production en relation avec la transformation et la commercialisation? Les Cahiers de la Recherche-Développement, 43: 7-19.

Raffaillac, J.P. 1998. Le Manioc et la Fertilité du Milieu. Montpellier, France, CNEARC-EITARC, 30 p.

Raffaillac, J.P. and Nedelec, G. 1988. Comportement du manioc en début de cycle en fonction de la durée de stockage de la bouture. In: VIIe Symposium ISTRC, June 1985. Paris, INRA.

Raffaillac, J.P. and Second, G. 1997. Le manioc. In: L'Amelioration des Plantes Tropicales. A. Charrier et al., eds., Montpellier, France, CIRAD-Orstom, collection Repères, pp. 429-455.

Roa, A.C., Chavarriaga, P., Duque, M.C., Maya, M.M., Bonierbale, M.W., Iglesias C., and Tohme, J., 2000. Cross-species amplification of cassava (Manihot esculenta Crantz) microsatellites: allelic polymorphism and degree of relationship. American Journal of Botany, 87(11): 1647-1655.

Roa, A.C., Maya M.M., Duque, M.C., Tohme J., Allem, A.C., and Bonierbale, M.W. 1997. AFLP analysis of relationships among cassava and other Manihot species. Theoretical and Applied Genetics, 95: 741-750.

Rogers D.J. and Appan, M. 1973. Manihot, Manihotoides (Euphorbiaceae). New York, Hafner Press, Flora Neotropica Monograph no. 13, 274 p.

Rogers D.J. and Fleming, H. 1973. A monograph of Manihot esculenta with an explanation of the taximetrics methods used. Economic Botany, 27: 1-113.

Rohlf, F.J. 1998. Ntsys-pc, numerical taxonomy and multivariate analysis system, version 2.0: user guide. New York, Exeter Software, 31 p.

Second, G. 1998. Manihot glaziovii contributed to the genetic make-up of cassava and represents an example of dynamic conservation and on-farm breeding of genetic resources. In: IVth International Scientific Meeting of the Casssava Biotechnology Network, Salavador de Bahia, Brazil.

Second, G., Allem, A.C., Emperaire, L., Ingram C., Colombo, C., Mendes, R.A., and Carvalho, J.C.B. 1997. Molecular markers (AFLP) based *Manihot* and cassava genetic structure analysis and numerical taxonomy in progress: implications for their dynamic conservation and genetic mapping. In: IIIrd International Scientific Meeting of the Cassava Biotechnology Network, A.M. Thro and M.O. Akoroda, eds., *African Journal of Root and Tuber Crops*, 2(1-2): 140-147.

Second, G., Colombo, C., Mendes, R.A. and Berthaud, J. 1998. A scheme for a dynamic conservation of the genetic resources of wild *Manihot* and cultivated Cassava in America and Africa and its extension to yam. In: Regional Workshop for the Conservation and Utilisation of Cassava, Sweetpotato and Yam Germplasm in Sub-Saharan Africa. Nairobi, Kenya, ILRI.

Second, G. and Iglesias, C. 2001. The state of use of cassava genetic diversity and a proposal to enhance it. In: *Broadening the Genetic Base of Crop Production*. H.D. Cooper, C. Spillane and T. Hodgkin, eds., IPGRI/FAO, pp. 203-231.

Serier, J.B. 1989. Historico da disseminação da maniçoba fora do Brasil. In: *Primeiro Encontro Nordestino da Maniçoba*. Recife, Brazil, IPA, pp. 89-95.

Webster, G.L. 1975. Conspectus of a new classification of the Euphorbiaceae. *Taxon*, 24: 593-601.

Citrus

Patrick Ollitrault, Camille Jacquemond,
Cécile Dubois and François Luro

Citrus fruits are the most extensively produced fruits in the world. About 90.9 million tonnes were produced in 1999/2000, of which 59.5 million tonnes were sweet oranges (FAO, 2000). The volume of fruit processed is increasing: concentrated and frozen orange juice for a large part of the processed fruit products in the United States and Brazil.

Citrus fruits were domesticated in Southeast Asia several thousand years ago and then spread throughout the world (Fig. 1). Citron (*C. medica* L.) was the first species cultivated in the Mediterranean basin, some centuries before the common era, while other species were introduced only during the second millennium. Citrus crops conquered America following the discovery of the New World during the 15th century. The area of citrus cultivation is today very wide, and it is located approximately between 40°N and 40°S latitude.

The cultivation of citrus faces increasing biotic and abiotic constraints in the major regions of production. Tristeza, a degenerating disease caused by the citrus tristeza virus, *Phytophthora* sp., and nematodes are found today throughout the cultivation areas. Other constraints are regional: cold and blight—which is a degenerating disease of still indeterminate origin—in the United States, citrus variegated chlorosis due to *Xilela fastidiosa* in Brazil, cercosporiosis caused by *Phaeoramularia angolensis* in Africa, and greening or citrus huanglongbing in Asia. Among the abiotic constraints, salinity and calcareous soils are major problems of the Mediterranean basin. The widespread use of grafted plants allows farmers to overcome soil-related constraints (calcareous soils, salinity, telluric parasites) to some extent, as well as tristeza. Scions are selected on the basis of qualitative aspects and, in some countries, characters of tolerance to citrus variegated chlorosis, to mal secco or to cercosporiosis (Ollitrault and Luro, 1997).

BOTANY AND GENETIC RESOURCES

Botany and Taxonomy

Partial apomixis by nucellar embryogenesis, associated with a wide sexual compatibility, has led to the production of clonal populations of interspecific

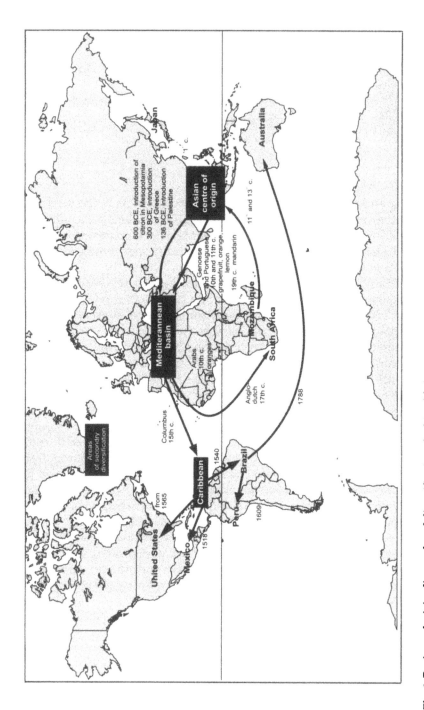

Fig. 1. Regions of origin, dispersal, and diversification of cultivated citrus.

hybrids, which have often been assimilated into new species by taxonomists. Botanic classifications are thus generally complicated. Tanaka (1961) identified 156 species, while Single and Reece (1967) distinguished only 16. The correspondence between these two classifications and the common names is given in Table 1 for the taxa studied in this chapter. In all the *Citrus* species and related genera, the base number of chromosomes (n) is equal to 9 (Krug, 1943). Almost all the *Citrus* are diploid and only a few natural polyploids have been identified, such as *Fortunella hindsii* or the Tahiti lime.

Genetic Resources

There are several collections of citrus throughout the world. They have two objectives, often divergent as to choice of plant material to be conserved: first, to preserve the diversity of *Citrus* and related genera over the long term, and second, to create orchards to provide grafts of valuable varieties. The collection of the Okitsu Branch (Fruit Tree Research Station) in Japan is the most important for cultivated material from the zones of origin, while the conservatory of the University of Malaysia is remarkable for its collection of Aurantioideae of Southeast Asia. The collections of the USDA (United States Department of Agriculture), IVIA (Instituto Valenciano de Investigaciones Agrarias) in Spain, and the University of Adana in Turkey contain certain Rutaceae related to the citrus but are regularly supplied for the most part by new varieties created throughout the world. The INRA and CIRAD station of San Giuliano, in France, has a unique status because of the favourable phytosanitary conditions of Corsica. It shelters a significant collection of healthy plant material, which includes numerous accessions of Southeast Asia and can be evaluated in the field. The Egid database management software, developed by CIRAD and INRA (Cottin et al., 1995) from the descriptors of the IPGRI (International Plant Genetic Resources Institute), has been adopted by the FAO to set up a global network to manage citrus genetic resources.

ORGANIZATION OF DIVERSITY

Agromorphological Variability

The agromorphological variability of citrus is considerable. It involves pomological and organoleptic characters as well as resistance to biotic and abiotic factors. The *Citrus* genus includes several sources of tolerance of biotic and abiotic stresses, which opens up interesting perspectives for the use of genetic resources in plant improvement.

Among the abiotic factors we can cite: cold tolerance in Satsuma mandarin trees; salinity tolerance in Rangpur lime trees and Cleopatra mandarins;

Table 1. Analysed Citrus accessions and genetic characteristics

Code	Cultivar	Swingle and Reece (1967)	Tanaka (1961)	Marker morpho.	iso.	Genome size (pg/2C)	ADH-1	AAT	IDH	LAP	MDH-1	MDH-2	PGI	PGM-1	PGM-2	PER	SKDH
Mandarins (M)																	
Mks	King of Siam	reticulata hybrid	nobilis	1	1	0.760	22	11	33	44	33	22	33	34	12	12	33
MSW*	Satsuma Wase	reticulata	unshiu	1	1	0.737	22	11	33	34	33	22	23	33	22	22	33
Mso	Satsuma Qwari	reticulata	unshiu	1	1	—	22	11	33	34	33	22	23	33	22	22	33
Mda	Dancy	reticulata	tangerina	1	1	0.736	22	11	33	44	33	22	23	23	22	12	33
Mte	Temple	reticulata hybrid	temple	1	1	0.748	22	11	33	24	33	22	23	22	22	22	33
Mcl	Cleopatra	reticulata	reshni	1	1	0.733	22	11	33	44	33	22	33	33	22	22	33
Mpo	Ponkan	reticulata	reticulata	1	1	0.744	22	11	33	44	33	22	33	23	22	11	22
Mco	Common	reticulata	deliciosa	1	1	0.730	22	11	33	45	33	22	34	23	22	12	23
M63	Clementine SRA63	reticulata	clementina	1	1	0.750	22	11	23	24	33	22	24	33	22	22	33
Mmu	Murcott	reticulata hybrid		1	1	0.746	22	11	33	44	33	22	33	33	22	22	33
Pemmelos (P)																	
Pme	Menara	grandis	sp.	1	1	0.751	22	12	33	22	33	22	22	13	11	12	23
Prk	Reinking	grandis	maxima	1	1	0.774	22	22	33	45	33	22	22	44	11	11	12
Pkp	Kao Pan	grandis	maxima	1	1	0.767	22	12	23	35	33	22	22	13	11	11	12
Psn	Sunshine	grandis	maxima	1	1	0.794	22	22	23	25	33	12	22	33	11	11	11
Ppi	Pink	grandis	maxima	1	1	0.779	22	11	33	55	33	22	22	13	11	11	11
Psp	Seedless	grandis	maxima	1	1	0.787	22	12	33	25	33	22	23	11	11	11	12
Pin	India	grandis	maxima	1	1	0.787	22	22	22	55	33	22	22	33	11	11	12
Pah	Tahiti	grandis	maxima	0	1	—	22	12	33	25	33	22	22	11	11	11	11
Pph	Philippines	grandis	maxima	0	1	—	22	22	33	55	33	22	22	11	11	11	11
Psu	Surinam	grandis	maxima	0	1	—	22	12	23	35	33	22	22	13	11	11	11
Pei	Eingedi	grandis	maxima	1	0	0.763	—	—	—	—	—	—	—	—	—	—	—
Pch	Chandler	grandis	maxima	1	0	0.764	—	—	—	—	—	—	—	—	—	—	—
Limes (L)																	
Lbs	Brazil Sweet	aurantifolia	limettioides	1	1	0.756	22	12	33	36	13	22	23	22	12	11	12

(Contd.)

(Table 1. Contd.)

Lga	Gallet	*aurantifolia*	*aurantifolia*	1	1	0.787	12	12	13	36	13	12	22	22	22	11	12
Lta	Tahiti	*aurantifolia*	*latifolia*	1	1	1.170											
Lme	Mexican	*aurantifolia*	*aurantifolia*	1	1	0.779											
Lel	Elkseur	*aurantifolia*	*latifolia*	1	1	1.170											
Lbe	Bears	*aurantifolia*	*latifolia*	1	1	1.170	22	22	13	36	13	12	22	23	12	11	22
Lca	Calédonie	*aurantifolia*	*aurantifolia*	1	1	0.784											
Lki	Kirk	*aurantifolia*	*aurantifolia*	1	1	0.779											
Lra	Rangpur	*aurantifolia*	*limonia*	1	1	0.772	22	12	13	36	13	22	23	22	22	11	13
Lka	Kanghzi	*aurantifolia*	*aurantifolia*	0	1	-	22	22	12	36	13	22	22	12	12	11	22
Lsr	IAC SRA618	*aurantifolia*	*aurantifolia*	1	0	1.170	-	-	-	-	-	-	-	-	-	-	-

● Lemons (C)

Cme	Meyer	*limon*	*meyeri*	1	1	0.772	22	12	23	46	13	22	23	23	12	12	12
Cfi	Fino	*limon*	*limon*	1	1	0.784											
Cve	Verna	*limon*	*limon*	1	1	-											
Cad	Adamapoulos	*limon*	*limon*	1	1	0.769											
Cdx	Doux	*limon*	*limon*	1	1	0.778											
Cli	Lisbon	*limon*	*limon*	1	1	0.786	22	12	13	46	13	22	24	23	12	12	12
Cvi	Villafranca	*limon*	*limon*	1	1	0.776											
Cmo	Monachello	*limon*	*limon*	1	1	0.787											
Ceu	Euréka	*limon*	*limon*	1	1	0.777											
Clu	Lunari	*limon*	*limon*	0	1	-											
Cst	Santa Teresa	*limon*	*limon*	1	0	0.786	-	-	-	-	-	-	-	-	-	-	-

● Sweet oranges (O)

Owa	Washington Navel	*sinensis*	*sinensis*	1	1	0.757											
Odf	Double Fine	*sinensis*	*sinensis*	1	1	0.778											
Ota	Tarocco	*sinensis*	*sinensis*	1	1	0.772											
Onh	New Hall	*sinensis*	*sinensis*	1	1	0.778											
Ona	Navelina	*sinensis*	*sinensis*	1	1	0.755	22	23	24	33	22	23	33	12	22	22	12

(Contd.)

(Table 1. Contd.)

| Code | Cultivar | Name of species | | Marker morpho. iso. | Genome size (pg/2C) | Enzymatic genotype | | | | | | | | | | |
		Swingle and Reece (1967)	Tanaka (1961)			ADH-1	AAT	IDH	LAP	MDH-1	MDH-2	PGI	PGM-1	PGM-2	PER	SKDH
Oha	Hamlin	*sinensis*	*sinensis*	1	0.749											
Osh	Shamouti	*sinensis*	*sinensis*	1	0.756											
Opb	Parson Brown	*sinensis*	*sinensis*	1	0.756											
Oca	Cadenera	*sinensis*	*sinensis*	1	0.751											
Ovl	Valencia Late	*sinensis*	*sinensis*	1	0.757											
● Sour oranges (B)																
Bfe	Ferrando	*aurantium*	*aurantium*	1	0.755											
Bfl	Florida	*aurantium*	*aurantium*	1	0.755											
Bse	Thornless	*aurantium*	*aurantium*	1	0.779											
Bma	Maroc	*aurantium*	*aurantium*	1	0.750											
Bqn	Nice (bouquetier)	*aurantium*	*aurantium*	1	0.756											
Bqf	Fleurs (bouquetier)	*aurantium*	*aurantium*	1	0.750	22	11	33	44	33	22	24	13	12	12	22
Bbs	Brazil Sour	*aurantium*	*aurantium*	1	-											
Bdd	Dai Dai	*aurantium*	*aurantium*	1	0.756											
Btu	Tuléar	*aurantium*	*aurantium*	1	-											
Bav	Avanito	*aurantium*	*aurantium*	0	-											
Bgr	Granito	*aurantium*	*aurantium*	1	0.753	-	-	-	-	-	-	-	-	-	-	-
● Citrons (K)																
Kdc	Corse	*medica*	*medica*	1	0.814											
Ket	Etrog	*medica*	*limonimedica*	1	0.821											
Kde	Digite	*medica*	*medica*	1	0.815	22	22	22	66	11	22	22	22	22	11	22
Kpc	Poncire	*medica*	*medica*	1	0.807											
Kdi	Diamante	*medica*	*medica*	0	-	-	-	-	-	-	-	-	-	-	-	-
● Grape fruits (G)																
Gsh	Shambar	*paradisi*	*paradisi*	1	0.749	-	-	-	-	-	-	-	-	-	-	-

Code						22	12	33	25	33	22	22	13	22	11	12	23
Gce	Cecily	*paradisi*	*paradisi*	1	0.778												
Gal	Alanoek	*paradisi*	*paradisi*	1	0.759												
Gre	Reed	*paradisi*	*paradisi*	1	0.772												
Gsr	Star Ruby	*paradisi*	*paradisi*	1	0.772												
Grb	Red Blush	*paradisi*	*paradisi*	1	0.788												
Glr	Little River	*paradisi*	*paradisi*	1	0.784												
Gth	Thomson	*paradisi*	*paradisi*	1	0.784												
Gma	Marsh	*paradisi*	*paradisi*	1	0.783												
Gru	Ruby	*paradisi*	*paradisi*	1	0.781	-	-	-	-	-	-	-	-	-	-	-	-
• Other Citrus																	
ROL	rough lemon	*limon*	*jambhiri*	1	0.777	12	12	23	46	13	22	23	22	22	22	11	22
PEC	*pectinifera*	*reticulata hybrid*	*depressa*	1	0.751	22	11	33	24	23	22	33	33	22	12	12	33
JUN	-	*ichang austera*	*junos*	1	0.810	22	12	33	24	23	22	33	13	12	11	11	22
GUL	-	*maxima*	*pseudogulgul*	1	0.745	22	12	33	24	33	22	23	11	23	11	11	22
ICH	ichangensis lemon	*ichangensis*	*ichangensis*	1	0.774	22	11	22	44	23	22	23	34	23	12	11	22
BGM	bergamot	*aurantifolia*	*bergamia*	1	0.771	22	12	13	44	13	22	24	13	24	12	12	12
PDC	commander pear	*limon*	*lumia*	1	-	22	12	33	24	33	22	22	34	22	12	12	22
COM	*combava*	*hystrix*	*hystrix*	1	0.803	22	12	33	14	12	22	22	33	22	11	11	12
INT	-	*paradisi*	*intermedia*	1	0.764	22	12	33	23	33	22	23	23	23	11	12	22
MAC	-	*aurantifolia*	*macrophylla*	1	0.798	22	22	23	44	12	22	12	23	12	12	11	22
PEN	-	*aurantifolia*	*pennivesiculata*	1	0.813	22	22	13	22	11	12	22	23	22	12	11	12
EXE	-	*aurantifolia*	*excelsa*	1	0.793	22	22	23	44	33	12	12	23	22	22	11	22
SIA	siamelo	*hybrid*	*hybrid*	1	0.745	22	11	33	24	33	22	23	33	12	12	11	12
KPA	khasi papeda	*latipes*	*latipes*	1	0.780	22	12	34	44	23	23	12	34	12	11	11	12
HAL	-	*halimii*	*halimii*	1	0.778	22	22	22	44	22	23	12	34	22	22	11	22
VOL	-	*limon*	*limonia*	1	0.764	12	12	13	46	13	22	23	22	23	22	11	12
NAS	*nasnaran*	*reticulata hybrid*	*amblycarpa*	0	-	12	12	33	44	33	22	13	33	13	11	11	12

*The codes in bold face represent the common enzymatic type in the analyses. All the individuals in a set of rows shaded in grey have the same enzymatic profile.

calcareous soil tolerance in *C. jambhiri*, *C. macrophylla*, *C. volkameriana*, *C. amblycarpa*, and sour oranges; and drought tolerance in Rangpur lime. Tolerance of the major pests and diseases has also been identified: tolerance of *Phytophthora* sp. in some pummelos, sour orange, *C. volkameriana*, and *C. amblycarpa*; African cercosporiosis tolerance in grapefruit, lemon, and Satsuma and Beauty mandarin; tristeza tolerance in Cleopatra mandarin, *C. amblycarpa*, Rangpur lime, *C. jambhiri*, and *C. volkameriana*; blight tolerance in orange; tolerance of citric canker due to *Xanthomonas campestris* in *C. junos* and some mandarins (Satsuma and Dancy, for example); and resistance to phytophagous acarids of Marsh pomelo and mandarins (Satsuma and Dancy). In view of these examples, there seems to be no link between the distributions of sources of resistance to biotic factors and the specific structure of the genus *Citrus*.

On the other hand, the morphophysiological variability is strongly marked between the species, even though certain characters selected by humans have a strong intraspecific diversity (precocity, calibre, colour of fruits). For example, within the genus *Citrus*, the diameter of fruits varies from a few centimetres for certain mandarins and limes to more than 30 cm for some grapefruits. Albedo is nearly non-existent in mandarins but is the essential characteristic of the fruit in the citron. The fruit pulp is green, orange, yellow, or red. Its acidity is nil in some sweet oranges and very high in limes and lemons. Although the leaves of all the species of *Citrus* are monofoliate, their size and shape as well as the shapes of the trees vary considerably according to the species.

A more refined study of the structure of morphological diversity in the genus *Citrus* has been done from 20 descriptors of the vegetative apparatus observed among 74 cultivars. It supports the analysis of relations between morphological diversity and molecular diversity presented in this chapter.

Biochemical and Molecular Variability

Essential oils and polyphenols were the first markers used to characterize varieties (Tatum et al., 1974) and to study the phylogenesis of citrus (Scora, 1988). Isozymes were used routinely to identify the zygotic or nucellar origin of seedlings (Soost et al., 1980; Khan and Roose, 1988; Ollitrault et al., 1992). They also make it possible to specify phylogenetic relations between species (Torres et al., 1982; Hirai et al., 1986; Ollitraul and Faure, 1992; Herrero et al., 1996, 1997). The techniques of direct analysis of DNA polymorphism—DNA, RFLP, RAPD, variable number of tandem repeats (VNTR)—were mainly applied in genome mapping programmes (Durham et al., 1992; Jarrel et al., 1992; Luro et al., 1994b; Fang et al., 1998; Moore et al., 2000) or programmes of varietal characterization and taxonomy (Luro et al., 1994a, 1995; Fang and Roose, 1996; Federici et al., 1998; Nicolosi et al., 2000). Nevertheless, the allelic determinism of these markers is sometimes difficult to clarify, so they have limited use in genetic studies of populations concerning heterozygosity and index of fixation or index of gametic inequilibrium.

Cytogenetic studies and flow cytometry analyses have demonstrated the existence of great variations between species as to chromosome size (Nair and Randhawa, 1969; Ollitrault et al., 1994). They also have proved many cases of structural heterozygosity (Raghuvanshi, 1969; Gmitter et al., 1992; Guerra, 1993; Miranda et al., 1997). These elements on the structure of genomes of different taxa are determinants for analysis of the organization of allelic diversity in evolutionary terms.

In order to study the parameters of population structure, the analysis of allelic diversity presented in this chapter relies on the polymorphism of 9 isozymic systems. The nuclear structural diversity is also examined by evaluation of genome size using flow cytometry. The varietal sampling for the cultivated forms is the same as for the study of morphological diversity. Seventeen non-cultivated *Citrus* spp. complete the analysis.

ISOZYMIC DIVERSITY

Thirty-five alleles were identified for 11 polymorphous loci. Only 5 of these alleles were not observed in cultivars. The null allele of the locus *LAP* (*LAP-6*), identified at the homozygous state in the citrons, was detected in the heterozygous state in a certain number of acid citrus (lemons, limes) when controlled hybrids were examined. Several cultivars of a single species presented identical profiles. This was particularly the case for orange, sour orange, pomelo, and lemon. The 74 cultivars were thus grouped into 30 isozymic genotypes (Table 1).

There appears to be widely varying intraspecific diversity among the edible species (Table 2). The citrons have nil allelic diversity due to a high homozygosity and the absence of polymorphism between the cultivars analysed. The grapefruit, sweet orange, and sour orange have similar intraspecific structures. The allelic diversity and heterozygosity in them are moderate and the intercultivar polymorphism is nonexistent. Lemons are

Table 2. Structure of intraspecific allelic diversity observed for 11 loci coding for isozymes

	No.	Mean no. of alleles per locus	Total diversity	Intercultivar diversity	Observed heterozygosity	Deviation of panmixia
Citron	4	1.00	0.00	0.00	0.00	—
Grapefruit	10	1.45	0.23	0.00	0.45	***(5 loci)
Sour orange	10	1.36	0.18	0.00	0.36	***(4 loci)
Sweet orange	10	1.45	0.23	0.00	0.45	***(5 loci)
Lemon	10	1.00	0.42	0.02	0.82	***(9 loci)
Lime	10	2.09	0.34	0.08	0.54	**(2 loci)
Pummelo	10	2.09	0.25	0.13	0.24	ns
Mandarin	10	2.00	0.19	0.10	0.17	ns

ns: non-significant at threshold of 5%.
**significant at threshold of 1%.
***significant at threshold of 1‰.

highly heterozygous but have very little intervarietal polymorphism. Indeed, only the cultivar "Meyer" can be differentiated from the other ones. The limes are also highly heterozygous and manifest a stronger intervarietal polymorphism than the lemons. The pummelos and mandarins have a very high allelic richness, mainly due to significant intervarietal polymorphism. The two species that have great intercultivar diversity—mandarins and pummelos—do not display a significant deviation to panmixis, which undoubtedly demonstrates an important genetic exchange within these taxa. All the other species, with the exception of citrons, which are totally fixed, have an excess of heterozygotes.

The total diversity of the sample of cultivated citruses, in the sense of Nei (1973), is 0.45. It is broken down in a balanced manner in terms of intraspecific diversity (0.23), and interspecific diversity (0.22), with a high value of the G_{ST} coefficient (0.49). This value indicates a marked allelic differentiation between the cultivated taxa. Indeed, it is significant for 10 of the 11 loci analysed. This differentiation between taxa, observed for nearly all the loci, is also found in the multilocus structure evaluated from the 30 genotypes of cultivated *Citrus*. The linkage disequilibrium thus involves 23 locus pairs out of 55 and 9 loci out of 11.

This strong structuration observed within the cultivars is confirmed when one looks at 47 enzymatic genotypes identified, which relates the 30 genotypes of cultivars to 17 other *Citrus* spp. Nine loci out of 11 present significant deviation to panmixis and a shortage of heterozygotes. This type of deviation is classically linked to structures in sub-populations (Walhund effect) and to systems of reproduction that limit gene flow.

The high level of genetic organization observed using genetic parameters of populations is found in the principal coordinates analysis (PCoA)done on the genotypes of cultivars, where 50.4% of the total variance is represented on the 1-2 plane (Fig. 2). The diversity of cultivated *Citrus* is structured around three gene pools: the first contains the mandarins, the second contains the grapefruit and pummelo, and the third is made up of the citrons, which show a marked relationship to the limes. The oranges and sour orange are close to the mandarins, with a probable introgression of pummelo. The lemons, highly heterozygous, may have evolved from a hybridization between the citron/lime group and the group made up of the mandarins, sweet oranges, and sour oranges. Factorial analysis allows us to identify the hybrid forms and their potential parents for this highly organized population.

This organization of cultivated forms around three pools is not called into question by the introduction of non-cultivated forms, as shown by the diversity tree that is constructed by NJ analysis of Dice dissimilarity (Fig. 3). Certain non-cultivated *Citrus* are associated with the groups formed by the cultivars: *C. pectinifera* with the mandarins; siamelo with the oranges; *C. pseudogulgul* and *C. intermedia* with the group of grapefruits and pummelos;

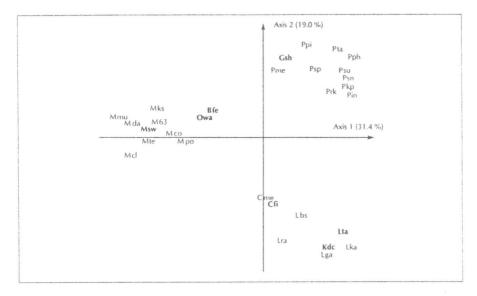

Fig. 2. Isozymic diversity of cultivated citrus on the basis of 11 loci: representation of the first factorial plane of PCoA done on a Dice matrix of dissimilarity between 30 different genotypes identified among 74 cultivars. The codes are the same as those used in Table 1.

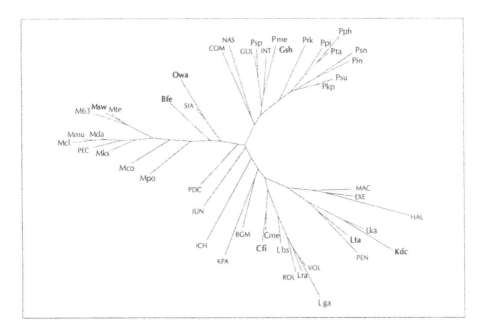

Fig. 3. Isozymic diversity of the genus *Citrus* on the basis of 11 loci: tree representation according to the NJ method, done on a Dice matrix of dissimilarity between 47 genotypes (30 cultivated genotypes and 17 other *Citrus*). The codes are the same as those used in Table 1.

C. *pennivesiculata, C. volkameriana* and *C. jhambivi* with the group of limes. The others are distinguished from these groups either because they carry alleles that are not observed in the cultivars—as with *C. macrophylla, C. excelsa, C. junos, C. ichangensis, C. latipes, C. hystrix,* and *C. amblycarpa*—or because they have original recombined allelic structures, such as *C. bergamia* or *C. lumia.*

GENOME SIZE

The size of nuclear genomes of individuals is given in Table 1. The diploid genotypes have relatively small genomes, between 0.73 and 0.82 pg of DNA per diploid genome (Fig. 4). The values of 1.17 pg correspond to triploid genotypes; they were observed for four cultivars of lime, Tahiti, Bears, Elkseur, and IAC SRA618. Among the edible species, the interspecific differences are

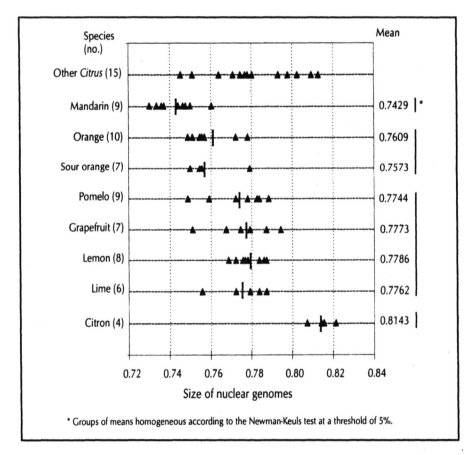

Fig. 4. Size of nuclear genomes of 75 individuals, of which 60 are edible cultivars grouped by species (mean of 3 measurements in picograms of DNA per diploid genome).

statistically significant and represent a deviation of 10% between the mandarins and the citrons (Fig. 5). The other species are divided into two groups of intermediate sizes. One comprises oranges and sour oranges, the other, corresponding to the larger sizes, comprises lemons, limes, grapefruits, and pummelos. The inedible types also have genome sizes between those of mandarins and citrons. Thus, two out of three taxa that structure the diversity, mandarin and citron, have genome sizes that are at the extremes observed in the genus *Citrus*. The other taxa have genome sizes that agree with the genetic affinities determined by isozymic analyses.

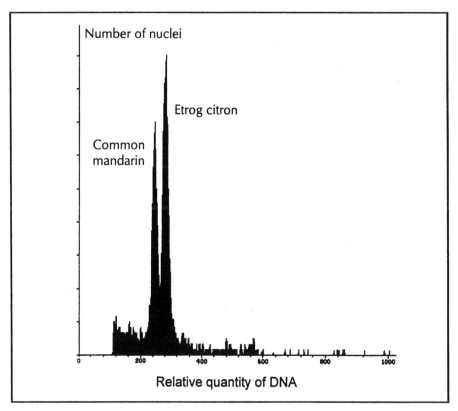

Fig. 5. Relative sizes of nuclear genomes of Etrog citron and common mandarin: flow cytometry of a mixture of nuclei stained with propidium idodide.

RELATIONS BETWEEN THE DIFFERENT LEVELS OF VARIABILITY

Analysis of morphological diversity from 20 vegetative descriptors allows us to find the overall organization around three gene pools previously identified according to the isozymic data (Fig. 6). The relative positions of cultivated species around these three axes are in the conserved set. On the other hand, the monomorphic species in the enzymatic sense present a morphological dispersal equivalent to that of polymorphic species in the molecular sense (Fig. 7). Two levels thus coexist in the organization of morphological diversity: one major level, which responds to the constraints affecting the evolution of the genome as a whole, and a secondary level, dissociated from the molecular evolution visualized by the isozymes.

Interspecific Organization

Except for the system of gametophytic self-incompatibility, there is no sexual incompatibility within the genus *Citrus*: hybrids are obtained easily for all the interspecific combinations. The notion of specific differentiation could thus be called into question. Nevertheless, this genus seems to be very highly organized to the extent that generalized gametic disequilibrium has been identified for the isozymes and to the extent that the major axes of molecular and morphological structuration appear similar. This indicates an organization into sub-populations between which the gene flows are limited, as confirmed by the deviations from the panmixia observed for almost all loci.

The organization of *Citrus* diversity around three taxa (*C. reticulata, C. medica*, and *C. maxima*) confirms the results of numerical taxonomy of Barret and Rhodes (1976), which have suggested that these taxa were the origin of the cultivated *Citrus* group. It is also in agreement with total protein analysis (Handa et al., 1986), isozyme analysis (Herrero et al., 1996, 1997), RFLP and RAPD analysis (Luro et al., 1994a; Federici et al., 1998; Nicolosi et al., 2000) and STMS analysis (Luro et al., in press). The differentiation between these sexually compatible taxa can be explained by foundation effect in three geographic zones and by an allopatric evolution. The pummelos originated in the Malay Archipelago and Indonesia, the citrons evolved in northeastern India and the nearby regions of Burma and China, and the mandarins were diversified over a region including Vietnam, southern China, and Japan (Webber, 1967; Scora, 1975).

The other cultivated species—sweet orange, sour orange, lemon, grapefruit, lime—appeared subsequently by recombinations among the basic taxa, which came into contact during the course of trade and migrations. The enzymatic data—generally high heterozygosity and absence of intervarietal polymorphism, confirmed recently with STMS (Luro et al., in press)—prove that there are typical cases of false species, in which varietal diversification is produced from an ancestral hybrid by accumulation of mutations without the intervention of sexual recombination. It is to be noted that all the cultivars of these species are polyembryonic, which allows us to fix the heterozygosity and to conserve the morphological and pomological type even without manual methods of vegetative propagation, such as layering, budding, or grafting.

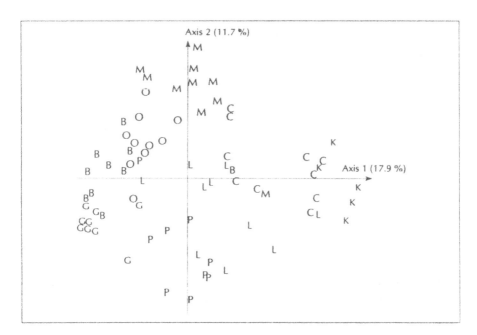

Fig. 6. Morphological diversity: representation of the primary factorial plane of PCoA done on a Sokal and Michener matrix of distance between 74 cultivars on the basis of 20 vegetative descriptors. The codes used are the same as those in Table 1.

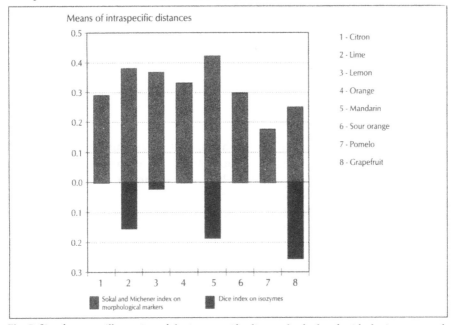

Fig. 7. Simultaneous illustration of the intraspecific dispersal calculated with the isozyme and morphological markers.

Our conclusions are in agreement with the ones obtained by isozyme analysis (Herrero et al., 1996), RFLP (Federici et al., 1998), RAPD and SCAR (Nicolosi et al., 2000), and STMS (Luro et al., in press). Sweet oranges and sour oranges are close to mandarins but have introgressed nuclear genomic fragments of pummelo. The last species also transmits its cytoplasmic genomes to sweet and sour oranges (Nicolosi et al., 2000; Ollitrault et al., 2000). Grapefruit is close to pummelo but includes nuclear genomic fragments of the mandarins/oranges group. It should have resulted from a hybridization between pummelo and sweet oranges introduced in the Caribbean islands after the discovery of the New World by Christopher Columbus. The genetic relationship between citron, limes, and lemons is clearly established by morphological and nuclear molecular markers. Synthesis of nuclear and cytoplasmic data (Ollitrault et al., 2000) indicated that mandarin and pummelo gene pools also contribute to lemon genesis. Nicolosi et al. (2000) suggested that it should result from a hybridization between citron and sour orange. Lime is the only cultivated species for which there is evidence of interspecific origin between cultivated and non-cultivated taxa; it should result from a hybridization between citron and C. *micrantha* (Nicolosi et al., 2000).

The strong organization, still observed today at the molecular rather than morphological scale indicates that the genetic exchanges between the three original groups are limited. The partial apomixis, linked to the polyembryony, has certainly been an essential element in the limitation of gene flows. Other factors, such as the structural differentiation of genomes, have also favoured the maintenance of gametic disequilibrium by limiting recombination on large portions of the genome. This differentiation in genome size is in agreement with the cytogenetic observations of Nair and Randhawa (1969) and of Raghuvanshi (1969). It testifies to the advanced state that the three basic taxa have reached on the way to real speciation.

Intraspecific Diversification

Intervarietal morphological polymorphism, relatively significant within sweet orange, sour orange, grapefruit, lemon, and lime, is explained largely by human selection. This is particularly marked for the pomological and phenological criteria. It can lead to a rapid morphophysiological evolution, independent of the molecular evolution analyses using isozymes. The most obvious example is that of the clementine. Appearing about a century ago in a seedling of common mandarin planted by Father Clement, it has since been considerably diversified. This diversification, result of a simple selection of bud mutations in the orchard, involves precocity—the period of production today extends from October to March—as well as pomological characters such as calibre, colour, and the presence of pips (Bono et al., 1982).

Over a much longer period, sweet oranges have diversified in the same way. This species, for which molecular studies with isozymes, RAPD

(Luro et al., 1994a), and microsatellites (Luro et al., 1995, in press) have not displayed any intervarietal polymorphism, is, however, highly polymorphic for morphological and phenological characters. Even though its introduction in the Mediterranean Basin is relatively recent (around the year 1000), this area constitutes the main centre of diversification, where all the main types of modern sweet oranges have been selected, such as common oranges, blood oranges, and navel oranges (Aubert, in press).

On the other hand, sexual recombination has also played a determining role in the diversification of pummelo, of which the cultivars are all monoembryonic, and of mandarin, certain cultivars of which are mono-embryonic. These two have high intervarietal isozymic polymorphism without significant difference in the panmixia.

GENETIC RESOURCE MANAGEMENT

The situation of citrus illustrates the uses and limitations of molecular markers in the construction of core collections. In the evolution of the genus *Citrus* we find factors that, on the global scale, show a good correlation between organization of the phenotypic diversity and organization of the molecular diversity (foundation effect, allopatric evolution, and limitation of gene flow that allow the maintenance of global gametic disequilibrium). For the secondary species, there are also, on the intraspecific scale, evolutionary mechanisms, such as somatic reproduction and strong selection pressures on the mutations affecting morphophysiological characters, which lead to dissociation of the two levels of evolution. In the case of citrus, the chief utility of the marking studies lies in the identification of sequences and evolutionary factors at the origin of taxa and their diversification. Studies on the constitution of a core collection must thus be based more on this general information than on the allelic constitution of individuals.

Among the three basic species, pummelos and mandarins have significant molecular polymorphism. Intraspecific varietal improvement can be done traditionally by sexual hybridization. The management of intraspecific genetic resources can thus be rationalized conventionally in the form of core collections. The results obtained from a collection of 100 mandarin trees indicate the existence of genetic organization on the intraspecific scale, which could help establish, among other things, a sampling strategy on the basis of molecular data.

The set of characters defining the other cultivated species—sweet orange, sour orange, grapefruit, lemon—relies on genotypes that have a relatively high heterozygosity but are stabilized by vegetative propagation. Conser-vation of the genetic resources of each of these species must be based on the constitution of genotype collections. This intraspecific diversity is difficult to recombine sexually for improvement of the 'species' because the characters defining the 'species' are thereby recombined. The genotype collections, which aim to conserve the widest adaptive diversity and morphological diversity within each 'species', help inform citrus farmers about cultivars best adapted to particular regions. Classical molecular markers (isozymes, STMS, RFLP, RAPD) offer no information at this level, given the mechanisms of intraspecific evolution described earlier; the stratification must be based mainly on geographic criteria and agromorphological data.

When we discuss citrus diversity in general, genetic resource management can be rationalized also in terms of gene conservation. The three taxa identified as being the origin of most of the cultivated forms thus constitute an essential reservoir since a large part of the allelic diversity exists at the intercultivar level. The mandarins and pummelos seem in this case to be more important in the conservatories. The limes group displaying important genotypic diversity as well as the evidence of the contribution of a fourth taxon (probably *C. micrantha*; Nicolosi et al., 2000) must also be preserved on a priority basis. Moreover, as our study has shown, certain non-cultivated citrus carry a rich allelic diversity. These taxa thus are not particular genotype combinations arising from hybridization between the three basic taxa of the cultivated forms. It seems essential to conserve them, particularly because they may contribute tolerances to biotic or abiotic factors in the process of stock improvement. Finally, the development of biotechnologies, particularly somatic hybridization, considerably enlarges the gene pool that can be used for the breeding (Grosser et al., 2000). It is thus advisable today to conserve the genetic resources of citrus at the level of the tribe Citreae.

APPENDIX

Plant Material

Seventy-four cultivars representing the 8 species cultivated for their fruits (Swingle and Reece, 1967) and 17 non-edible types, some of which are used as stock, served as the basis of the enzymatic study (Table 1). To the extent possible, 10 cultivars were retained for each species cultivated, with the exception of citron, for which we had only 4 genotypes available in the collection. The trees, protected from any viral or viroidal disease, were cultivated at the agronomic research station of INRA and CIRAD of San Giuliano, in Corsica. Ninety of these genotypes were the subject of a morphological description.

Enzymatic Analyses

Nine enzymatic systems were analysed by electrophoresis on starch gel or polyacrylamide gel (Ollitrault et al., 1992): alcohol dehydrogenase (ADH), malate dehydrogenase (MDH), isocitrate dehydrogenase (IDH), shikimate dehydrogenase (SKDH), phosphoglucomutase (PGM), phosphogluco-isomerase (PGI), peroxydases (PER), leucine aminopeptidase (LAP), and aspartate aminotransferase (AAT). For the locus $PGM-2$, only two allele positions were retained. For the other systems, the interpretation and allelic nomenclature were the same as those of Ollitrault et al. (1992) and were in accordance with the interpretation given by Torres et al. (1978, 1982) for MDH, IDH, PGI, and LAP.

Flow Cytometry Analysis

The nuclear genome size of each of the diploid genotypes was estimated by the mean of three measurements relative to that of a triploid cultivar (Tahiti lime), used as an internal control. Leaf pieces of the sample and of the control were prepared in mixtures and coloured with propidium iodide according to the protocol described by Ollitrault et al. (1994). Two thousand nuclei were then analysed on a Fascan cytometer. The nuclear genome size of each genotype was estimated in picograms per diploid genome from the mean of relative values multiplied by 1.17 pg, which corresponds to the genome size of Tahiti lime estimated by Ollitrault et al. (1994).

Morphological Studies

Twenty qualitative descriptors of the vegetative parts (Table 3) were studied. The set of data on the morphology of citrus was managed by the computerized database system for the citrus germplasm network EGID (Cottin et al., 1995).

Table 3. The twenty qualitative morphological descriptors

A. Shape of tree
   1. Erect
   2. Spheroid
   3. Flat ellipsoid

B. Position of branches
   1. Erect
   2. Spread out
   3. Drooping
   4. Weeping

C. Density of foliage
   1. Sparse
   2. Dense

D. Surface of trunk
   1. Smooth
   2. Rough

E. Colour of leaf surface
   1. Light green
   2. Green
   3. Dark green

F. Colour of underside of leaf in
   relation to leaf surface
   1. Identical
   2. Lighter

G. Nerves on leaf surface
   1. Prominent
   2. Not prominent

H. Angle of leaf base
   1. Acute
   2. Obtuse

I. Angle of leaf tip
   1. Acute
   2. Obtuse

J. Articulation of leaf
   1. Present
   2. Absent

K. Attachment of petiole to branch
   1. Straight
   2. Angled

L. Density of spines
   1. Nil
   2. Low
   3. Moderate
   4. High

M. Length of spines
   1. Nil
   2. Very short (0 to 5 mm)
   3. Short (5 to 15 mm)
   4. Medium (15 to 40 mm)
   5. Long (> 40 mm)

N. Shape of section of young branches
   1. Angular
   2. Round

O. Leaf edge
   1. Crenellate
   2. Dentate
   3. Entire
   4. Undulate

P. Leaf form
   1. Elliptical
   2. Oval
   3. Inverse oval
   4. Lanceolate
   5. Orbiculate

Q. Length of petiole
   1. Nil
   2. Short (0 to 10 mm)
   3. Medium (10 to 15 mm)
   4. Long (15 to 35 m)
   5. Very long (>35 mm)

R. Shape of lamina
   1. Absent
   2. Cordiform
   3. Deltoid
   4. Oval

S. Size of lamina
   1. Insignificant
   2. Small
   3. Medium
   4. Large
   5. Very large (equal to the limb)

T. Colour of young shoots
   1. Anthocyanate
   2. Green

Statistical Analyses

The parameters of genetic structuration were studied using Genepop software for analysis of deviations at panmixia, differentiation between cultivated taxa (study of allele distribution in the species by the exact test of Fisher), and gametic disequilibrium. The descriptive parameters of the diversity—total diversity, diversity between taxa, diversity between individuals, G_{ST}—are those proposed by Nei (1973). The tree representations and PCoA were done on the basis of the Dice matrix of distance for the enzymatic data and the Sokal and Michener matrix of distance for the morphological data. The trees were constructed by the neighbour-joining method with the help of Darwin software (Perrier et al., 1999).

REFERENCES

Aubert, B. Text of presentation of the 2000 revision. In: *Histoire Naturelle des Orangers*. A. Risso and A. Poiteau, eds., Connaissance et Memoires Europeenne. (In press).

Barret, H.C. and Rhodes, A.M. 1976. A numerical taxonomic study of affinity relationships in cultivated *Citrus* and its close relatives. *Systematic Botany*, 1: 105-136.

Bono, R., Fernandez de Cordova, L., and Soler, J. 1982. Arrufatina, Esbal and Guillermina, three Clementine mandarin mutations recently appearing in Spain. Proceedings of the International Society of Citriculture, 1: 94-96.

Cottin, R., Allent, V., and Jacquemond, C. 1995. Gestion informatique des ressources génétiques: Egid. In: Symposium Méditerranéen sur les Mandarines. San Giuliano, France, INRA, p. 2.

Durham, R.E., Liou, P.C., Gmitter, R.G., and Moore, G.A. 1992. Linkage map of restriction fragment length polymorphisms and isozymes in *Citrus*. *Theoretical and Applied Genetics*, 84: 39-48.

Fang D. and Roose M.L. 1996. Fingerprinting citrus cultivars with inter-SSR markers. Proceedings of the International Society of Citriculture. 185-188.

Fang, D.Q., Federici, C.T., and Roose, M.L. 1998. A high resolution linkage map of the citrus tristeza virus resistance gene in *Poncirus trifoliata* (L.) Raf. *Genetics*, 150: 883-890.

FAO, 1997. Annuaire production: 1996. Rome, FAO.

Federici, C.T., Fang, D.Q., Scora, R.W., and Roose, M.L. 1998. Phylogenetic relationships within the genus *Citrus* (Rutaceae) and related genera as revealed by RFLP and RAPD analysis. *Theoretical and Applied Genetics*, 96: 812-822.

Gmitter, F.G., Deng, X.X., and Hearn, C.J. 1992. Cytogenetic mechanism underlying reduced fertility and seedlessness in *Citrus*. In: Vllth International Citrus Congress, pp. 113-116.

Green R.M., Vardi, A., and Galun, E. 1986. The plastome of *Citrus*: physical map, variation among *Citrus* cultivars and species and comparison with related genera. *Theoretical and Applied Genetics*, 72: 170-177.

Grosser, J.W., Mourao-Fo, A.A., Gmitter, F.G. JR., Louzada, E.S., Jiang, J., Baergen, K., Quiros, A., Cabasson, C., Schell, J.L., and Chandler, J.L. 1996. Allotetraploid hybrids between *Citrus* and seven related genera produced by somatic hybridization. *Theoretical and Applied Genetics*, 92: 577-582.

Grosser, J., Ollitrault, P., and Olivares, O. 2000. Somatic hybridization in *Citrus*: an effective tool to facilitate variety improvement. *In Vitro Cell Development Biology-Plants*, 36: 434-449.

Guerra, M.S. 1993. Cytogenetics of Rutaceae. 5. High chromosomal variability in *Citrus* species revealed by CMA/DAPI staining. *Heredity*, 71: 234-241.

Handa, T., Ishizawa, Y., and Oogaki, C. 1986. Phylogenetic study of fraction I protein of *Citrus* and its close related genera. *Journal of Genetics*, 61: 15-24.

Herrero, R., Asins, M.J., Carbonell, E.A., and Navarro, L. 1996. Genetic diversity in the orange subfamily Aurantiodeae. 1. Intraspecies and intragenus genetic variability. *Theoretical and Applied Genetics*, 92: 599-609.

Herrero, R., Asins, M.J., Pina, J.A., Carbonell, E.A., and Navarro, L., 1997. Genetic diversity in the orange subfamily Aurantiodeae. 2. Genetic relationships among genera and species. *Theoretical and Applied Genetics*, 93: 1327-1 334.

Hirai, M., Kozaki, I., and Kajiura, I. 1986. Isozyme analysis and phylogenic relationship of *Citrus*. *Japanese Journal of Breeding*, 36: 377-389.

Jarrel, D.C., Roose, M.L., Traugh, S.N., and Kupper, R.S. 1992. A genetic map of *Citrus* based on the segregation of isozymes and RFLPs in an intergeneric cross. *Theoretical and Applied Genetics*, 84: 49-56.

Khan, I.A. and Roose, M.L. 1988. Frequency and characteristics of nucellar and zygotic seedlings in three cultivars of trifoliate orange. *Journal of the American Society for Horticultural Science*, 113: 105-110.

Krug, C.A. 1943. Chromosome numbers in the subfamily Arantioideae, with special reference in the genus *Citrus*. *Citrus Botanical Gazette*, 104: 602-611.

Luro, F., Laigret F., Bove, J.M., and Ollitrault, P. 1994a. Application of RAPD to *Citrus* genetics and taxonomy. In : VIIth International Citrus Congress, pp. 225-228.

Luro F., Laigret F., Ollitrault, P., and Bove, J.M. 1995. DNA amplified fingerprinting (DAF), an useful tool for determination of genetic origin and diversity analysis in *Citrus*. *HortScience*, 30: 1063-1067.

Luro, F., Lorieux, M., Laigret, F., Bove, J.M., and Ollitrault, P. 1994b. Genetic mapping of an intergeneric *Citrus* hybrid using molecular markers. *Fruits*, 49: 404-408.

Luro, F., Ris, D., and Ollitrault, P. Evaluation of genetic relationships in *Citrus* genus by means of sequence tagged microsatellites. *Acta Horticulturae*. (In press).

Miranda, M., Ikeda, F., Endo, T., Moriguchi, T., and Omura, M. 1997. Chromosome markers and alterations in mitotic cells from interspecific *Citrus* somatic hybrids analysed by fluorochrome staining. *Plant Cell Reports*, 16: 807-812.

Moore, G.A., Tozlu, I., Weber, C.A., and Guy, C.L. 2000. Mapping quantitative trait loci for salt tolerance and cold tolerance in *Citrus grandis* (L.) Osb. × *Poncirus trifoliata* (L.) Raf. hybrid populations. *Acta Horticulturae, 535*, ISHS, 37-45.

Nair, P.K.R. and Randhawa, G.S. 1969. Chromosome morphology of the pachytene stage with respect to different *Citrus* types. In: Ist International Citrus Symposium, pp. 215-223.

Nei, M. 1973. Analysis of gene diversity in subdivided population. Proceedings of the National Academy of Science of the United States of America, 70: 3321-3323.

Nicolosi, E., Deng, Z.N., Gentile, A., La Malfa, S., Continella, G., and Tribulato, E. 2000. Citrus phylogeny and genetic origin of important species as investigated by molecular markers. *Theoretical and Applied Genetics, 100:* 1155-1166.

Ollitrault, P., Dambier, D., Luro, F., and Duperray, C. 1994. Nuclear genome size variations in *Citrus. Fruits, 49:* 390-393.

Ollitrault, P. and Faure, X. 1992. Système de reproduction et organisation de la diversité génétique dans le genre *Citrus.* In: *Complexes D'Espèces, Flux de Génes et Ressources Génétiques des Plantes.* Paris, BRG, pp. 133-151.

Ollitrault, P., Faure, X. and Normand, F. 1992. Citrus rootstocks characterization with dark and leaf isozymes: application for distinguishing nucellar from zygotic trees. In: VIIth International Citrus Congress, pp. 338-341.

Ollitrault, P. and Luro, F. 1997. Les agrumes. In: *L'Amélioration des Plantes Tropicales.* A. Charrier et al., eds., Montpellier, France, CIRAD-Orstom, pp. 13-36.

Ollitrault, P., Dambier, D., Froelicher, Y., Luro, F., and Cottin, R. 2000. La diversite des agrumes: structuration et exploitation par hybridation somatique. *Comptes rendus de l'Academie d'Agriculture, 86-88:* 197-221.

Perrier, X., Flori, A., and Bonnot, F. 1999. Les méthodes d'analyse des données. In: *Diversité Génétique des Plantes Tropicales Cultivées.* P. Hamon et al., eds., Montpellier, France, CIRAD, collection Repéres, pp. 43-76.

Raghuvanshi, S.S. 1969. Cytological evidence bearing on evolution in *Citrus.* In: Ist International Citrus Symposium, pp. 207-214.

Scora, R.W. 1975. On the history and origin of citrus. In: Symposium on the Biochemical Systematics, Genetics and Origin of Cultivated Plants. *Bulletin of the Torrey Botanical Club,* 102: 369-375.

Scora, R.W. 1988. Biochemistry, taxonomy and evolution of modern cultivated citrus. In: VIth International Citrus Congress, pp. 277-289.

Soost, R.K., Williams, T.E., and Torres, A.M. 1980. Identification of nucellar and zygotic seedlings with leaf isozymes. *HortScience,* 15: 728-729.

Swingle, W.T. and Reece, P.C. 1967. The botany of *Citrus* and its wild relatives. In: *The Citrus Industry. 1. History, World Distribution, Botany and Varieties.* W. Reuther et al., eds., Berkeley, University of California Press, pp. 190-430.

Tanaka, T. 1961. Citrologia: semi centennial commemoration papers on *Citrus* studies. Osaka, Citrologia Supporting Foundation, 114 p.

Tatum, J.H., Berry, R.E., and Hearn, C.I. 1974. Characterization of citrus cultivars and separation by thin layer chromatography. Proceedings of the Florida State Horticultural Society, 87: 75-81.

Torres, A.M., Soost, R.K., and Diedenhofen, U. 1978. Leaf isozymes as genetic markers in *Citrus*. American Journal of Botany, 65: 869-881 .

Torres, A.M., Soost, R.K., and Mau-Lastovicka, T. 1982. Citrus isozymes: genetic and distinguishing nucellar from zygotic seedlings. *Journal of Heredity*, 73: 335-339.

Webber, H.J. 1967. History and development of the citrus industry. In: *The Citrus Industry. 1. History, World Distribution, Botany and Varieties.* W. Reuther et al., eds., Berkeley, University of California Press, pp. 1-39.

Coconut

Patricia Lebrun, Yavo-Pierre N'cho, Roland
Bourdeix and Luc Baudouin

The coconut is an emblematic plant of tropical coastal countries, but it is
also a vital resource for many populations of these regions. It is cultivated
over around 11 million ha, 94% of which is located in Asia and the Pacific
(Bourdeix et al., 1997). The major producer countries are the Philippines and
Indonesia. It is essentially a smallholder crop, and the large plantations
represent less than 10% of the total production.

Almost all parts of the coconut tree are exploited in numerous ways
(Persley, 1992). The wood, although difficult to work with, is of excellent
quality. The leaves are used to make roofs. The leaf midribs are used to make
brooms. The sap tapped from inflorescences yields sugar and fermented
beverages. The roots are used as dyes and in traditional medicines.

But coconut is best known for its fruit: the epidermis of this drupe covers
a thick husk, the fibres of which are widely used as coir. Inside this is a
voluminous seed—the coconut—comprising a brown, lignified shell and an
albumen, the peripheral part of which is solidified at maturity. The remaining
cavity encloses the liquid part of the albumen: the coconut water. The
immature tender coconut provides a sweet and refreshing drink. When it is
mature, it is mostly valued for its solid albumen, which can be consumed
directly or after various transforming processes. The dried albumen, called
copra, is an important item of international trade. It is the source of one of
the principal oils of the lauric type, particularly useful in soap-making. This
oil is also used in foods and in cosmetics.

TAXONOMY AND GENETIC RESOURCES

Botany and Taxonomy

The coconut (*Cocos nucifera* L.) is a diploid arborescent monocotyledon (2n =
2x = 32) of the family Arecaceae. It is a monospecific genus, without any
closely related wild species. It is a palm tree, the unbranched trunk of which
bears a crown of fronds produced at the rate of about one a month. At the

base of each frond, a ramified inflorescence emerges. Each branch of the inflorescence has some female flowers at the base and a large number of male flowers at the summit.

The flowering is protandrous and the development of its cycles explains the reproductive behaviour of the two main types of coconut, between which are ranged the various cultivars (Rognon, 1976). The 'Tall' coconuts are mainly allogamous: the female flowering begins after the end of the male flowering. A certain rate of autogamy is, however, possible when the male flowering overlaps the female flowering of the preceding inflorescence. The Tall coconuts are also characterized by rapid growth, the presence of a voluminous bole at the base of the trunk, and widely spaced leaf scars. The Dwarf coconuts are most often autogamous: the female flowering occurs entirely (or mostly, as for the Brazilian Green Dwarf) before the male flowering ends. They are a small part of the world population and are generally located close to habitations. Apart from their autogamy, the Dwarfs are distinguished by slower growth, closely spaced leaf scars, greater precocity, and nuts that are smaller and often have a vividly coloured epidermis. There is also a variety of coconut similar to the Dwarfs in size, but allogamous: the Dwarf *Niu Leka*.

Genetic Resources

ORIGIN AND AREA OF DISTRIBUTION

Even though an American origin has been suggested, most authors now agree that the coconut is related to the Indo-Malayan centre of origin, as Vavilov has defined it (Child, 1964; Zohary, 1970). It is at present distributed in all the tropical coastal zones. The seeds were disseminated by flotation to the South Pacific, as well as by human intervention, which has long been the major means of dissemination. There are occasional spontaneous populations, but the overwhelming majority of coconut trees have been planted.

THE DIVERSITY OF POPULATIONS

In the collections conserved around the world, more than 300 local varieties ('cultivars' or 'ecotypes') of coconut have been numbered, according to their geographic provenance and the plant and fruit morphology. Apart from the distinction between Dwarfs and Talls, Harries (1978) proposed that coconuts be grouped into two subtypes: The *Niu Kafa* type, with a slender trunk and elongated fruits that are triangular in section and rich in fibre, represent the wild type, adapted to dissemination by ocean currents. The *Niu Vai* type, with a thicker trunk, more rigid shape, and rounded fruits rich in water, have been domesticated since ancient times. A certain number of populations that have intermediate characters come from an introgression into 'wild' populations by domesticated cultivars. Both types are good for the production of copra and were widely planted from the end of the 19th century.

Various methods have been used to characterize the cultivars of coconut and study the relations between populations. The evaluation of morphological and agronomic characters, for which N'cho et al. (1993) contribute valuable information on the diversity as well as the potential use of cultivars, has a limited efficacy because of the influence of the environment. The enzymatic markers studied by Benoit and Ghesquiere (1984) reveal low polymorphism and give little information. They were taken up later in Indonesia (Hartana et al., 1993) and in Sri Lanka (Fernando and Gamini, 1997) with greater success, but the number of usable systems remains small. The polyphenols have provided promising results (Jay et al., 1989), but they proved to be not reproducible from one environment to another. Molecular markers have been the subject of active research using various methods: ISTR (Rohde et al., 1995), RAPD (Ashburner et al., 1997), AFLP (Perera et al., 1998), RFLP (Lebrun et al., 1998a, b), and microsatellites.

USE IN VARIETAL IMPROVEMENT

Hybridization between Dwarf coconuts began in Fiji (Marechal, 1928), and Patel (1938) created the first Dwarf × Tall hybrids. In the 1940s to 1960s, comparative tests of Dwarf × Tall and Tall × Tall hybrids enabled the demonstration of the superiority of hybrids in terms of potential production (de Nuce de Lamothe and Bernard, 1985). Thammes (1955) proposed interplanting of two varieties in isolation, one used as male and the other as female, which made possible the mass production of hybrid varieties. Ultimately, this method was replaced by assisted pollination, which consists of applying a massive input of exogenous pollen on emasculated inflorescences (de Nuce de Lamothe and Rognon, 1972).

A strategy for coconut improvement was proposed by Bourdeix et al. (1990, 1991a, b). It was inspired from a recurrent reciprocal selection scheme that was designed by Comstock et al. (1949) for maize. Its originality lies in the simultaneous operation of two selection axes, one oriented towards the Tall × Tall hybrids and the other towards the Dwarf × Tall hybrids (Fig. 1).

The characterization of many available cultivars is essential to their use in improvement and to the search for heterosis, as well as to their conservation as a source of variability. This chapter summarizes the results of the study of Lebrun et al. (1998b) and attempts to find out what the tools of molecular markers can contribute to the knowledge and improvement of coconut in addition to the methodologies used earlier.

ORGANIZATION OF DIVERSITY

Nuclear and Mitochondrial Diversity Revealed by RFLP

For the nuclear RFLP, 25 probe-enzyme combinations used reveal 60 polymorphic bands among the 289 trees analysed. The Cox 1 probe reveals two mitochondrial profiles.

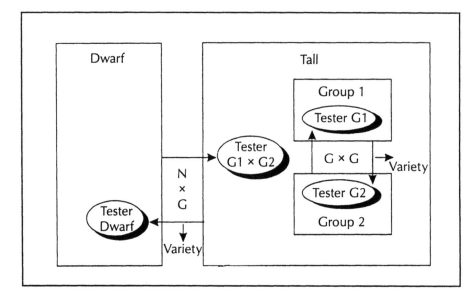

Fig. 1. Relationship between the two axes of recurrent reciprocal selection in coconut (Bourdeix et al., 1991a).

Correspondence analysis (CA) and geographic information can be used to constitute groups of relatively homogeneous cultivars and thus to identify the distinctive markers of each group.

THE TWO MAJOR ZONES OF DIVERSITY OF TALL COCONUTS

On the right of the CA (Fig. 2) can be distinguished a primary group that corresponds to the set of Tall cultivars from the ecogeographic zone of the Pacific, comprising Southeast Asia and the South Pacific, to which are added all the Dwarf cultivars and the Panama coconuts: this primary group is called the 'Pacific group'. Another group, located at left, corresponds to the coconuts of India, Sri Lanka, and West Africa: this is the 'Indo-Atlantic group'. These two groups are separated by a third group of three cultivars along the Indian Ocean, called the 'Indian Ocean group'.

In terms of nuclear markers, the Indo-Atlantic group is distinguished by the predominant presence of five markers, which are nearly absent otherwise, and of three others, which they share out with the single populations of Southeast Asia. Twenty-one markers are absent or have a low frequency in this group, while they are well represented in most of the other Talls. Finally, the 'rapid' allele of the cytoplasmic marker Cox 1 seems characteristic of this group: out of 45 individuals that it is made up of, 35 have this allele, while all the individuals of the Pacific group, except the individual Tonga Tall, have the 'slow' allele. The Kappadam Tall is a particular case since the five

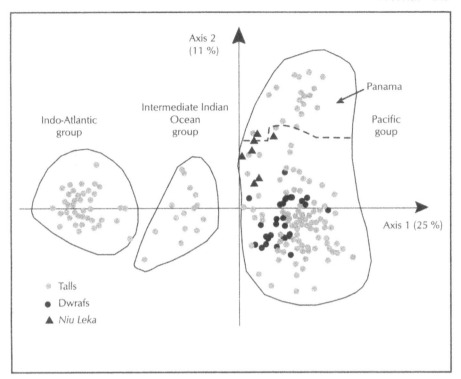

Fig. 2. Synthesis of RFLP data: the major groups of cultivars.

individuals that represent it have the 'slow' allele. In addition, even though its nuclear markers are essentially characteristic of the Indo-Atlantic group, a higher frequency is observed than that expected for some nuclear markers generally found in the Pacific group. It will be seen later that these two apparent anomalies could constitute the key of the origin of the Kappadam Tall, the fruit characters of which are very peculiar for this region.

These two groups are also distinguished by their degree of polymorphism: the Pacific group has more markers common to different cultivars—42 against 33 for the Indo-Atlantic group—and more markers of intermediate frequency, between 10% and 90%, 20 to 38 bands depending on the cultivar, as against 11 to 20 for the Indo-Atlantic group.

The intermediate Indian Ocean group has almost all the nuclear markers of the two primary groups and thus contains the widest molecular diversity. The cytoplasmic allele is the 'slow' allele, except for two individuals of the Mozambique Tall.

DIVERSITY IN THE PACIFIC GROUP

In order to characterize further the diversity of the Pacific group, a second CA was done (Fig. 3). In the first place, a geographic gradation can be seen

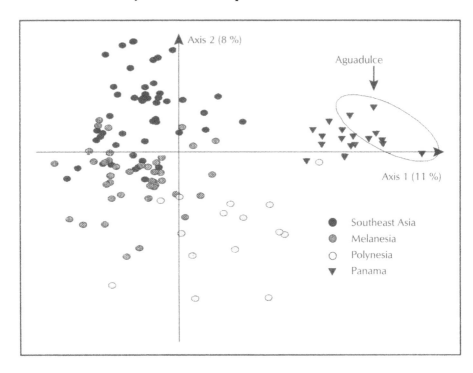

Fig. 3. Synthesis of RFLP data: diversity within the Pacific group.

between Southeast Asia and the South Pacific according to the second axis: the coconut trees of Polynesia appear at the lower part of the figure, those of Southeast Asia at the upper part, and those of Melanesia in the intermediate position. Contrary to the divergences revealed between the groups, the differences here appear to lie less in the existence of specific alleles than in frequency variations. There is also some overlapping between neighbouring subregions.

In the second place, it can be seen very clearly that the three populations of Panama are individualized on the first axis. They are related to the other cultivars of the Pacific group as to their RFLP profile, to the extent that all the markers frequent in Panama are present in this group. However, 19 bands of this group are absent or rare in Panama. Certain very frequent markers in the populations of Panama are more frequent in Polynesia than in the rest of the Pacific group. For the others, it is the reverse. The markers of intermediate frequency are few (12 to 17). The Aguadulce population is slightly distinct from the two other Panamanian populations, especially because of the presence of some bands specific to the Indo-Atlantic, with a low frequency. On the other hand, the two other populations, Monagre and Bowden (Jamaica), seem indistinguishable.

Dwarf coconut

The Dwarf coconuts are essentially autogamous. This characteristic is expressed in the near absence of bands of intermediate frequency and by a very low rate of heterozygosity. The two exceptions are *Niu Leka*, the only clearly allogamous Dwarf cultivar, and Malayan Green Dwarf, known to be partly allogamous. All the bands common to Dwarfs are found in the Pacific group, but 13 bands present in this group are absent from most of the Dwarfs. Among these latter, several correspond to the fixation of alleles of intermediate frequency in the Talls, which strongly suggests a common origin.

Four Dwarf cultivars of distinct origin and colour have a nearly identical profile: the Green Dwarfs from Sri Lanka and Kiribati, and the Brown Dwarfs from Ternate and Madang. On the other hand, the Malayan Dwarfs form a homogeneous group distinguished by the fixation of five alleles different from those found in the other Dwarfs. The seven Ghana Yellow Dwarfs and the 15 Malayan Yellow Dwarfs have exactly the same profile, which confirms the identity of these two cultivars. The three Dwarf cultivars of the Philippines, as well as the Brazilian Green Dwarf, also have common traits, while the profile of two Red Dwarfs of the South Pacific and that of the Cameroon Red Dwarf (a cultivar taken from Cameroon but probably exotic) appear clearly divergent even though certain traits are close to the Dwarf group.

Finally, the *Niu Leka* Dwarf has a profile that recalls that of the Talls of its region of origin (Tonga Tall and Rotuma Tall) and has the 'rapid' allele of Cox 1, rare in the region but present in a Tonga Tall individual. The presence of an allele typical of the Indo-Atlantic group in this region cannot be explained through historic data or through the morphoagronomic characters. It may be that the two regions have two different but indistinguishable alleles. Whatever the case, the *Niu Leka* is distinct from the other Dwarfs, which seem to be closer to the Southeast Asian or Melanesian coconut.

The Polyphenols

Figure 4 summarizes the principal results of the study on polyphenols conducted by Jay et al. (1989). It represents a discriminant analysis, done on 32 cultivars (or ecotypes), each of them being represented on average by five individuals. The cultivars studied are included in the RFLP analysis, with the exception of Thailand Green Dwarf and the Tahiti Tall. Contrary to the CA done on the molecular markers, the origin of the individuals figures explicitly in the data provided for the statistical analysis. Even when this difference is taken into account, the major conclusions of this study converge with those of the molecular data. For example, we note that the Tall cultivars are distributed around three major zones: Africa, the South Pacific, and the Far East, these latter two being partly overlapped. The Indian coconuts are distributed between several groups. Those of Lakshadweep, an archipelago situated southwest of India, are close to the African coconuts, while those of

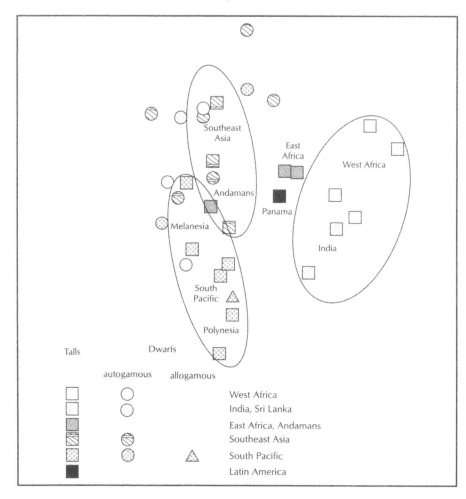

Fig. 4. Polyphenolic data: discriminant analysis of individuals according to the cultivar criterion (1-2 plane) (Jay et al., 1989).

the Andamans fall between the populations of the Far East and those of the Pacific. The East African cultivars are also found towards the centre of the figure in relation to the other African cultivars. Moreover, all of the Dwarfs seem close to the Far East and Melanesian Talls. However, there is no apparent link between the geographic origin of a Dwarf and its polyphenolic profile. This is particularly the case with those collected in Sri Lanka and in Africa, the exotic origin of which is clearly confirmed.

Thus, despite the differences in the method of statistical analysis and in the mode of presentation, the results of the two approaches are overall convergent. The most notable difference lies in the situation of the Panama Tall, which here appears intermediate and relatively close to the African cultivars.

Agromorphological Data

Figure 5 shows the two primary axes of a discriminant analysis made on morphological and production data for 17 Tall cultivars, each represented by 30 individuals (N'cho et al., 1993). The first axis represents 34% of the total variability and comprises variables linked to the general vigour of the plant and, more particularly, that of the trunk. The second (21% of the variability) is positively linked to the size of the fruit and of the cavity left by the albumen (corresponding to the volume of water in the nearly mature fruit). It is also associated negatively with the number of nuts produced.

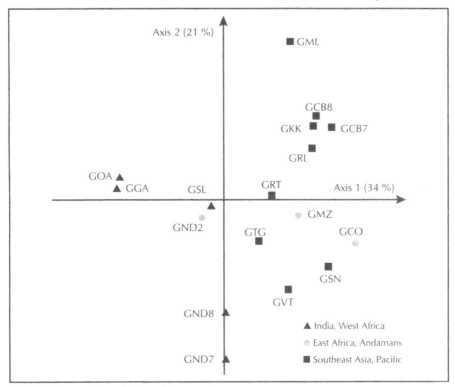

Fig. 5. Synthesis of morphological data (N'cho et al., 1993).

Following this representation, the cultivars of India and West Africa are characterized by a relatively slender trunk and by several small fruits, enclosing a small volume of water. Those of East Africa and the Melanesian islands are distinguished by a more robust trunk, with small fruits (except the Rennell Tall), while those of Southeast Asia and Papua New Guinea are robust and have large nuts. The cultivars of Polynesia occupy a median position. A third axis, which is not represented (13% of the variability) and is associated with the length of the different parts of the inflorescence, contains the Rennell Tall alone.

As with polyphenolic markers and RFLP, we find an east-west zonation, however less obvious. The intermediate position of the East African cultivars is not observed here.

DISCUSSION AND CONCLUSION

Contribution of Molecular Markers to the History of Coconut Diversification

The molecular analyses presented here are overall in agreement with the results of earlier studies. The major differences pertain to the better reproducibility of results and the independence of RFLP with respect to the environment, which results in greater precision. The RFLP markers thus contribute to a better understanding of the major events that marked the diversification of coconut (Lebrun et al., 1998b). In order to highlight the specific contributions of molecular markers, we place them in the context of a brief history of the coconut (Fig. 6).

The most probable region of origin is between Southeast Asia and Papua New Guinea (Zohary, 1970; Child, 1964). The coconut is adapted to dispersal by flotation, and the distribution of the islands in this region allows the nut to easily reach a coast favourable to its germination. Presumably, there was a noticeable variation in the composition of the fruit from the beginning. Gradually, the coconut tree migrated spontaneously by flotation towards the Pacific and, eventually, closer towards India. In these long-distance

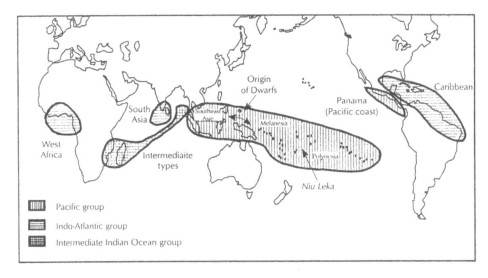

Fig. 6. Geographic location of the major coconut groups.

migrations, it was undoubtedly the form best adapted to marine transport that was favoured, the *Niu Kafa* type of Harries (1978).

At a very distant and uncertain date, the populations typical of the Indian subcontinent were isolated from the Pacific group. Even though it is not considered an indigenous plant, the coconut has grown in India for at least 3000 years (Child, 1964). The fact that the *Niu Kafa* type is predominant in India could suggest transportation by flotation, although it does not prove it. The later imports from Southeast Asia were limited and could not greatly modify the genetic structure.

● The distribution of cultivars into two major groups is not new. On the other hand, several RFLP markers have a great disequilibrium of frequency, which could go as far as specificity, between the Pacific and Indo-Atlantic groups. This indicates an ancient divergence probably due to prolonged geographic isolation.

From about 1300 BCE, colonization of the South Pacific consolidated the plantation of coconut. The fibre provided rope needed for construction of supple and durable boats. The immature fruits (for water) and mature fruits (for albumen) served as food supplies. The movement was gradual and uninterrupted. The two fruit forms, *Niu Vai* and *Niu Kafa*, coexisted and were probably selected for their own qualities. They were selected from an initially polymorphic stock, the genetic basis of which is similar to that of Southeast Asia and Papua New Guinea.

The populations of the Pacific group have the same markers, and the differences between the subregions pertain mostly to frequencies.

From 600 BCE, Malayan-Polynesian peoples migrated from Southeast Asia to Madagascar. The immigrants imported the coconuts with their costumes and their language, including the name of the plant (*voanio*). From there, the coconut reached East Africa and the Comoro Islands.

From 800 or a little earlier, the Arabs traded between India and East Africa (Ibn Battuta, 1351). They transported coconuts typical of these regions. Populations with intermediate genetic structure developed in East Africa by hybridization between the imported and local coconuts.

● The intermediate Indian Ocean group accumulates markers in its genome that characterize the Indo-Atlantic and the Pacific groups. From its distribution in two distinct geographic zones and the historical data, we can consider this intermediate group the result of a fusion between populations arising from the two original groups.

At an indeterminate period, flows were established from India towards Southeast Asia (perhaps through the Arab travellers). The populations of the Andamans, Indian islands to the northwest of Sumatra, thus have a genetic structure close to that of West Africa. Further east, from Indonesia to the Philippines, very limited exchanges occurred. An inverse exchange was the origin of the Kappadam Tall, which is distinguished from populations of the same region by its voluminous and round fruits.

- Three characteristic markers of the Indo-Atlantic populations were observed at a low frequency in Southeast Asia, but absent further east. They indicate a low genetic input of Indian origin.
- The Kappadam Tall has an exotic cytoplasmic marker and, in the nuclear markers, shows traces of introgression from the east, on a genetic basis that is typically Indo-Atlantic. The combination of these peculiarities may be explained as follows: this cultivar must have arisen from an introduction originating in Southeast Asia, followed by a selection of mother trees with round fruits during several successive generations. The open pollination, ensured for the most part by local populations, must in the long term have given it a nuclear genotype close to that of the Indo-Atlantic group. Towards 1498, in the wake of Vasco da Gama, the Portuguese circled the African continent and reached India by the sea route. The coconut was taken on the return voyage and established in West Africa.

  Between 1525 and 1550, directly or perhaps from West Africa (Cape Verde), it reached the Caribbean and the Atlantic coast of America. Strong historical arguments and the morphological resemblance between the coconuts of Brazil, the Caribbean (not included in this study), and Africa suggest that they belong to the same group.
- The similarity of RFLP profiles of the West African and Indian cultivars seems to confirm the results obtained by other methods.

  During the pre-Colombian era, the coconut was present on the Pacific coast of Panama, as related by Oviedo in 1535 in *La historia general y natural de las Indias, Islas y Tierra Firme del Mar Océano* (cited by Zizumbo and Quero, 1997), but absent in Mexico. The present-day populations could come from these coconuts. The distance between the Latin American coast and the Pacific islands, as well as the typically *Niu Vai* character of the local material, seem to exclude transportation by flotation. On the other hand, human transport seems possible, either by the Polynesians or by Southeast Asian populations, of which archaeological traces have been found as far as Ecuador (Langdon, 1995).

  During the 16th century, imports from Panama, the Philippines, and the Solomon Islands were reported in Mexico.
- The RFLP markers underline the specificities of the coconuts from Panama. All the evidence points to their origin in the Pacific group, although at present it cannot be decided whether they came through the Polynesian or the Southeast Asian route. Their considerable homogeneity and homozygosity are evidence of a strong foundation effect.
- Within the Panama origin, the Aguadulce population has, at low frequency, alleles specific to the Indo-Atlantic group, which is the highly probable signature of an introgression of genes coming from populations of the Atlantic coast.

  At a date that is difficult to fix, but after the separation of Pacific and Indo-Atlantic stocks, the Dwarfs appeared between the Southeast Asia and Papua New Guinea populations, similar to those of the Philippines with respect to the number and diversity of Dwarfs found there. The underlying differentiation operates on a regional basis during the course of diffusion of this material, probably by intercrossing with the local Talls.

● Reliable indexes, linked to the number of alleles that they fix, are drawn from RFLP in favour of a single origin for Dwarfs, with the exception of *Niu Leka*. The Tall cultivars with an allelic composition most compatible with that of the Dwarfs are located in a zone ranging from Southeast Asia to Papua New Guinea.

At an indeterminate date, the *Niu Leka*, an allogamous Dwarf coconut, appeared in the region of Fiji, the Samoas, and Tonga.

● The resemblance between the RFLP profile of *Niu Leka* and that of the Tonga and Rotuma Talls originating from the same region testifies in favour of a local origin for this cultivar, and thus one independent of the origin of the other Dwarfs.

An Approach to the Relation between Genetic Distance and Heterosis

In many plants, it can be observed that crosses between genetically distant populations are better than those with a narrow genetic base. This 'interpopulation heterosis' is widely exploited in recurrent reciprocal selection programmes. It is useful to verify whether such a phenomenon is observed in coconut. Table 1 shows the results of hybrid tests of Tall × Tall as a function of the origin of parents. The intermediate Indian Ocean populations are represented by the Mozambique Tall.

In five trials comparing the Tall × Tall hybrids, all the hybrids are superior to the West African Tall control. Among them, crosses between cultivars of

Table 1. Production of copra of some Tall × Tall hybrids classified according to the groups defined by RFLP: results of five assays

Assay Year of plantation	PBGC 1 1965	PBGC 3 1969-70	PBGC 7 1971	PBGC 8 1972-73	PBGC 9 1971
Pacific × Indo-Atlantic	(2)* 130%	(2) 182%	(2) 174%	(3) 212%	(3) 182%
Indian Ocean × Pacific	(2) 124%	(2) 151%	(2) 153%		
Indian Ocean × Indo-Atlantic	(1) 136%				
Pacific × Pacific		(2) 138%	(1) 137%	(3) 142%	
Indo-Atlantic × Indo-Atlantic					(3) 142%
Control West African Tall (%)	(1) 100%	(1) 100%	(1) 100%	(1) 100%	(1) 100%
Production (kg/tree/year)	23.0	15.8	11.5	8.2	11.5
Self-fertilized control	(1) 88%				

*For each assay, the number of crosses per type of recombination is given in parentheses.

the same group had the lowest yields, while crosses between the Indo-Atlantic and Pacific cultivars generally had the best production. The Indian Ocean group was represented by the Mozambique Tall. Its crosses with the cultivars of the two preceding groups had a performance between that of intragroup crosses and that of intergroup crosses.

On the other hand, even though the molecular data strongly suggest that the Dwarfs originate from the Pacific group, the latter may give excellent hybrids with the partners of either of the major groups defined above. This is the case, for example, of Malayan Yellow Dwarf × West African Tall (PB121) and Malayan Red Dwarf × Rennell Tall. Whatever the factors responsible for the 'interpopulation heterosis' mentioned above, the regime of autogamous reproduction that prevails in the Dwarfs seems to have induced a sufficiently significant genetic divergence in relation to their group of origin—due to the fixation of one allele per locus, rather than the appearance of new alleles— for heterosis to take place with these as well as with the Indo-African group.

CONCLUSION

The results obtained with molecular markers largely agree with the results found earlier with other methods. However, they bring greater precision to the study of genetic relationships between populations. Studies of agronomic and morphological criteria are essential for characterization of the variability of a species such as coconut, to the extent that they provide elements essential for the use of cultivars studied in varietal improvement. The image of the genetic organization of populations drawn from only these criteria, however, is less precise than that made possible by molecular markers. On the one hand, environmental effects bias the comparisons when the studies are done in varying conditions of place or periods; on the other hand, natural or human selection may in some cases lead to similar phenotypes from populations of distinct origin. The similarity of phenotypes may conceal actual genetic complementarities. Finally, expression of the genetic value may be affected by the greater or lesser consanguinity of populations.

Polyphenol markers have proved to be highly effective in a preliminary study. However, the results obtained could not be reproduced in different environments. In plants, polyphenols are implicated in reactions to different stresses. It can thus be expected that the profiles obtained vary as a function of external stimuli. Finally, their implication in defence against stress suggests that they could not remain neutral with respect to selection. On the other hand, RFLP markers are chosen independently of the activity of the sequence concerned. Even though it is difficult to prove rigorously, the hypothesis of neutrality is more easily maintained.

Molecular markers offer several advantages for the improvement of coconut. They are precious tools in managing collections. In a perennial plant such as the coconut, it is important to collect the widest possible genetic variability in a limited area. In certain situations, RFLP allows us to identify the varieties precisely. It is even possible, in certain populations, to detect the presence of genes of foreign origin and to explain the probable cause of them. Considering the small numbers used, this technique has proved to be very effective.

Moreover, the distribution of genetic stock of the species into two major geographical groups that have been subject to prolonged genetic isolation contributes a solid basis for the choice of a mechanism of recurrent reciprocal selection for the production of Tall × Tall hybrids. It allows the prior assignment of a place to most of the cultivars in this scheme, in order to maximize the heterosis. The few intermediate cultivars located in East Africa and in South Asia and the Far East can be used to enrich the variability of one of two groups, especially the Indo-Atlantic, in which the genetic basis and phenotype variability is the least. Within each of the heterotic groups, it is possible to choose the cultivars that should be recombined on a priority basis to maximize the selectable variability.

Finally, molecular markers in some cases provide precise indexes on the history of plant material, which could be useful in the search for new sources of particular characters. The study of populations of the Pacific coast of Latin America with respect to tolerance to lethal yellowing in the Caribbean zone illustrates this field of application. The populations of Panama seem to have factors of tolerance to this disease. The RFLP profile of three of these populations suggests that they arise from a small initial population. Thus, the diversity and dynamics of populations in the Talls of the Pacific coast can be studied from a wider sample to enable more precise characterization of the nature of initial inputs and more certain identification of populations that can transmit disease tolerance. This broadening of the research could prove to be particularly useful in light of the fact that the Panama Tall is sensitive to *Phytophthora*, which is rampant in the same region.

APPENDIX

Plant Material

The leaf samples collected from 289 trees represent 26 Tall cultivars and 16 Dwarf cultivars from the collections of Côte d'Ivoire (Marc Delorme station), Vanuatu (Saraoutu station), and Jamaica. They cover most coconut cultivation zones, except the Caribbean zone and the east coast of America. The list of cultivars, with their geographic origin, is given in Table 2.

RFLP Analyses

The extracts of total DNA are taken from lyophilised leaflet, taken from leaf no. 1 (youngest green leaf). The method used is that of CTAB (cetyl-trimethyl ammonium bromide), adapted on maize by Hoisington (1992). The restrictions are done using four enzymes: EcoR1, EcoRV, Bgl2, and Sst1. The restricted DNA migrates in a 0.8% agarose gel in TAE buffer (tri-acetate EDTA), then it is transferred on a nylon membrane. The probes, used for the molecular hybridization following the protocol of Hoisington (1992), are marked with ^{32}P. The results are detected from autoradiograms.

Origin of Probes

This study was done using 20 cDNA probes and a mitochondrial probe (Cox 1). Among the nuclear probes is a cDNA of coconut, the others being heterologous probes of rice, oil palm, and maize, the origin of which is cited in Lebrun et al. (1998b).

Data Analysis

Each band is coded as a dominant marker: 10 for its presence and 01 for its absence. The binary matrix of bands × individuals thus obtained makes it possible to do a CA (Benzecri, 1973) using Addad software (Addad, 1983). The graphic representations of these multivariate analyses are used to describe the genetic structuration of the material.

Table 2. Number and geographic distribution of cultivars sampled

Origin	Talls	Code	No.*	Dwarfs	Code	No.*
West Africa	4 cultivars			2 cultivars		
● Côte d'Ivoire	West African Tall	GOA	5			
	Mensah West African Tall	GOA04	10			
	Ouidah West African Tall	GOA06	10			
● Benin						
● Cameroon	Kribi Cameroon Tall	GCA	5	Cameroon Red Dwarf	NRC	5
● Ghana				Ghana Yellow Dwarf	NJG	7
East Africa	2 cultivars					
● Comoro Islands	Moheli Comoro Tall	GCO	5			
● Mozambique	Mozambique Tall	GMZ	5			
South Asia	4 cultivars			1 cultivar		
● India	Micro Laccadives Tall	GND07	5			
	Kappadam Tall	GND05	5			
	Ordinary Andaman Tall	GND02	4			
● Sri Lanka	Sri Lanka Tall	GSL	5	Sri Lanka Green Dwarf	NVS	5
Southeast Asia	8 cultivars			7 cultivars		
● Thailand	Thailand Tall	GTH	5	Catigan Green Dwarf	NVP02	5
● Philippines	Baybay Tall	GPH04	5	Pilipog Green Dwarf	NVP05	5
	Tagnanan Tall	GTN	5	Tacunan Green Dwarf	NVP03	5
● Cambodia	Cambodia Tall	GCB	10			
● Indonesia	Tenga Tall	GDO02	5	Ternate Brown Dwarf	NBO	5
	Palu Tall	GDO03	5			
	Takome Tall	GDO04	5			

(Contd.)

(Table 2. Contd.)

Origin	Talls	Code	No.*	Dwarfs	Code	No.*
● Malaysia	Malaysia Talls	GML	11	Malayan Yellow Dwarf	NJM	15
				Malayan Green Dwarf	NVM	5
				Malayan Red Dwarf	NRM	5 + 5**
South Pacific						
● Papua New Guinea	9 cultivars			5 cultivars		
	Karkar Tall	GNG01	5	Madang Brown Dwarf	NBN	5
● New Guinea	Markham Valley Tall	GNG03	5			
● Solomon Islands	Gazelle Tall	GNG04	5 + 5**			
	Rennell Tall	GRL	7 + 5**			
	Solomon Tall	GSL	6			
● French Polynesia	Polynesia Rangiroa Tall	GPY01	5	Polynesian Red Dwarf	NRY	5
● Fiji	Rotuma Tall	GRT	5	*Niu Leka* Dwarf	NNL	7
● Tonga	Tonga Tall	GTG	5			
● Vanuatu	Vanuatu Tall	GVT	5	Vanuatu Red Dwarf	NRV	5**
				Kiribati Green Dwarf	NVT	5**
Latin America						
● Panama	3 cultivars			1 cultivar		
	Panama Tall	GPA	10***			
	Panama Tall (Aguedulce)	GPA01	6			
	Panama Tall (Monagre)	GPA02	6			
● Brazil				Brazilian Green Dwarf	NVB	5

*All samples from Côte d'Ivoire, unless otherwise mentioned.
**Sampled at Vanuatu.
***Sampled at Jamaica.

REFERENCES

ADDAD, 1983. *Manuel de référence*. Paris, Association pour le développement et la diffusion de l'analyse des données.

Ashburner, G.R., Thompson, W.K., and Halloran, G.M. 1997. RAPD analysis of South Pacific coconut palm populations. *Crop Science*, 37: 992-997.

Benoit, H. and Ghesquiere, M. 1984. Electrophorèse, compte rendu cocotier. IV. Déterminisme génétique. Montpellier, France, CIRAD-Irho, 11 p. (internal document).

Benzecri, J.P. 1973. *L'Analyse des Données, vol. II. L'Analyse dès Correspondances*. Paris, Dunod, 616 p.

Bourdeix, R., Baudouin, L., Billotte, N., Labouisse, J.P., and Noiret, J.M. 1997. Le cocotier. In: *L'Amélioration des Plantes Tropicales*. A. Charrier et al., eds., Montpellier, France, CIRAD-Orstom, pp. 217-239.

Bourdeix, R., Meunier, J., and N'cho, Y.P. 1991a. Une stratégie de sélection du cocotier *Cocos nucifera* L. 2. Amélioration des hybrides Grand × Grand. *Oléagineux*, 46(7): 267-282.

Bourdeix, R., Meunier, J., and N'cho, Y.P. 1991b. Une stratégie de sélection du cocotier *Cocos nucifera* L. 3. Amélioration des hybrides Nain × Grand. *Oléagineux*, 46(10): 361 -374.

Bourdeix, R., N'cho, Y.P., and Le Saint, J.P. 1990. Une stratégie de sélection du cocotier. 1 . Synthèse des acquis. *Oléagineux*, 45(8-9): 359-371 .

Child, R. 1964. *Coconuts*. 2nd ed. London, Longman, 335 p.

Comstock, R.E., Robinson, H.F., and Harvey, P.H. 1949. A breeding procedure to make maximum use of both general and specific combining ability. *Agronomy Journal*, 41: 360-367.

de Nuce De Lamothe, M. and Benard, G. 1985. L'hybride de cocotier PB121 (ou Mawa) (NJM × GOA). *Oléagineux*, 40(5): 261-266.

de Nuce de Lamothe, M. and Rognon, F. 1972. La production de semences hybrides chez le cocotier par pollinisation assistée. *Oléagineux*, 27(10): 539-544.

Fernando, W.M.U. and Gamini, G. 1997. Profil des variations isoenzymatiques chez les populations de cocotier (*Cocos nucifera* L.) utilisées pour la sélection des variétés améliorées. *Plantations, Recherche, Développement*, 4(4): 256-263.

Harries, H.C. 1978. The evolution, dissemination and classification of *Cocos nucifera* L. *The Botanical Review*, 44: 265-320.

Hartana, A., Hengky, and Dwi Asmono. 1993. Analisis Keragaman dan pewarisan pola pita isozim tanaman kelapa. *Journal Matematika dan Sains*, 1 (supplement D): 63-76.

Hoisington, D. 1992. *Laboratory Protocols*. Mexico, CIMMY. CIMMYT Applied Molecular Genetics Laboratory.

Ibn Battuta 1351 . Voyages et périples. In: *Les Voyageurs Arabes*. Paris, Gallimard, collection La Pléiade, 1408 p.

Jay M., Bourdeix, R., Potier, F., and Sanlaville, C. 1989. Premiers résultats de l'étude des polyphénols foliaires du cocotier. *Oléagineux*, 44(3): 151-161 .

Langdon, R. 1995. The banana as a key to early American and Polynesian history. *Journal of Pacific History*, 28(1): 15-35.

Lebrun, P., Grivet, L., and Baudouin, L. 1998b. Dissémination et domestication du cocotier à la lumière des marqueurs RFLP. *Plantations, Recherche, Développement*, 5(4): 233-245.

Lebrun, P., N'cho, Y.P., Seguin, M., Grivet, L., and Baudouin, L. 1998a. Genetic diversity in coconut (*Cocos nucifera* L.) revealed by restriction fragment length polymorphism (RFLP) markers. *Euphytica*, 101: 103-108.

Marechal, H. 1928. Observation and preliminary experiments on the coconut palm with a view to developing improved seed for Fiji. *Fiji Agricultural Journal*, 1: 16-45.

N'cho, Y.P., Sangare, A., Bourdeix, R., Bonnot, F., and Baudouin, L. 1993. Evaluation de quelques écotypes de cocotier par une approche biométrique. 1. Etude des populations de Grands. *Oléagineux*, 48(3): 121-132.

Patel, J.S. 1938. *The Coconut: A Monograph*. Madras, Government Press, 350 p.

Perera, L., Russel, J.R., Provan, J., McNicol, J.W., and Powell, W. 1998. Evaluating genetic relationships between indigenous coconut (*Cocos nucifera* L.) accessions from Sri Lanka by means of AFLP profiling. *Theoretical and Applied Genetics*, 96: 545-550.

Persley, G.J. 1992. *Replanting the Tree of Life*. Wallingford, UK, CAB International, 156 p.

Rognon, F. 1976. Biologie florale du cocotier. *Oléagineux*, 31(1): 13-18.

Rohde, W., Kullaya, A., Rodriguez, J., and Ritter, E. 1995. Genome analysis of *Cocos nucifera* L. by PCR amplification of spacer sequences separating a subset of *copia*-like *Eco*RI repetitive elements. *Journal of Genetics and Breeding*, 49: 179-186.

Thammes, P.L.M. 1955. Review of coconut selection in Indonesia. *Euphytica*, 4: 17-24.

Zizumbo, V.D. and Quero, H.J. 1997. Re-evaluation of early observations on coconut in the New World. *Economic Botany*, 52(1): 68-77.

Zohary, D. 1970. Centers of diversity and centers of origin. In: *Genetic Resources in Plants, Their Exploration and Conservation*. O.H. Frankel and E. Bennett, eds., Oxford, Blackwell, 547 p.

Coffee
(*Coffea canephora*)

Stéphane Dussert, Philippe Lashermes,
François Anthony, Christophe Montagnon,
Pierre Trouslot, Marie-Christine Combes,
Julien Berthaud, Michel Noirot and Serge Hamon

Coffee is the primary agricultural export product (Charrier and Eskes, 1997). It is produced from two species: *Coffea arabica* L. and *C. canephora* Pierre. *Coffea arabica* is known for its gustatory qualities. It is cultivated on the high humid tropical plateaux, essentially in Latin America and East Africa. *Coffea canephora* is renowned for its agronomic hardiness, whence its common name of Robusta. It is cultivated mainly in humid tropical zones of low altitude and represents 30% of the world production of coffee. It comes mostly from Brazil, Indonesia, and Côte d'Ivoire. It is now widely produced in Southeast Asia—the Philippines and Vietnam—and in India.

During the 18th and 19th centuries, only Arabica was produced and that mainly in tropical America, the Caribbean, and Asia (Charrier and Eskes, 1997). However, this species appeared to be highly sensitive to parasitic threats, especially orange rust. That is why, in Africa, during the 19th century, the spontaneous forms of other species of coffee, especially *C. canephora*, were cultivated locally. For *C. canephora*, it was mostly in the Belgian Congo (now the Democratic Republic of Congo) and Uganda that coffee plants from local forest populations, of the Robusta type, were cultivated. They were transferred to Java, a major breeding centre of *C. canephora* from 1900 to 1930 (Montagnon et al., 1998). At the same time, in Africa, the diversity of material cultivated was extended with the use of local spontaneous forms: Kouilou in Côte d'Ivoire, Niaouli in Togo and Benin, and Nana in the Central African Republic. The material selected in Java was reintroduced in the Belgian Congo around 1916 at INEAC (Institut National pour l'Étude Agronomique du Congo Belge), which has become the major breeding centre of *C. canephora* from 1930 to 1960 (Montagnon et al., 1998). However, although the overall performance of cultivated trees has increased noticeably after a few breeding cycles at

Java and the Belgian Congo, the cultivars nonetheless have remained genetically very close to individuals of the original natural populations. Moreover, in the African countries where the species originated and where *C. canephora* is cultivated, local spontaneous forms could be crossed with the introductions and the cultivated plants could revert to the wild forms.

BOTANY AND GENETIC RESOURCES

Botany and Mode of Reproduction

Coffea canephora belongs to the family Rubiaceae, genus *Coffea* L., subgenus *Coffea* Bridson. The genus *Coffea* has an area of distribution limited to the African continent, Madagascar, and the Mascarene Islands. It contains close to 80 species, of which 25 are endemic to Africa (Bridson and Verdcourt, 1988). The species of the genus *Coffea* that are closest genetically to *C. canephora* are *C. congensis* and *C. brevipes* (Lashermes et al., 1997). Moreover, *C. canephora*, or an ancestral form of it, is one of the two parental diploid species of *C. arabica* (Lashermes et al., 1997). All the species of the genus are diploid, with the exception of *C. arabica*, which is allotetraploid. Similarly, they all have a mode of reproduction that is strictly allogamous, with the exception of *C. arabica*, which is autogamous. Studies on *C. canephora* have indicated a system of gametophytic self-incompatibility (Berthaud, 1980).

For the diploid species of the genus, the quantity of DNA per genome, measured by flow cytometry, varies from 0.95 to 1.78 pg (Cros et al., 1995). It is 1.54 pg for *C. canephora* and 2.61 pg for *C. arabica*.

Coffea canephora has one of the widest areas of distribution of the subgenus *Coffea*: it extends west to east from Guinea to Sudan, and north to south from Cameroon to Angola (Berthaud, 1986).

The growth of coffee plants of this species is dimorphic. The main stems (orthotropic axes) grow vertically and the branches (plagiotropic axes) grow horizontally. Horticultural propagation is relatively easy. The plant may flower once or twice a year, after a rainfall of at least 10 mm, which follows a period of water stress. Berries mature at 8 to 12 months depending on the variety and environment.

The seeds of *C. canephora* do not behave in an orthodox manner (Roberts, 1973) when dehydrated or stored at low temperature (Couturon, 1980). Their longevity is only one to two years in the hydrated state at ambient temperatures.

Genetic Resources

Given the behaviour of *C. canephora* seeds, the long-term conservation of genetic resources of this species is done in the field.

Collections that are more or less representative of the most widespread introductions, of material taken from plantations and of local forms, are conserved in Côte d'Ivoire, Cameroon, Uganda, India, Indonesia, and Brazil. But the only reference collection for wild forms of *C. canephora* is the Divo collection, in Côte d'Ivoire. It contains more than 700 wild genotypes collected by ORSTOM (now the IRD, Institut de Recherche pour le Développement, France) in collaboration with CIRAD, the FAO (Food and Agriculture Organization, Italy), the IPGRI (International Plant Genetic Resources Institute, Italy), and the MNHN (Muséum national d'histoire naturelle, France) between 1975 and 1987, in five African countries: Côte d'Ivoire and Guinea, in West Africa; and Cameroon, Congo, and Central African Republic, in Central Africa. Management of this collection relies on clonal duplication of each genotype in the field. Dead trees are replaced by horticultural propagation from the other representative of the same genotype. In parallel, CIRAD constituted a significant collection of cultivated material, also conserved in the Divo experimental station. This collection contains more than 600 accessions of diverse origin: local varieties and populations, forms taken from village plantations, and selected material.

STRUCTURE OF GENETIC DIVERSITY

Isozymic Variability

Primary analysis of the genetic diversity of *C. canephora* from enzymatic polymorphism was done by Berthaud (1986). Fifteen samples were classified using genetic distances calculated from allelic frequencies of each sample. Twelve out of 15 samples corresponded to the forest populations: nine populations studied in Côte d'Ivoire and three in the Central African Republic. For the three other samples, it was necessary to group individuals of different populations or origins. One sample was made up of material from Cameroon; the second combined all the genotypes cultivated in the working collection of CIRAD involved in the agronomic trials; and the third combined cultivated coffee plants of the Ebobo type, which originated in Côte d'Ivoire (today there are no more representatives of the Ebobo type in collection). This study indicated, for the first time, a genetic structure in the species *C. canephora*. Two groups were identified: the 'Guinean' group, composed of wild populations of Côte d'Ivoire, and the 'Congolese' group, which comprises the wild material of the Central African Republic and of Cameroon and the cultivated material. Subsequently, by increasing the number of genotypes analysed and classifying the collection of cultivated material into 11 samples, Montagnon et al. (1992) identified two subgroups within the Congolese group: SG1 and SG2.

In our study, we took into account individuals—60 wild and 50 cultivated—and not 'populations' (Tables 1 and 2). In total, 29 alleles were

Table 1. Origin of wild material studied: country, year of collection, number of forest populations sampled, and number of genotypes analysed

Country	Year of collection	No. of populations	No. of genotypes	Reference
Cameroon	1983	10	15	Anthony et al., 1985
Congo	1985	7	13	De Namur et al., 1988
Côte d'Ivoire	1975-1986	21	36	Berthaud 1983
				Le Pierres et al., 1989
Guinea	1987	1	2	Le Pierres et al., 1989
Central African Republic	1975	6	11	Berthaud and Guillamet, 1978
Total		45	77	

Table 2. Origin of cultivated material studied: type of introduction, denomination in collection, donor institute or reference of the collection, country (of origin for donations, of cultivation for plantation samples) and number of genotypes analysed

Type of introduction	Name	Donor or collector	Country	No.
Donation	Aboisso	Aboisso,[1] Côte d'Ivoire	Gabon	6
	Niaouli	Bingerville,[2] Côte d'Ivoire	Togo	3
	Kouilou of Madagascar	Bingerville,[3] Côte d'Ivoire	Gabon	4
	C10 Man	—	Rep. of Congo	2
	INEAC	INEAC,[4] Rep. of Congo	Rep. of Congo	12
Plantation sample	Côte d'Ivoire	Berthaud, 1983, Le Pierres et al., 1989	Côte d'Ivoire	7
	Guinea	Le Pierres et al., 1989	Guinea	9
	Togo		Togo	2
	Hybrids		Côte d'Ivoire	6
Unknown	Robusta A1	Unknown	Unknown	4
Total				55

[1] Introduction at Aboisso (Côte d'Ivoire) by Beynis in 1910, of material cultivated in Gabon (Cordier, 1961).
[2] Introduction at the trial garden at Bingerville (Côte d'Ivoire), in 1914, of material cultivated in Togo (Cordier, 1961).
[3] Introduction at Bingerville (Côte d'Ivoire), in 1951, of material selected at Madagascar and originating in Gabon (Cordier, 1961).
[4] Introduction in Côte d'Ivoire, in 1935, of material selected at INEAC in the Belgian Congo (Cordier, 1961).

detected for the 8 polymorphic loci, with 2 to 6 alleles per locus and an average of 3.6 alleles per locus. There is no significant difference between the wild and cultivated individuals for mean number of alleles per locus.

The classification of the 60 wild genotypes indicates a structure in two groups (Fig. 1a). Group 1 contains only individuals collected in West Africa (Côte d'Ivoire and Guinea). Group 2 combines all the individuals originating from Central Africa (Cameroon, Congo, and Central African Republic) and

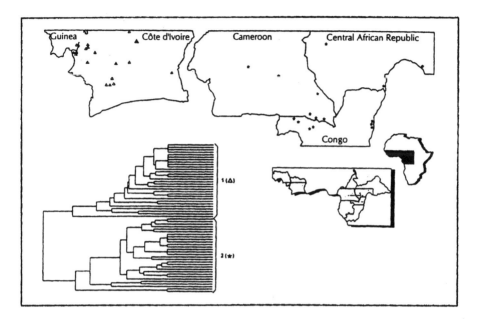

Fig. 1a. Classification of 60 wild genotypes according to the Dice similarity index and the UPGMA method of aggregation from data observed for 29 isozyme markers.

two genotypes from Côte d'Ivoire. These two groups correspond, in their composition, to the Guinean and Congolese groups of Berthaud (1986).

The dendrogram obtained with all the genotypes studied, wild and cultivated, is structured in three groups (Fig. 1b). No separation appears between wild and cultivated forms. The group most distant from the other two (group 1) contains all the individuals of the wild group 1 (West Africa) and genotypes taken from plantations in Côte d'Ivoire. The second group (group 2) comprises the genotypes of the wild group 2 (with the exception of three genotypes), all the cultivated material originating in the Republic of Congo, the individuals of the 'hybrid' groups, and those that were taken from the plantations in Guinea. The cultivated material originating in Gabon (except three genotypes), the individuals of the Robusta A1 group, two genotypes taken from the plantations of Côte d'Ivoire, a genotype taken from Togo, and three individuals of the wild group 2 form the third group (group 3). With regard to the origin of the material they contain, our group 2 corresponds to subgroup SG2 of Montagnon et al. (1992) and our group 3 corresponds to their subgroup SG1.

Molecular Variability

The use of molecular markers for the study of genetic diversity of the genus *Coffea* is recent. The first studies covered the analysis of the diversity of

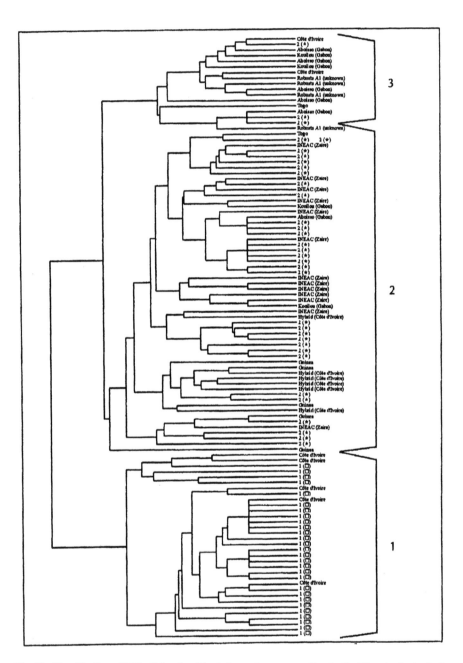

Fig. 1b. Classification of 110 wild and cultivated genotypes according to the Dice similarity index and the UPGMA method of aggregation from data observed for 29 isozyme markers.

C. *arabica* using RAPD markers (Lashermes et al., 1996). The results that we present are the first data on the molecular variability of C. *canephora*.

Out of the 26 homologous probes tested, 10 were found to be mono-locus and polymorphic. They allowed the detection of 2 to 14 alleles per locus or 66 alleles in total and an average of 6.6 alleles per polymorphic locus. The total number of alleles observed for the wild genotypes (62) is significantly higher (χ^2 = 4.55; P = 0.0329) than that of cultivated genotypes (54).

The dendrogram obtained from molecular data indicates a structure of wild material into five groups (Fig. 2a). The genotypes of a population of northwest Congo and a population of southwest Cameroon make up group A. Group B comprises all the genotypes collected along the southern frontier of the Central African Republic. The individuals of group C are distributed in the three countries of Central Africa: northwest Congo, southwest Cameroon, and southwest Central African Republic. Group D is made up of all the genotypes collected in Guinea and Côte d'Ivoire, with the exception of four individuals of western Côte d'Ivoire. Group E contains the genotypes collected in northeast Congo, those that belong to populations of northwest

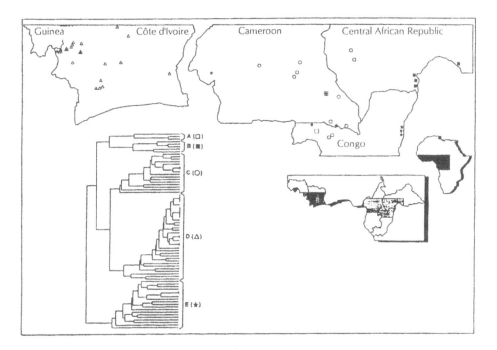

Fig. 2a. Classification of 77 wild genotypes according to the Dice similarity index and the UPGMA method of aggregation from data observed for 66 RFLP markers.

Congo and southern Cameroon, and individuals from three populations in western Côte d'Ivoire.

A global analysis of the dendrogram indicates that the wild material originating from West Africa is classified in a single group, while the material collected in Central Africa is structured into four groups. Group E is the most distant from the other four groups. The group of Central Africa (group C) closest to that of West Africa (group D) has the widest geographic distribution.

When the cultivated material is taken into account in the analysis, the structure of the species in five groups is conserved (Fig. 2b). Similarly, the position of groups with respect to each other remains unchanged. For each group, the composition of wild material remains identical to that defined previously. However, no cultivated material is present in the wild groups B and C. The individuals from the Republic of Congo, with the exception of one individual, are included in group E, as are individuals collected in the plantations of Guinea and a hybrid taken from the plantations of Côte d'Ivoire. In addition to the wild individuals of group A, most of the cultivated genotypes originating from Gabon, those of Togo (sampled from plantations and the Niaouli group), and individuals of the Robusta A1 group are classified in group A. Finally, almost all of the hybrid individuals, the genotypes taken from the plantations of Côte d'Ivoire, and most of the genotypes from the Guinea plantations are included in group D.

Agromorphological Variability

To our knowledge, only two studies have been done on the analysis of agromorphological diversity of *C. canephora* (Montagnon et al., 1992; Leroy et al., 1993). In the two cases, the principal components analysis did not reveal a high level of organization of the species. On the other hand, subsequent comparisons between the groups established on the basis of isozymes were done in several studies and for various combinations of agromorphological variables (Berthaud, 1986; Montagnon et al., 1992, 1993; Leroy et al., 1993; Montagnon and Leroy, 1993; Moschetto et al., 1996). Significant differences between the means of isozyme groups have been shown with some traits: leaf morphology, length of internodes, ramification, drought-sensitivity, phenology of fructification, and sensitivity to orange rust due to *Hemileia vastatrix*. On the other hand, in a plantation it is not possible to determine what genetic group a coffee plant belongs to on the basis of its morphology.

In our study, analysis of agromorphological data leads to a classification of wild genotypes into two major groups and a third group comprising two individuals relatively close to each other but very distant from other wild genotypes (Fig. 3a). This classification is not geographical: all the countries studied have representatives of group I and group II. Moreover, for around one third of the populations, the individuals from a single population are

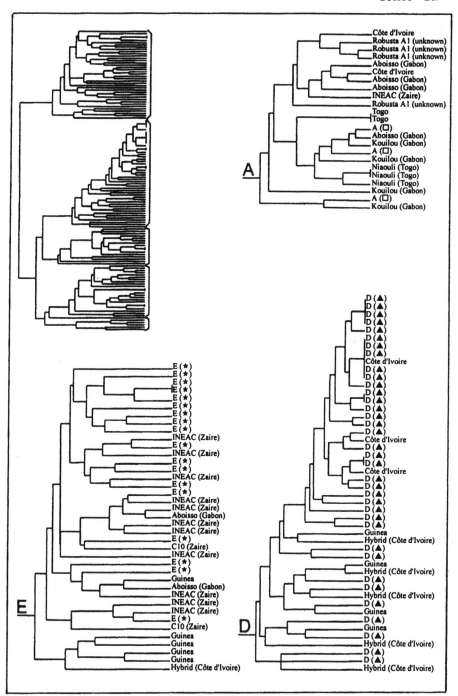

Fig. 2b. Classification of 132 wild genotypes according to the Dice similarity index and the UPGMA method of aggregation from data observed for 66 RFLP markers.

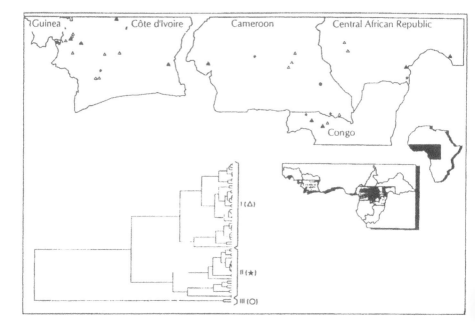

Fig. 3a. Classification of 61 wild genotypes according to the Euclidean distance and the UPGMA method of aggregation from data observed for 11 agromorphological markers.

distributed between the two major groups I and II. Analyses of variance for each of the agromorphological markers studied indicate a significant difference between the means of groups I and II for the berry development duration, bean weight, and leaf morphology.

When the cultivated forms are taken into account, the structure of the diversity is overall similar to that obtained when only the wild genotypes are analysed (Fig. 3b). The distribution of wild individuals in the groups remains unchanged with the exception of two genotypes of the wild group II, which are included within group III. No cultivated form is found in group III. All the genotypes originating in Gabon, with the exception of a genotype of the Aboisso group, as well as the individuals of groups Robusta A1 and hybrids are included in group I. The individuals of the Republic of Congo, except one genotype, and the genotypes taken from plantations in Côte d'Ivoire are found in group II.

Relationships between Various Levels of Diversity

The distribution of individuals of the biochemical groups within molecular and agromorphological groups (Table 3) allows us to compare the structures

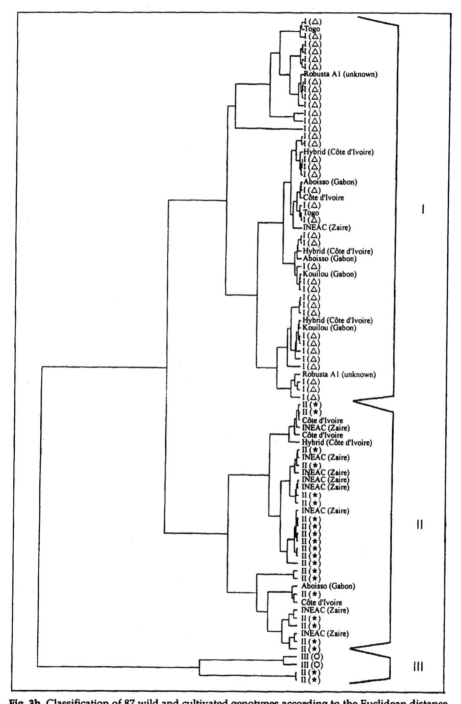

Fig. 3b. Classification of 87 wild and cultivated genotypes according to the Euclidean distance and the UPGMA method of aggregation from data observed for 11 agromorphological markers.

Table 3. Distributions of wild genotypes, on the one hand, from wild and cultivated genotypes, on the other, classified according to their relationship to biochemical groups within molecular groups and agromorphological groups. For each comparison, only the genotypes common to analyses corresponding to two compared markers have been taken into account

Biochemical groups		Molecular groups					Agromorphological groups		
		A	B	C	D	E	I	II	III
Wild	1	0	0	0	30	0	14	7	0
genotypes	2	2	4	12	1	12	7	7	1
Wild and	1	0	0	0	32	1	17	9	0
cultivated	2	6	4	12	8	29	16	20	2
genotypes	3	13	0	0	0	1	8	1	0

observed. The biochemical groups present different levels of diversity. Biochemical group I, relatively homogeneous in molecular terms, corresponds only to molecular group D. Biochemical group 2 has wide diversity and comprises representatives of the five molecular groups. When only wild genotypes are taken into account, a nearly perfect agreement (excepting one individual) is observed between biochemical group 1 and molecular group D. The individuals of biochemical group 2 are distributed among molecular groups A, B, C, and E. When the wild and cultivated genotypes are considered simultaneously, the agreements indicated with the wild genotypes alone are not modified in their essentials. The biochemical variability of cultivated genotypes is greater than that of wild genotypes. Biochemical group 3, containing mainly cultivated genotypes, corresponds essentially to molecular group A.

The agromorphological markers indicate a relatively strong organization of the diversity of *C. canephora* in two major groups. However, no agreement could be established between the two biochemical groups and the three agromorphological groups. Similarly, the agromorphological structure does not coincide with that drawn from RFLP markers. The fact that each morphological group includes representatives of different biochemical and molecular groups makes it impossible to distinguish the Guinean and Congolese groups according to their morphological characteristics. Similar selection pressures are probably exerted in West Africa and Central Africa, which has led to morphologically undifferentiated forms.

The high agreement of structures observed with the two types of neutral marker used, isozymes and RFLP, conforms to the classification within *C. canephora* (Table 4). Moreover, the results of our study prove the utility of molecular markers in the analysis of the genetic structure of *C. canephora*. On the one hand, for an equivalent number of loci, RFLP markers allowed detection of a higher number of alleles per polymorphic locus than did biochemical markers (6.6 against 3.6). On the other hand, RFLP markers

Table 4. Agreement of structures of the wild and cultivated diversity observed using isozyme markers and RFLP markers during three successive studies and composition of wild material, taken from plantations and selected, from observed groups

Berthaud, 1986	Montagnon et al., 1992	Our study				
		Markers			Material	
Isozymes	Isozymes	Isozymes	RFLP	Wild	Plantation	Selected
Guinea	Guinea	1	D	Côte d'Ivoire Guinea	Côte d'Ivoire Guinea	Hybrid (Côte d'Ivoire)
			B	Central African E		
	SG2	2	C	Central African O Cameroon Congo NO		
Congo			E	Congo Cameroon S	Guinea	INEAC (Rep. of Congo) C10 (Rep. of Congo)
	SG1	3	A	Congo NO Cameroon SE	Togo Côte d'Ivoire	Kouilou (Gabon) Aboisso (Gabon) Niaouli (Togo) Robusta A1 (unknown)

allowed us to refine the analysis by indicating an intragroup structure: the Congolese group defined by Berthaud (1986) corresponds to four molecular groups.

DEVELOPMENT AND MANAGEMENT OF GENETIC RESOURCES

Structure of Diversity and Use of Genetic Resources

Using isozymic loci that discriminate between the Guinean and Congolese groups, Berthaud (1986) identified intergroup hybrids in the cultivated clones. Among the 12 intergroup clones identified, 6 are the highest-yielding cultivars. From this observation, Berthaud proposed a reciprocal recurrent selection procedure for *C. canephora* based on the use of the Guinean and Congolese groups. The efficiency of this procedure was demonstrated subsequently (Leroy et al., 1993) and confirmed that it is important to know the structure of the diversity to best exploit the genetic resources.

The results of our study with molecular markers could be used to improve this selection procedure. Indeed, it should first be verified that there are no combinations within the Congolese group presenting a heterosis higher than the mean heterosis between the Guinean and Congolese groups. Second, if the hybrids between Guinean and Congolese groups remain the most promising, the preliminary results seem to indicate that the value of the heterosis between the two groups depends on the molecular group to which the individual of the Congolese group belongs. Thus, based only on the genetic distances established using RFLP markers, the heterosis between individuals in groups D and C could be lower than that between the individuals in groups E and C.

Structure of the Diversity and Management of Genetic Resources

From the results of this study we can propose a certain number of recommendations to ensure better *ex situ* management of genetic resources of *C. canephora*.

Most collections of *C. canephora* contain cultivated material and are highly redundant. Our study shows that a large part of the diversity existing in the wild material is not represented in the cultivated material. Moreover, the cultivated material does not contain an original diversity in relation to the wild material. Consequently, the wild material collected in many expeditions and conserved in Côte d'Ivoire presently constitutes the largest source of variability available for this species. Efforts must thus be made to preserve this collection, either by duplicating plant material or by other means of conservation.

In Côte d'Ivoire, the two collections of cultivated and wild coffee trees have so far been managed independently, which increases the task of management. Our results indicate that it is possible to consolidate the collection and hierarchize it on the basis of molecular groups. Moreover, the use of algorithms of sampling that maximize the intragroup diversity, such as that proposed by Noirot et al. (1996), would, by defining a core collection, enable optimal management of the global collection and help establish priorities for conservation as well as for evaluation, use, and diffusion of the genetic resources.

At present, the plant material is maintained only in the form of field genotypes. The creation of stratified, small core collections could allow conservation of genes rather than genotypes by rationally constituting bulks of seeds within each of the genetic groups indicated in our study. *In vitro* conservation of microcuttings established from such seed bulks has shown its limits: some genetic groups are quickly lost, especially within *C. canephora* (Dussert et al., 1997). On the other hand, the cryopreservation of seeds, already attempted with *C. arabica* (Dussert et al., 1998), is a promising alternative to field conservation.

CONCLUSION

Our study shows that molecular markers of the RFLP type can be used to increase our knowledge of the organization of the diversity of coffee C. *canephora*. This organization agrees to a great extent with that obtained with biochemical markers. However, the molecular markers show a higher differentiation than other markers.

Even though our analysis was done on a reduced sample of the genotypes conserved in Côte d'Ivoire, it is interesting to observe that the cultivated material is not generally differentiated from the wild material and that only part of the diversity of this wild material has so far been exploited in C. *canephora* cultivation. Moreover, there is a differentiation within the Congolese group, the material originating from Central Africa being structured in several molecular groups. On the other hand, the material originating in West Africa, of the Guinean group and classified in a single molecular group, is not more distant from the Central African groups than the latter are among themselves.

Thus, the present results enable us to consider new strategies for varietal selection as well as for rational management of genetic resources of C. *canephora*.

APPENDIX

Plant Material

A sampling of 132 genotypes was done within collections of wild and cultivated material conserved in Côte d'Ivoire (Tables 1 and 2). The 77 wild genotypes were sampled in order to have a representation of each of 45 forest populations studied (Table 1). For the cultivated material, a random proportional sampling was done for each of the 10 principal origins identified in collection (Table 2). The grouping is highly heterogeneous. For the material that was donated, the denomination of the groups corresponds to the name, the location, and the donor experimental station or to the varietal type of the material (Robusta or Kouilou). For these groups, the country of origin could correspond to the cultivation zone or the breeding centre. For the material collected in the plantations, the country mentioned is that in which the material was collected. The group called 'hybrids' includes genotypes for which it was later shown that they are hybrids between the forms originating in West Africa and those originating in Central Africa (Berthaud, 1983). Finally, the history of the introduction of the Robusta A1 group could not be traced. The origin of this group thus remains unknown.

RFLP Analysis

The total genomic DNA was extracted according to the method described by Agwanda et al. (1997). The technique of molecular marker analysis used is that described by Lashermes et al. (1995). Two restriction enzymes were used: EcoRI and HindIII. The 26 probes tested come from a genome bank of C. arabica. Among these, 10 were retained for their polymorphic and mono-locus characteristics. Each probe was used after restriction by one or the other restriction enzyme. The presence and absence of 66 bands corresponding to 66 alleles were coded 1 and 0, respectively.

Enzymatic Analysis

Within the total sampling of 132 individuals (Tables 1 and 2), the analysis of isozymic polymorphism was done on 60 wild individuals and 50 cultivated individuals. Among the 60 wild individuals, 48 were common to analysis done with agromorphological markers. For the cultivated material, the number of individuals common to isozyme and agromorphological analyses was 26. The techniques of extraction, electrophoresis, and detection of isozymes are those of Berthaud (1986). The analyses were done on 5 enzymatic systems revealing 8 loci: esterases a and b (3 loci), 6-phosphogluconate dehydrogenase (2 loci), isocitrate dehydrogenase (1 locus), phospho-glucomutase (1 locus), and phosphoglucoisomerase (1 locus). The 29 alleles identified were coded as present or absent (1 or 0).

Agromorphological Study

Within the132 genotypes studied (Tables 1 and 2), 61 wild genotypes and 26 cultivated genotypes were evaluated for 11 agromorphological markers. Four classes of markers could be distinguished: morphological (length, width, area and shape of leaves, length of acumen, length of petiole); technological (100 bean weight, percentage of peaberries, outturn or bean weight to berry weight ratio, percentage of empty loges, for the wild material only); phenological (for the wild genotypes only, berry development duration and extension of maturation of berries); and agronomic (yield). For a detailed description of markers, see Anthony (1992).

Statistical Analyses of Classification

For the RFLP and isozyme markers the distance matrixes between individuals were calculated using the Dice similarity index (1945). The Euclidean distance was used for agromorphological markers. For the three types of markers, the method of aggregation used to construct the dendrograms was the UPGMA method.

REFERENCES

Agwanda, O.A., Lashermes, P., Trouslot, P., Combes, M.C., and Charrier, A. 1997. Identification of RAPD markers for resistance to coffee berry disease, *Colletotrichum kahawae*, in arabica coffee. *Euphytica*, 97: 241-248.

Anthony, A. 1992. Les ressources génétiques des caféiers: collecte, gestion d'un conservatoire et évaluation de la diversité génétique. Montpellier, France, Orstom, collection Travaux et documents, 320 p.

Anthony, F., Couturon, E., and de Namur, C. 1985. Les caféiers sauvages du Cameroun: résultats d'une mission de prospection effectuée par l'Orstom en 1983. In: *XIe Colloque Scientifique International sur le Café*. Paris, ASIC, pp. 495-501 .

Berthaud, J. 1980. L'incompatibilité chez *Coffea canephora*: méthode de test et déterminisme génétique. *Café, Cacao, Thé*, 24: 267-274.

Berthaud, J. 1983. Liste du matériel provenant des prospections de Côte d'Ivoire. Paris, Orstom (document interne).

Berthaud, J. 1986. Les ressources génétiques pour l'amélioration des caféiers africains diploïdes. Montpellier, France, Orstom, collection Travaux et documents, 379 p.

Berthaud, J. and Guillaumet, J.L. 1978. Les caféiers sauvages en Centrafrique: résultats d'une mission de prospection (janvier-février 1975). *Café, Cacao, Thé*, 3: 171-186.

Bridson, D.M. and Verdcourt, B. 1988. Rubiaceae (Part 2). In: *Flora of Tropical East Africa*. R.M. Polhill, eds., Rotterdam, Balkema, 727 p.

Charrier, A. and Eskes B. 1997. Les caféiers. In: *L'Amélioration des Plantes Tropicales*. A. Charrier et al., eds., Montpellier, France, CIRAD-Orstom, collection Repères, pp. 171-196.

Cordier, L. 1961 . Les objectifs de la sélection caféière en Côte d'Ivoire. *Café, Cacao, Thé*, 5: 147-159.

Couturon, E. 1980. Le maintien de la viabilité des graines de caféiers par le contrôle de leur teneur en eau et de la température de stockage. *Café, cacao, thé*, 24: 27-32.

Cros, J., Combes, M.C., Chabrillange, N., Duperray, C., Monnot des Angles, A., and Hamon, S. 1995. Nuclear DNA content in the subgenus *Coffea*: inter- and intra-specific variation in African species. *Canadian Journal of Botany*, 73: 14-20.

de Namur, C., Couturon, E., Sita, P., and Anthony, F. 1988. Résultats d'une mission de prospection des caféiers sauvages du Congo. In: XIIe Colloque Scientifique International sur le Café. Paris, ASIC, pp. 397-404.

Dice, L.R. 1945. Measures of the amount of ecologic association between species. *Ecology*, 26: 297-302.

Dussert, S., Chabrillange, N., Anthony, F., Engelmann, F., Recalt C., and Hamon, S. 1997. Variability in storage response within a coffee (*Coffea* spp.) core collection under slow growth conditions. *Plant Cell Report*, 16: 344-348.

Dussert, S., Chabrillange, N., Engelmann, F,, Anthony, F., Louarn, J. and Hamon, S. 1998. Cryopreservation of seeds of four coffee species (*Coffea arabica, C. costatifructa, C. racemosa* and *C. sessiliflora*): importance of water content and cooling rate. *Seed Science Research*, 8: 9-15.

Lashermes, P., Combes, M.C., and Cros, J. 1995. Use of non-radioactive digixigenin-labelled DNA probes for RFLP analysis in coffee. In: *Techniques et Utilisations des Marqueurs Moléculaires*. A. Berville and M. Tressac, eds., Paris, INRA, pp. 21-25.

Lashermes, P., Combes, M.C., Trouslot, P., and Charrier, A. 1997. Phylogenetic relationships of coffee-tree species (*Coffea* L.) as inferred from ITS sequences of nuclear ribosomal DNA. *Theoretical and Applied Genetics*, 94: 947-955.

Lashermes, P., Trouslot, P., Anthony, F., Combes, M.C., and Charrier, A. 1996. Genetic diversity for RAPD markers between cultivated and wild accessions of *C. arabica*. *Euphytica*, 87: 59-64.

Le Pierres, D., Charmetant, P., Yapo, A., Leroy, T., Couturon, E., Bontems, S., and Tehe, H. 1989. Les caféiers sauvages de Côte d'Ivoire et de Guinée: bilan des missions de prospection effectuées de 1984 à 1987. In: XIIIᵉ Colloque Scientifique International sur le Café. Paris, ASIC, pp. 420-428.

Leroy, T., Montagnon, C., Charrier, A., and Eskes, A. 1993. Reciprocal recurrent Selection applied to *Coffea canephora* Pierre. 1. Characterization and evaluation of breeding populations and value of intergroup hybrids. *Euphytica*, 67: 113-125.

Montagnon, C. and Leroy, T. 1993. Réaction à la secheresse de jeunes caféiers *Coffea canephora* de Côte d'Ivoire appartenant à différents groupes génétiques. *Café, cacao, thé*, 37: 1 79-190.

Montagnon, C., Leroy, T., Cilas, C., and Eskes, A.B. 1993. Differences among clones of *Coffea canephora* in resistance to the scolytid coffee twig-borer. *International Journal of Pest Management*, 39: 204-209.

Montagnon, C., Leroy, T., and Eskes, A.B. 1998. Amélioration variétale de *Coffea canephora*. 2. Les programmes de sélection et leurs résultats. *Plantations, recherche, développement*, 5: 89-98.

Montagnon, C., Leroy, T., and Yapo, A. 1992. Etude complémentaire de la diversité génotypique et phénotypique des caféiers de l'espèce *C. canephora* en collection en Côte d'Ivoire. In: XIVᵉ Colloque Scientifique International sur le Café. Paris, ASIC, pp. 444-450.

Moschetto, D., Montagnon, C., Guyot, B., Perriot J.J., Leroy T., and Eskes, A.B. 1996. Studies on the effect of genotype on cup quality of *Coffea canephora*. *Tropical Science*, 36: 18-31.

Noirot, M., Anthony, F., and Hamon, S. 1996. The principal component scoring: a new method of constituting a core collection using quantitative data. *Genetic Resources and Crop Evolution*, 43: 1-6.

Roberts, E.H. 1973. Predicting the storage life of seeds. *Seed Science and Technology*, 1: 499-514.

Pearl Millet

Gilles Bezançon

Pearl millet is, along with sorghum, the staple food for a significant part of the population in Africa and the Indian subcontinent. It is the cereal most tolerant to drought; it is cultivated in the Sahel, in zones where rainfall is no more than 200 mm. For these reasons, the conservation, evaluation, and commercialization of genetic resources of pearl millet constitute a considerable task. Pearl millet is mostly cultivated in arid and semi-arid areas in Africa, where it covers 11.5 million ha, and in India, where it covers 14.7 million ha (FAO, 1996).

Three quarters of the African production comes from the western part of the continent, and the major producer countries, in decreasing order, are Nigeria, Niger, Burkina Faso, Chad, Mali, and Mauritania. In East Africa, the Sudan and Uganda are the primary producers, while in southern Africa this traditional crop has nearly disappeared. In Africa, the local varieties, early or late, are tall and put out many suckers. They are not highly productive: planted in poor soils, they do not benefit from tilling and only rarely get inputs in the form of organic manure from livestock. Seeds are planted during the beginning of the rains and must be planted several times in case of poor rainfall.

In India, pearl millet is cultivated on a large scale in Rajasthan, Gujarat, and Haryana. The use of animals for ploughing allows the soil to be worked and seeding is done by means of traditional seeders. The traditional varieties are late-yielding and short-strawed, and they put out few suckers. But here also, because chemical fertilizers are rarely used, grain yields are not high. Yields remain low, 0.6 to 0.8 t/ha, compared with those of other cereals cultivated in the tropics, such as rice and maize. Pearl millet can be intercropped, with cowpea, for example. Finally, in the United States, pearl millet is cultivated on more than 150,000 ha to produce fodder and grain: the yields are over 1.2 t/ha (FAO, 1996).

The primary product of the plant is the grain, which has greater nutritive value than wheat and rice. It is consumed in the form of dough, porridge, couscous, or pancakes, and in certain regions is used to make alcoholic beverages (pearl millet beer). The straw can also be used as forage or in construction of roofs and fences for traditional houses.

TAXONOMY AND GENETIC RESOURCES

Botany and Taxonomy of the Genus *Pennisetum*

Pearl millet, *Pennisetum glaucum* (L.) R. Br., belongs to the genus *Pennisetum* (family Poaceae, subfamily Panicoideae, tribe Paniceae). The species of *Pennisetum* (numbering about 60) are distributed in tropical and subtropical zones across the world. The genus is divided into five sections and pearl millet belongs to the section *Penicillaria*, which is characterized by the presence of a tuft of hairs on the apex of the stamen. In this section, van der Zon (1992) recognized three subspecies within the species *P. glaucum* (x = 7, 2n = 2x = 14): *P. glaucum* ssp. *glaucum*, cultivated pearl millet, found in Africa; *P. glaucum* ssp. *violaceum*, the wild form, widely distributed in Africa in the Sahelian and sub-desert zone, in a discontinuous manner from the Atlantic to the Red Sea and in varied ecological situations; and *P. glaucum* ssp. *sieberianum*, which combines the intermediate forms resulting from natural hybridizations between the cultivated and wild forms.

On the model of Harlan and de Wet (1971), the species complex of the genus *Pennisetum* can be structured in three gene pools. The primary pool, monospecific, combines the three subspecies of *P. glaucum*. In certain regions of cultivation, the importance of intermediate forms has led local people to give them a particular name: *shibra* in Haoussa in Niger and *n'doul* in Wolof in Senegal. The secondary pool is composed of two species, *P. purpureum* and *P. squamulatum*, which can easily hybridize with *P. glaucum* (Hanna, 1987). *Pennisetum purpureum* is a perennial species, allotetraploid (x = 7, 2n = 4x = 28), sexual, and allogamous. *Pennisetum squamulatum* is a perennial species, tetraploid (x = 9, 2n = 4x = 36) and apomictic. These two species are used in improvement of pearl millet by apomixis (Hanna, 1987, 1990). The tertiary pool comprises the other species of the genus, including species of the section *Brevivalvula*. Widely represented in Africa, where they occupy various ecological niches, *Brevivalvula* are either annual or perennial with varied systems of reproduction (sexual, apomictic, vegetative). For these reasons, they can be used in genetic improvement of cultivated pearl millet.

Genetic Resources

Around 1960 the Rockefeller Foundation and the Indian Agricultural Research Institute undertook the establishment of a world collection of pearl millet cultivars. This first attempt fell through. A relatively large number of traditional cultivars disappeared partly because of desertification in zones particularly exposed to climatic fluctuations and partly because of socioeconomic transformations in some countries. This led the Food and Agriculture Organisation of the United Nations and then the IBPGR (International Board for Plant Genetic Resources) to undertake, around the

1970s, the collection and conservation of genetic resources of pearl millet and sorghum in the Sahelian zone. The International Crops Research Institute for the Semi-Arid Tropics (ICRISAT), mandated by the CGIAR (Consultative Group for International Agricultural Research), coordinated these collections in collaboration with IBPGR, ORSTOM (Institut Français de Recherche Scientifique pour le Développement en Coopération), now IRD (Institut de Recherche pour le Développement) and national institutes for agricultural research. A world collection of close to 24,000 samples—traditional cultivars and wild forms together—originating from 44 countries is presently conserved, for the long and medium term, in the Indian centre of ICRISAT at Patancheru. Other sites of conservation are, notably, the GeneTrop laboratory of the IRD at Montpellier, in France—for medium-term conservation of around 3200 samples, 11% of them are wild species—and the Sahelian centre of ICRISAT of Sadore, in Niger. Doubles of these collections were established at the Centre for Conservation of Genetic Resources of Ottawa, in Canada, as well as at the National Seed Storage Laboratory in Fort Collins, United States, for long-term conservation.

For each of these collections the samples were accompanied by a profile comprising 11 descriptors, which gives information on the original situation of the sample.

ORGANIZATION OF GENETIC DIVERSITY

Agromorphological Variability

The large size of the collections makes it difficult to evaluate the diversity they contain. Several evaluation studies conducted by ICRISAT have been done on specific collections of some countries, including Ghana, Malawi, Cameroon, and the Central African Republic, and of several states in India. The results for 1939 and 2458 samples from the world collection, evaluated for 20 and 18 characters respectively, are presented in a catalogue published by the NBPGR (National Bureau of Plant Genetic Resources, India) and ICRISAT (NBPGR and ICRISAT, 1993). These data are reported only for cultivated forms. There are no studies that include the wild forms, which would make it possible to describe the genetic diversity of the entire collection and to compare the data with enzymatic electrophoresis data obtained elsewhere.

Apart from these purely descriptive studies conducted in India, various analyses have been done. They have resulted in classifications, notably for the cultivated forms, that are consistent with the classification of cultivars into large regional groups established from observations during various expeditions (Clément, 1985). Portéres (1950, 1976) recognized 16 species distributed in four groups on the basis of their geographic distribution: far West Africa group, West and Central Africa group, Nile-Sudan group, and

East Africa and Angola group. For francophone West Africa, Bono (1973) constituted two groups essentially on the basis of characters of false ears: group I, divided into two subgroups (the forms found in Mali, Côte d'Ivoire, and Mauritania, and some forms of Niger, on the one hand, and the forms found in Burkina Faso, on the other), and group II, also divided into two subgroups (pearl millets of Niger and of Senegal). On the basis of grain shape, Brunken et al. (1977) defined four races: *typhoides*, *nigritarum*, *globosum*, and *leonis*. The notations made on the length of the cycle as well as on about 10 characters of the ears and grain for some 1500 ears collected in Mali and Senegal have indicated a regional differentiation in each of these countries (Marchais, 1982).

Marchais et al. (1993) studied 267 accessions of cultivated pearl millet and 118 wild pearl millet using 14 morphological characters. The results obtained (Fig. 1) show a marked distinction between cultivated and wild pearl millets on the basis of just a few characters: length of ears, density of ears, length of pedicel of the involucre, grain size, stem diameter, tiller number, and leaf width.

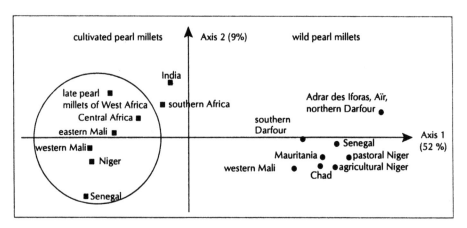

Fig. 1. Position of means of major morphological families of pearl millet on the 1-2 plane of a principal components analysis of 14 botanical characters with high heritability (Marchais et al., 1993). The pearl millets cultivated in West Africa in the wider sense are circled.

Within the cultivated pearl millets, there are five groups for West Africa, one for Central Africa, one for India, and one for southern Africa, the two last being more homogeneous. Discrimination among the wild pearl millets is less marked because of their greater morphological homogeneity.

Phenological Diversity of Cultivated Pearl Millets

Data on the flowering phenology of more than 12,000 cultivars obtained by ICRISAT (Marchais et al., personal communication) indicated two phenomena that concur with the triggering of flowering in pearl millet: the need for a minimum number of degree-days and sensitivity to day length. Figure 2 compiles the results observed for two cultivation cycles, one completed in long days and the other in short days. It was observed that, for each of the regions studied, there is a range of varieties that, during long days, flower between 45 and 140 days. This diversity can be imputed to differences in the duration of the rainy season in each region.

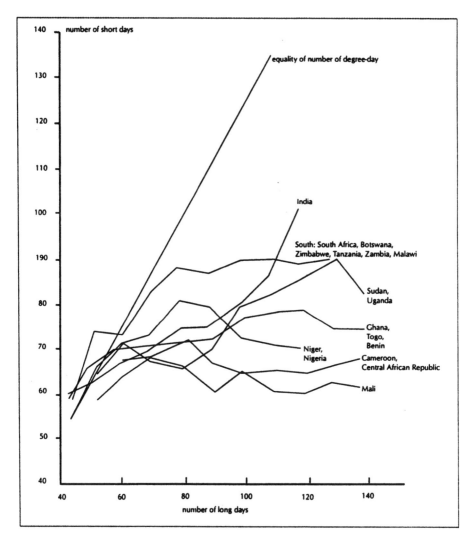

Fig. 2. Comparative evolution of flowering in long and short days according to the geographic origin of the pearl millets observed (Marchais et al., personal communication).

On the other hand, during short days, the cycles do not go beyond 90 days, except for cultivars of southern India. The cultivars that flower between 40 and 60 days for a long-day cultivation flower during a period that is equivalent in number of degree-days when cultivated during short days. These cultivars can thus be considered heat-sensitive, but not photosensitive. In cultivars that flower beyond 60 days for a long-day cultivation, behaviour during short days varies according to the region.

Overall, there are three possible behaviours. The first pertains to cultivars of the South group: up to about 80 days, the increase in the cycle of long days is accompanied by an increase in the cycle of short days, then the length of the cycle of short days does not increase further, which could be explained by a maximal thermoperiodic demand with the appearance of photosensitivity. The second is represented by cultivars of West Africa: the traditional early pearl millets that flower in less than 80 days and can be called thermoperiodic (but with a lower demand in degree-days than pearl millets of the South group) are distinguished from the traditional late pearl millets that flower after 80 days, which can be called photosensitive and have a minimal thermoperiodic demand. The third behaviour pertains to the Indian varieties, which represent an intermediate situation, in which the elongation of the long-day cycle is accompanied by an elongation of the short-day cycle, in a continuous fashion. There is a north-south gradient here, the longer cycles characterizing the cultivars of southern India. In this situation, it is difficult to talk of photosensitive and non-photosensitive pearl millets.

The other groups shown in the figure are more or less close to one of the major groups that we have just described.

Overall, in this phenological analysis of flowering of cultivated pearl millets, we find groups close to those that have been observed with other descriptors of diversity. The distinction between early and late pearl millets is obvious only in West Africa. The most characteristic are very early and non-photosensitive cultivars from Ghana, Togo, and Benin, the photosensitive and long-cycle pearl millets of Mali, and the peculiar pearl millets of Sierra Leone that have a very long cycle.

Biochemical and Molecular Variability

Most of the results on enzymatic polymorphism of pearl millets come from the studies of Tostain (1994) on 549 accessions, of which 361 were cultivated pearl millets and 188 were wild pearl millets. For the eight enzymatic systems studied, 12 loci proved polymorphic, for a total of 46 alleles, or an average of 3.8 alleles per locus.

The results obtained show that, in their zone of origin, wild and cultivated pearl millet cannot be differentiated at the loci studied by alleles that would be fixed in one group and absent from the other. The two groups differ only by the allelic frequencies at these loci.

The analyses done on the wild pearl millets indicate five groups, which correspond to well-defined geographic entities. The 'west' group (I) comprises the populations of Senegal, Mauritania, and western Mali; the 'centre' group (II) corresponds to the populations of Adrar des Iforas and from Gourma to Mali, Oudalan in Burkina Faso, the Azawak and Ader Doutchi valleys to Niger; the 'Aïr' group (III) is made up of the population from the Aïr massif to Niger; the 'Chad-west' group (IV) comprises the populations located south of Lake Chad; and the 'Darfour' group (V) comprises populations east of Lake Chad and from Darfour to Sudan.

For the cultivated pearl millets, various partial enzymatic analyses have led the author to distribute the set of samples into seven groups, on a geographical as well as physiological basis (length of cycle): group A (early pearl millets presently found in Senegal and western Niger), group B (early pearl millets of western Mali), group C (early pearl millets found from Niger to Sudan), group D (early pearl millets of Togo and Ghana), group E (late pearl millets of West Africa), group F (early and late pearl millets of East Africa and southern Africa), and group G (early and late pearl millets of India). The multivariate analyses on the whole do not confirm the entire classification, since certain groups, notably A, C, and E, partly overlap. Discriminant analysis gives 72% to 76% individuals well classified for these three groups, against 83% to 94% for the other groups.

A global analysis was done on 188 samples of the wild form from the entire area of distribution and 123 samples of the cultivated form belonging to four groups A, B, C, and D. It shows, for the wild pearl millets, a genetic diversity comparable to that of cultivated pearl millets: the Nei index of diversity is equal to 0.249 for the wild pearl millets and 0.256 for the cultivated pearl millets. On the other hand, the structuration of wild pearl millets is weaker than that of cultivated pearl millets (G_{ST} = 0.13 against 0.17). It is the cultivated pearl millets of groups A and B (early cultivars of West Africa) that are the closest to the wild pearl millets (Fig. 3a), while the pearl millets of group C (early forms found from Niger to Sudan) are in a less central position. In sum, two major sets can be distinguished: the first contains the wild and cultivated forms of groups A, B, and D; the second contains the rest of the cultivated pearl millets (late pearl millets of West Africa, pearl millets of East and southern Africa, pearl millets of India). A peculiar case is the Tiotande variety, originating in Senegal and cultivated in the off season, which stands apart from the rest (Fig. 3b).

Samples that represent only a limited part of the genetic diversity of wild and cultivated pearl millets were analysed using other molecular descriptors of polymorphism (sequencing, RFLP, RAPD). The comparison of some samples of pearl millet by RFLP showed that chloroplast DNA probe is very poorly polymorphic and that ribosomal DNA is polymorphic only in wild pearl millet (Gepts and Clegg, 1989). An RFLP analysis on the *ADH-1* region indicates genetic variability within and between populations as well

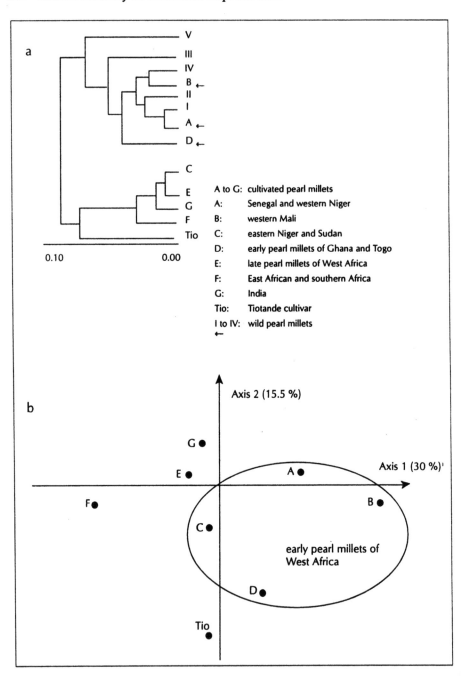

Fig. 3. (a) Dendrogram of Nei distances between groups of cultivated and wild pearl millets (Tostain, 1998). (b) Principal components analysis of 361 accessions of cultivated pearl millet. Only the projections from the centres of gravity of 7 groups of cultivated pearl millets and the Tiotande cultivar are represented (Tostain, 1998).

as the specificity of certain restriction profiles of wild forms (Pilate-Andre et al., 1993). The sequencing of alleles of the *ADH-1* gene of these samples showed no significant difference between wild and cultivated pearl millet (Gaut and Clegg, 1993). Tostain (1996) studied the diversity of 14 populations of pearl millet (4 wild, 9 cultivated, and 1 accession of Tanzania that is intermediate between wild and cultivated forms) using RAPD markers. The results confirm overall the classifications obtained with enzymatic markers but, there also, the RAPD markers reveal a very low diversity within populations and do not enable a clear separation between wild and cultivated forms. All the results obtained using different descriptors of polymorphism, whether biochemical (isozymes) or molecular (RFLP, RAPD sequencing), are consistent in terms of difficulties in differentiating the wild and cultivated pearl millet.

Relationships between Levels of Variability

Classification of wild and cultivated pearl millets obtained from data of enzymatic polymorphism was similar to classifications established on the basis of morphological characters. A good correspondence was observed with the classification of Marchais et al. (1993).

In the particular case of Niger pearl millets, a comparative analysis was done on a sample of 21 cultivars representative of the cultivated pearl millets in Niger using 12 quantitative characters and 3 qualitative characters (Siaka et al., 1996). The results (Fig. 4a) show a structuration of the diversity into three groups: cultivars of group 1 originate in the desert zone, cultivars of group 2 are cultivated in the eastern part of countries between longitudes 8°E and 13°E, and cultivars of group 3 are cultivated in the western part of countries between longitudes 1°E and 8°E. This structuration corresponds to three groups (Fig. 4b) indicated by Tostain (1994) on the basis of enzymatic polymorphism: late pearl millets (group 1), early pearl millets with short ears (group 2), and early pearl millets with long ears (group 3).

Other criteria have also been used to describe the diversity of pearl millets. L. Marchais (personal communication) tested close to 200 cultivars for their capacity to restore male fertility on the male-sterile cytoplasm described by Marchais and Pernes (1985). The results indicate a geographic differentiation: the pearl millets of India and southern Africa constitute two homogeneous groups, which have a very low rate of restoration, while in West Africa early pearl millets, which show a high variability for restorative ability, are distinguished from late pearl millets, which for the most part maintain sterility.

The results taken together—morphological, phenological, biochemical, and cytoplasmic data—tend to confirm the information collected by the researchers with respect to the geographic differentiation of pearl millets. This differentiation results first in natural selection and adaptation of wild pearl millets and cultivars to various environments and, second, in

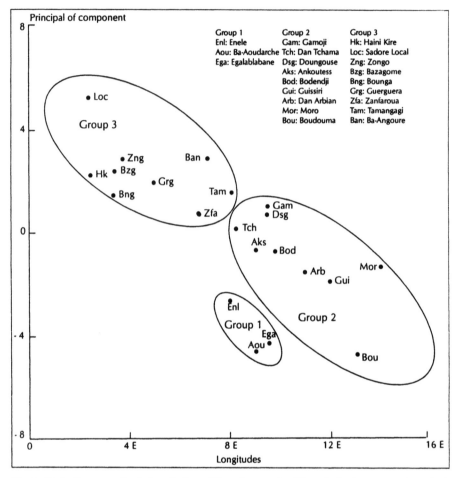

Fig. 4a. Projection according to longitude and the primary axis of the principal components analysis of 21 local cultivars of Niger (Siaka et al., 1996)

domestication, a process in which the role of the farmer is critical. This geographic structuration of the diversity of wild and cultivated pearl millets must thus be taken into account on a priority basis in the elaboration of new methods of conservation, management, and commercialization of genetic resources.

The origin of pearl millet domestication can be more precisely defined only through new information gained from supplementary studies based on descriptors of genetic polymorphism that are less sensitive than isozymes to the actual gene flows, such as chloroplast DNA. Moreover, it would be desirable to extend the sample to the entire area of distribution of wild pearl millet, up to Sudan and Ethiopia. This region is extremely important for understanding the domestication of pearl millet. It corresponds to the centre of origin of cowpea and sorghum, which, like pearl millet, are also cultivated

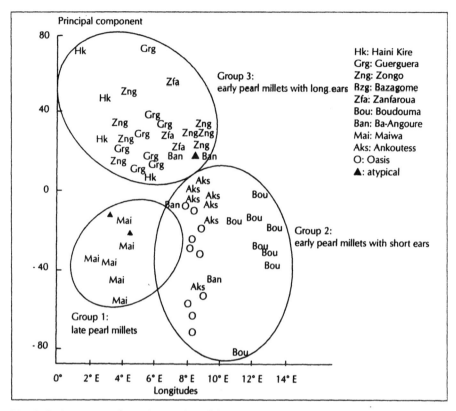

Fig. 4b. Projection according to longitude and the primary axis of the principal components analysis of 66 accessions of Niger (Tostain, 1994).

in India, unlike species such as rice *Oryza glaberrima* and yams of the *Dioscorea cayenensis-D. rotundata*, which were domesticated in West Africa.

DOMESTICATION AND MANAGEMENT OF GENETIC RESOURCES

The geographic distribution of wild pearl millet, limited to Sahelian Africa, suggests that it was domesticated here to give us the cereal we know today. The oldest vestiges of cultivated pearl millet in contact with wild pearl millet were found in Mauritania and are more than 3000 years old (Amblard and Pernes, 1989), while imprints of wild pearl millet were discovered on pottery dating around 5000 years ago in the centre of the Sudan (Stemler, 1990). Various genetic analyses have shown that the phenomenon of domestication depends on a small number of linked genes—length of pedicel and involucre, senility of ear at maturity, size of seed envelope, and presence or absence of hairs—the recessive alleles of which enable the expression of the cultivated phenotype (Pernes et al., 1984), which could explain a rapid domestication.

In pearl millet, an allogamous species, there are gene flows between cultivated and wild forms, which are more or less controlled by barriers to reproduction before and after the zygote stage. They give rise to viable hybrid forms that the farmer must eliminate during domestication to maintain the cultivated phenotype, as is done even today. But such selection remains flexible, and the hybrids can be harvested if needed. These constant exchanges blur the traces of a possible genetic structuration arising directly from domestication. In these conditions, the calculations of genetic distance from enzymatic polymorphism data are difficult to use to locate the centre of domestication of the cultivated form. The differences in genetic distance could reflect mainly variations in the efficacy of barriers to reproduction between wild and cultivated forms. The hypothesis of domestication accompanied by reinforcement of these barriers to gene exchanges between the ancestral and cultivated forms can not be ruled out and thus goes against the association of a centre of domestication with the shortest genetic distance between populations of the present wild form and those of the cultivated form (Bezancon et al., 1997). This is why, rather than look for a centre of domestication of pearl millet in a region delimited by Mauritania, Senegal, and western Mali (Tostain, 1992), it is preferable to conserve the idea of a non-centre in the sense of Harlan (1971). The domestication of pearl millet, associated with migratory phenomena, could have occurred in different geographic zones, simultaneously or in different eras, each agrarian civilization selecting a specific morphotype particularly well adapted to the local conditions—climate, soil, and cultivation practices. Long-distance migrations, on the continental scale in southern Africa and on the intercontinental scale in India, led to the constitution of groups more homogeneous and well-differentiated from the original forms.

At ICRISAT, several techniques have been recommended to maintain the diversity of the world collection. The technique of intercrosses between a defined number of plants (100) of the same cultivar (Appa-Rao, 1980) presents constraints related to the mechanisms needed to isolate breeding fields. These problems can be overcome by making controlled crosses with pollen mixtures, which increases the work involved. The technique of self-fertilization (Burton, 1983) saves time and labour. Self-fertilization is done on plants that serve to characterize the cultivars and enables conservation of dominant and recessive genes that are used to classify the cultivars. In this case, however, the plants produced are smaller and less vigorous and produce smaller and fewer grains. The technique of pools favours gene exchanges and can be used to increase adaptability. It relies on the constitution of pools from a mixture of grains from closely related cultivars—those from the same region or those having identical characteristics—the mixture being cultivated in conditions of isolation.

The technique of intercrosses and that of pools seem particularly suitable for pearl millet. In contrast, the technique of self-fertilization has major disadvantages, particularly problems of genetic drift.

The objective of ICRISAT is to produce enough grains for evaluation, selection, exchange, and conservation. There are two ways of conserving grains: long-term conservation (temperature –18°C and 5% relative humidity) designed for basic collections and medium-term conservation (temperature +5°C and 20% relative humidity) in collections for evaluation, multiplication, and distribution. It is clear that such methods applied to pearl millet collections are costly and labour-intensive.

The structuration of the genetic diversity of pearl millets, such as it has been indicated, serves as a basis for other projects of genetic resource conservation. In this context, the constitution of core collections is advisable.

Usually, to constitute a core collection, after stratification on the basis of taxonomic, geographic, sometimes morphological, enzymatic, or agronomic data, it is recommended that a sampling of 10% of the initial collection be drawn by random selection in each group of a number of samples proportionate to the logarithm of the group size (Brown, 1989). In cases where intragroup data are available, Marchais (1996), in the framework of a study on relative parental contributions that maximize the average value of certain characters such as grain yield for creation of a synthetic variety, proposed a method that enables the exact—rather than random—identification of samples that will constitute the core collection as well as their proportions. Such selection is done by maximizing the total Nei diversity (1973) of weighted bulk.

The algorithm follows the rationale used in population genetics for panmictic populations over several generations, with a constant selective value (Crow and Kimura, 1970). It establishes a relationship between allelic frequencies to generations n and n + 1 and the selective value of different genotypes. The method has been applied to a collection of 95 samples of wild pearl millet collected in Niger, which were described for 8 enzymatic systems—12 loci and 46 alleles—by Tostain (1992). This set presents a relatively homogeneous enzymatic diversity in relation to other geographic groups. A state of equilibrium was reached after 700 iterations and the results showed that the Nei diversity for the entire collection is equal to 3.58 in cases where all the samples are represented in equal proportions, while it is 6.48 after application of the algorithm. This level of diversity is obtained with only eight accessions.

The method has thus proved highly effective in increasing diversity while eliminating redundancies in this group of wild pearl millets of Niger. The same procedure could be applied to the average of allelic frequencies of each of the groups, in order to maximize the diversity of a core collection for the entire collection. Finally, this procedure of creating core collections can be used for qualitative as well as for quantitative characters, the latter being divided into classes.

The main objection that has been raised against the method is that, in most cases, the accessions are not sufficiently analysed before the constitution of a core collection. The procedure could thus be put into place on samples taken randomly according to the principles of Brown (1989), after the samples are analysed.

The application of this methodology to the world collection of pearl millet described on the basis of enzymatic and molecular markers would enable us to uncover the numerous redundancies, notably among the cultivars, while increasing the overall level of diversity.

On the regional scale, perfection of a methodology of *in situ* conservation, associated with management and participatory improvement of genetic resources of pearl millet would also be a solution. The IRD presently conducts research of this type in the southwestern region of Niger. The dynamic aspect of *in situ* conservation contrasts with all the other methodologies, which enable the conservation of rigid structures. Here, the evolution of traditional cultivars is not prevented: the gene flows with the wild forms, if they are present, continue, exchanges continue to occur between traditional cultivars (exchanges between farmers, migrations) and with genetically improved cultivars sometimes provided by agricultural research institutes. The overall diversity evolves under the influence of environmental factors, physical and human, and closely approaches the method used by farmers to maintain their traditional cultivars. The knowledge and control of the mechanisms that govern these different factors could enable better management of genetic resources.

APPENDIX

Agromorphological Studies

The latest evaluations of cultivated pearl millets were done on 1938 samples originating from 33 countries observed for 20 descriptors during the first round, and then on 2458 samples originating from 38 countries observed for 18 descriptors during the second round (NBPGR and ICRISAT, 1993).

The study of Marchais et al. (1993) was done on 267 cultivated plants and 118 wild plants from samples representative of the whole collection. They were grown in Niger during the rainy season (planted 25 June). Each sample was represented by a line of 10 plants. The notations were done on 14 morphological characters: 12 quantitative characters (height of main stem, length and width of third leaf below the ear of the main stem, length of ear on the main stem, diameter of rachis of the ear of the main stem, length of the pedicel of the involucre, length of hairs on the involucre, length of internodes, length of glumules of hairs on the involucre, number of ears per involucre, length and width of grain) and two qualitative characters (pilosity of foliar limb and vitreousness of albumen).

Phenological Analyses

More than 12,000 cultivars were evaluated on their flowering (time from seed to flowering) in two conditions of cultivation: rainy season (planted 25 June, long days) and off season (planted 25 November, short days). Each sample was cultivated on a line of 4 m (around 20 plants) and the flowering was noted when 50% of the plants had produced an ear in female flowering (L. Marchais et al., personal communication). This experiment was done at the ICRISAT station in Patancheru, India.

Biochemical and Molecular Analyses

In total, 549 samples were analysed: 361 represented cultivated forms originating from 29 African countries and India and 188 were collected in 8 sub-Saharan African countries, representing populations of the wild form in almost its entire area of geographic distribution. Eight enzymatic systems were used: ADH, CAT, EST, GOT, MDH, PGD, PGI, and PGM (Tostain, 1994).

Data Analysis

The methods of statistical analysis used were principal components analysis and discriminant analysis, done with Stat-ITCF software. Parameters of biochemical polymorphism were calculated using Biosys 1-7 software. Different genetic distances were also calculated: Rogers, Cavalli-Sforza, Wright.

REFERENCES

Amblard, S. and Pernes, J. 1989. The identification of cultivated pearl millet (*Pennisetum*) amongst plant impressions on pottery from Oued Chebbi (Dhar Oualata, Mauritania). *African Archeological Review*, 7: 117-126.

Appa-Rao, S. 1980. Progress and problems of pearl millet germplasm maintenance. In: *Trends in Genetical Research on* Pennisetum. V.P. Gupta and J.L. Minocha, eds., Ludhiana, India, Punjab Agricultural University, pp. 279-282.

Bezançon, G., Renno, J.F., and Anand Kumar, K. 1997. Le mil. In: *L'Amelioration des Plantes Tropicales*. A. Charrier et al., eds., Montpellier, France, CIRAD-Orstom, collection Repères, pp. 457-482.

Bono, M. 1973. Contribution à la morphosystématique des *Pennisetum* annuels cultivés pour leur grain en Afrique occidentale francophone. *L'Agronomie Tropicale*, 28(3): 229-356.

Brown, A.H.D. 1989. Core collections: a practical approach to genetic resources management. *Genome*, 31: 818-824.

Brunken, J.N. de Wet, J.M.J., and Harlan, J.R. 1977. The morphology and domestication of pearl millet. *Economic Botany*, 31: 163-174.

Burton, G.W. 1983. Breeding pearl millet. In: *Plant Breeding Reviews*, vol. 1, J. Janik, ed., Westport, USA, AVI Publishing, pp. 162-182.

Clement, J.C. 1985. Les mils pénicillaires de l'Afrique de l'Ouest: prospections et collectes. Rome, IBPGR, 231 p.

Crow, J.F. and Kimura, M. 1970. *An Introduction to Population Genetics Theory*. New York, Harper and Row, 591 p.

FAO, 1996. *Annuaire Production: 1995*. Rome, FAO.

Gaut, B.S. and Clegg, M.T. 1993. Nucleotide polymorphism in the *ADH-1* locus of pearl millet (*Pennisetum glaucum*, Poaceae). *Genetics*, 135(4): 1091-1097.

Gepts, P. and Clegg, M.T. 1989. Genetic diversity in pearl millet, *Pennisetum glaucum* (L.) R. Br., at the DNA sequence level. *Journal of Heredity*, 80: 203-208.

Hanna, W.W. 1987. Utilization of wild relatives of pearl millet. In: International Pearl Millet Workshop. J.R., Witcombe ed., Patancheru, India, ICRISAT, pp. 33-42.

Hanna, W.W. 1990. Transfer of germplasm from the secondary to the primary gene pool in *Pennisetum*. *Theoretical and Applied Genetics*, 80: 303-308.

Harlan, J.R. 1971. Agricultural origins: centers and non-centers. *Science*, 14: 468-474.

Harlan, J.R. and de Wet, J.M.J. 1971. Toward a rational classification of cultivated plants. *Taxon*, 20(4): 509-517.

Marchais, L. 1982. La diversité phénotypique des mils pénicillaires cultivés au Sénégal et au Mali. *L'Agronomie Tropicale*, 37: 68-80.

Marchais, L. 1996. Parental proportion maximizing the mean value of a parameter in a panmictic population can be useful in plant breeding. *Agronomie*, 16: 257-264.

Marchais, L. and Pernes, J. 1985. Genetic divergence between wild and cultivated pearl millets (*Pennisetum typhoides*). 1. Male sterility. *Zeitschrift für Pflanzenzüchtung*, 95: 103-112.

Marchais, L., Tostain, S., and Amoukou, I. 1993. Signification taxonomique et évolutive de la structure génétique des mils pénicillaires. In: *Le Mil en Afrique*. S. Hamon, ed., Paris, Orstom, Colloques et séminaires, pp. 119-128.

NBPGR and ICRISAT, 1993. *Evaluation of Pearl Millet Germplasm:* part 1 and part 2.

Nei, M., 1973. Analysis of gene diversity in subdivided populations. *Proceedings of the National Academy of Sciences of the United States of America*, 70: 3321-3323.

Pernes, J., Combes, D., and Leblanc, J.M. 1984. Le mil. In: *Gestion des Ressources Génétiques des Plantes*. 1. Monographies, J. Pernès, ed., Paris, ACCT, pp. 159-210.

Pilate-Andre, S., Lamy, F., and Sarr, A. 1993. Diversité génétique des mils détectable par RFLP au niveau de la région du géne *ADH-1*. In: *Le Mil en Afrique*. S. Hamon, ed., Paris, Orstom, Colloques et séminaires, pp. 67-75.

Porteres, R. 1950. Vieilles agricultures de l'Afrique intertropicale: centres d'origine et de diversification variétale primaire et berceaux de l'agriculture antérieurs au XVIe siècle. *L'Agronomie Tropicale*, 5(9-10): 489-507.

Porteres, R. 1976. African cereals: *Eleusine*, fonio, black fonio, teff, *Bracchiaria*, *Paspalum*, *Pennisetum* and African rice. In: *Origins of African Plant Domestication*. J.R. Harlan et al. eds., The Hague, Mouton, pp. 409-452.

Siaka, S., Ouendeba, B., and Anand Kumar, K. 1996. Caractérisation des cultivars locaux du Niger. In: *Premières Journées Biologiques et Agronomiques du Niger*, 20-25 sept. 1996.

Stemler, A. 1990. A scanning electron microscopic analysis of plant impressions in pottery from the sites of Kadero, El Zakiab, Um Direiwa and El Kadada. *Archéologie du Nil Moyen*, 4: 87-105.

Tostain S. 1992. Enzyme diversity in pearl millet (*Pennisetum glaucum* L.). 3. Wild millet. *Theoretical and Applied Genetics*, 83: 733-742.

Tostain, S. 1994. Evaluation de la diversité génétique des mils pénicillaires diploïdes, *Pennisetum glaucum* (L.) R. Br., au moyen de marqueurs enzymatiques: étude des relations entre formes sauvages et cultivées. Paris, Orstom, Travaux et documents microédites no. 124, 331 p.

Tostain, S. 1996. Genetic diversity of pearl millet (*Pennisetum glaucum*) estimated by random amplified DNA markers and compared with isoenzymatic diversity. In: *Réunion sur les Plantes Tropicales*. Montpellier, France, EUCARPIA-CIRAD, p. 255.

Tostain, S. 1998. Le mil, une longue histoire: hypothéses sur sa domestication et ses migrations. In: *Plantes et Paysages d'Afrique: une Histoire à Explorer*. M. Chastanet, eds., Paris, Karthala-CRA, pp. 461-490.

van der Zon, A.P.M. 1992. *Graminées du Cameroun, 2. Flore*. Wageningen, Netherlands, Agricultural University, Papers no. 92-1, 557 p.

Rubber Tree
(*Hevea brasiliensis*)

Marc Seguin, Albert Flori, Hyacinthe Legnaté
and André Clément-Demange

The taxonomic distribution of latex-producing plants is wide. Among the 12,500 latex species, which belong to 900 genera, 100 species distributed in 76 families produce rubber, or cispolyisoprene (Polhamus, 1962). Within the genus *Hevea*, which offers favourable traits for commercial exploitation, the species *H. brasiliensis* is almost exclusively exploited for natural rubber because of its productivity and the quality of its rubber.

Natural rubber, in the form of latex or coagulate, is collected and transported to a drying factory, which produces smoked sheets or dry granulate blocks of rubber. It is most often exported to places where it is transformed into consumer products. The world production has risen to nearly 7 million tonnes a year and comes from the tropical zone, where hevea is cultivated over around 8 million hectares. Almost 91% of the cultivation takes place in Southeast Asia and it is dominated by four countries: Thailand, Indonesia, India, and Malaysia (http://apps.fao.org). Africa contributes 6.5% of the production. In South America, cultivation is limited by a South American leaf disease caused by the fungus *Microcyclus ulei*.

Natural rubber represents a third of the total production of rubber. Compared to synthetic rubber made from petroleum, natural rubber has specific qualities, such as heat-resistance. The aviation tyre industry, for example, uses only natural rubber, and the tyre industry generally consumes 70% of the production, the remaining 30% being used for a variety of products.

BOTANY AND GENETIC RESOURCES

Botany and Taxonomy

Hevea is a tree with rhythmic growth and orthotropic branching. It may grow 30 m high and reach a circumference of 3 m in its natural environment, the Amazon basin. In plantations, grafted trees do not reach such dimensions. It is a tropical species that requires insolation and humidity and does not tolerate altitudes higher than 600 m. The leaves have three leaflets of ovoid shape, arranged at the tip of a long petiole. They are renewed each year by a natural process of defoliation and refoliation during the dry season. Hevea is a monoecious plant that is preferentially allogamous, with unisexual yellow flowers of a few millimetres grouped in racemes. The fruit of hevea, the size of which ranges from 0.5 to 5 cm depending on the species, has a trilocular structure characteristic of Euphorbiaceae. It is made up of three carpels, each containing a seed.

Within the genus *Hevea* (Willd.), which is well-defined, the species are difficult to differentiate and have been the subject of much confusion, disagreement, and fluctuations between authors. Except for the species *H. brasiliensis*, the observations have been made in often difficult field conditions and are generally quite brief. Schultes (1990) presented the most recent synthesis and distinguished ten species, of which three are subdivided into four varieties. The natural hybridizations seem limited in an undisturbed forest environment, but several interspecific morphological variations are observed. We thus speak of a species complex, according to the definition of Pernes (1984). The nine related wild species are part of the primary gene pool of the cultivated species *H. brasiliensis*. They are thus of great use as genetic resources of cultivated hevea, in so far as they have useful characteristics such as genetic resistance to certain diseases (Schultes, 1977, 1990).

The Genome Structure

With the exception of a triploid clone of *H. guianensis* and the possible existence of a race of *H. pauciflora* with 18 chromosomes (Baldwin, 1947), all the species of the genus have a chromosome number of 2n = 36.

Strong similarities between different chromosome pairs suggest an ancient duplication of the chromosome stock. At meiosis, the formation of multivalents, especially of quadrivalents, is observed from time to time, which indicates some affinity between pairs of homologous chromosomes. Hevea was thus considered an amphidiploid genus, that is, a tetraploid that behaves like a diploid at meiosis, resulting from a crossing of two unidentified, diploid ancestral wild species (Bouharmont, 1960; Ong, 1985). The basic chromosome number would thus equal 9, which seems closest to the usual chromosome number of Euphorbiaceae, which ranges from 6 to 11 (Ong, 1985).

However, the analysis of segregations of isozymic or RFLP markers reveals a majority of non-duplicated loci (Seguin et al., 1996; Lespinasse et al., 2000b). Thus, hevea could be an amphidiploid in which the two homologous genomes have diverged considerably. It cannot be affirmed that the rare duplications observed in the hevea genome (Seguin et al., 1998) indicate a polyploid origin. Such duplications could also come from chromosome remnants after polyploidization. The same situation is observed in cassava or manioc (*Manihot esculenta*), another Euphorbiacea with 2n = 36 chromosomes (Fregene et al., 1997).

A preliminary approach for global characterization of the nuclear genome of *H. brasiliensis* was proposed (Low and Bonner, 1985). The study of the kinetics of reassociation of short fragments of marked DNA (300 nucleotides) indicates the presence of 43% of single copy DNA and 32% of moderately repeated sequences (frequency of repetition around 1000); the remaining DNA are highly repeated or palindromic. From the kinetics of reassociation of single copy DNA, the overall genome size can be estimated at 6×10^8 base pairs. However, from an occasional study by flow cytometry, the size of the haploid genome of hevea has been estimated at 2×10^9 base pairs (2.1 pg/1C), identical in five species studied—*H. brasiliensis*, *H. benthamiana*, *H. guianensis*, *H. pauciflora*, and *H. spruceana*. This value has also been obtained by microdensitometry in *H. brasiliensis* and *H. camargoana* by Bennett and Leitch (1997).

GENETIC RESOURCES

Prospections, Introductions and Collections

In 1747, Fresneau gave the first description of the 'rubber' tree, which was later called hevea. First gathered by the *seringueros* in South America, natural rubber was produced and developed rapidly once its chewing and waterproofing qualities were discovered, the vulcanization process was developed in 1830, and plantations were established (Serier, 1993).

In 1876, on the insistence of Marckham, who had already successfully transferred cinchona from South America to Asia, Wickham, a British planter settled in Brazil, harvested 70,000 seeds close to Boim, at the mouth of the Amazon and Tapajos, upstream of Santarem (state of Pará, whence the initial name of the Para tree for the species *H. brasiliensis*). He shipped them to Kew Garden in the United Kingdom, where only 4% of the seeds germinated three weeks after they arrived. Several plantlets were then sent to Ceylon and Singapore, with varying success. The 22 trees planted in the botanical garden at Singapore in 1877 are thus considered the origin of almost all the hevea plantations in the world. The varieties (grafted clones) that resulted from selection in Asia from these plants introduced at the end of the 19th century were thus called Wickham clones. Even though it is probable that other genotypes were introduced subsequently, only the Wickham example has been recorded and the preponderant role of this material in the genetic composition of present cultivars has been acknowledged.

The Wickham clones proved to be very sensitive to *Microcyclus ulei* in South America. Important breeding efforts were made by the Ford and

Firestone companies in Brazil and Guatemala and subsequently by Brazilian research centres to obtain productive and resistant clones (clones F, FB, FX, FDR, MDF, MDX, IAN, CNSAM). The introgression of totally resistant genes from other species of the genus into *H. brasiliensis* was not successful, this type of resistance being easily overcome by the pathogen. The present strategy aims to associate several components of partial resistance through crosses to attain a sufficient level of long-lasting genetic resistance to the disease (Rivano, 1992; Lespinasse et al., 2000a).

The supposed narrowness of the genetic base of the Wickham material introduced and then selected in Asia and Africa, which is illustrated by the high general sensitivity of this material to *Microcyclus*, has led to the organization of prospecting efforts within the distribution area of *H. brasiliensis*.

In Malaysia, the Rubber Research Institute of Malaysia (RRIM) imported seedlings from wild trees of seven species of *Hevea* from Brazil in 1951-52 and in 1966. It also imported clones bred by Ford (F, FB, FX) and by the Instituto Agronomico do Norte (IAN) in Brazil (Ong, 1987).

In 1974, the Insitut de recherches sur le caoutchouc (IRCA, France), in co-operation with the Empresa Brasileira de Pesquisa Agropecuaria (EMBRAPA, Brazil), made a preliminary prospection in the Amazon region, in the states of Acre (22 clones) and Rondonia (20 clones), collecting graft wood in the forest from trees judged to be exceptional. To this study were added 18 clones from the Firestone prospection in the Peruvian Madre de Dios.

The international prospection organized in 1981 by the International Rubber Research and Development Board (IRRDB) for *H. brasiliensis*, in 16 districts and 60 localities of the Brazilian states of Acre (Lins et al., 1981), Rondonia (Gonçalves, 1981), and Mato Grosso (de Paiva, 1981), resulted in the collection of 9800 wild genotypes at the Asian centre of Malaysia (Ong et al., 1995) and 1500 genotypes at the African centre of Côte d'Ivoire (Chapuset et al., 1995). The geographic location of districts prospected is given in Table 1. This expedition collected essentially seeds, but also 130 genotypes from graft wood of trees judged to be exceptional (called ortet clones). Genetic material was exchanged between the two centres and distributed to other member countries of the IRRDB. Most of the genotypes were grafted and are conserved in the form of clones in living collection.

After 1945, Schultes put together, among others on the sites of Calima and Palmira, significant collections of material found in Colombia. In 1985, with the permission of the Colombian government, IRCA transferred 341 SCH genotypes from these two sites to Guadeloupe and then to Côte d'Ivoire (Nicolas, 1985).

The IRCA also received 23 CNSAM clones from the EMBRAPA at Manaus, comprising *H. pauciflora* genotypes and the only *H. brasiliensis* genotypes collected in the Brazilian state of the Amazonas, and introduced them in Côte d'Ivoire (Gonçalves et al., 1983).

Table 1. Code and origin of the major populations of *H. brasiliensis* in collection

Code	Collection	State and country of origin	Place of origin (districts)	River basin (tributaries of Amazon)
AC/T	IRRDB 1981	Acre, Brazil	Tarauaca	Jurua
AC/F	IRRDB 1981	Acre, Brazil	Feijo	Jurua
AC/S	IRRDB 1981	Acre, Brazil	Sena Madureira	Purus
AC/B	IRRDB 1981	Acre, Brazil	Basileia	Purus
AC/X	IRRDB 1981	Acre, Brazil	Xapuri	Purus
RO/A	IRRDB 1981	Rondonia, Brazil	Ariquemes	Madeira
RO/C	IRRDB 1981	Rondonia, Brazil	Calama	Madeira
RO/CM	IRRDB 1981	Rondonia, Brazil	Costa Marques	Madeira
RO/J	IRRDB 1981	Rondonia, Brazil	Jaru	Madeira
RO/JP	IRRDB 1981	Rondonia, Brazil	Jiparana	Madeira
RO/OP	IRRDB 1981	Rondonia, Brazil	Ouro Preto	Madeira
RO/PB	IRRDB 1981	Rondonia, Brazil	Pimenta Bueno	Madeira
MT/A	IRRDB 1981	Mato Grosso, Brazil	Aracatuba	Tapajos
MT/C	IRRDB 1981	Mato Grosso, Brazil	Juruena	Tapajos
MT/IT	IRRDB 1981	Mato Grosso, Brazil	Itauba	Tapajos
MT/VB	IRRDB 1981	Mato Grosso, Brazil	Vila Bella	Madeira
MDF	Firestone	Madre de Dios, Peru	?	Madeira
SCH	Schultes coll. Calima and Palmira	Colombia	?	?
W	Wickham	Para, Brazil	Santarem	Tapajos

In conclusion, *H. brasiliensis* can be considered to be well represented in the living collections, in Malaysia and Côte d'Ivoire, by several thousands of genotypes. However, the area of distribution of this species has been sampled in a highly heterogeneous fashion. The states of Amazonas and Para in Brazil, which are the major part of that area, were hardly prospected. It is also necessary to search for other species of *Hevea* in order to create a centre of conservation and study of the entire genus. The observations available on the Schultes and CNSAM collections show that there are genotypes of species other than *H. brasiliensis* in these collections.

CONSERVATION, MULTIPLICATION AND EXCHANGE OF ACCESSIONS

In the natural state, hevea reproduces by seed. Once techniques of grafting were introduced by van Helten in 1918, clones could be bred and genotypes conserved in the form of living collections, which were durable and relatively inexpensive to create and maintain in gardens of grafted trees. Propagation was done in nurseries by grafting on stocks grown from seed.

The viability is one week for graft wood and three weeks for seeds (the seeds are recalcitrant: the duration of germination capacity is short). At the experimental stage, the pollen is conserved for no more than a month and the quantities that can be harvested are very small. Genetic transfer in this form is thus not considered feasible.

ORGANIZATION OF GENETIC DIVERSITY

This chapter summarizes the present understanding of the organization of diversity of hevea. It is based on the conclusions of earlier studies and presents complementary original results from isozymic analyses on a very large number of genotypes. These supplementary data allowed us to develop, on a sufficient number of genotypes, an original outline of comparison between morphoagronomic variability and genetic diversity of neutral markers, by means of a multiple factorial analysis.

Research on genetic diversity of hevea aims to better understand the extent and organization of the diversity and the relationship between agronomic variability and genetic diversity. These studies aim to respond to the problems of breeders, who wish to define an optimal strategy of genetic resource conservation, population sampling for agronomic evaluation, crossing scheme design, and progenitor choice.

Agromorphological Variability

AGRONOMIC EVALUATION

Agronomic evaluation is done on prospected material in order to constitute two small collections before genetic recombination. One, called the working collection, is intended to quickly concentrate the alleles of agronomic interest in order to increase the possibilities of getting a usable clone, especially after crossing with the Wickham material. This objective could be met through simple agronomic evaluation, but genetic mapping allows us to assess the reduction of variability resulting from intensive and rigorous selection. The other collection must be constructed according to the concept of core collection, that is, in trying to maintain the widest possible diversity in a small sample. This task requires the use of all methods of evaluation: agronomic, morphological, isozymic, and molecular, for the nuclear and cytoplasmic genomes. Methodological research is in progress, especially on hevea, to optimize the selection of genotypes while integrating these different types of information (Hamon et al., 1998).

A preliminary assay on 2500 genotypes from the IRRDB prospection in the states of Acre (AC), Rondonia (RO), and Mato Grosso (MT) in Brazil, at the rate of one tree per genotype, allowed a preliminary overall evaluation of the production and growth of trees. The mean production of this wild material is very low, representing only 12% of that of the control Wickham clone, GT1. Similarly, its rate of growth is on average lower than that of GT1. The architectural aspect of the trees is characterized by a predominance of clones growing tall, with few or no branches; this character corresponds to a natural adaptation to the forest environment that allows young trees to quickly reach the light. However, there is a small proportion of clones that have production close to that of GT1, good vigour, or abundant ramification (Chapuset et al., 1995; Clement-Demange et al., 1997).

From the agronomic point of view, an east-west gradient appears in the area of distribution of the genotypes studied, as illustrated in the first plane of the principal components analysis (PCA) done on the mean values of production, growth, and architecture, by locality (Vi Cao, personal communication, Fig. 1). The genotypes of Mato Grosso are distinguished, especially from those of Acre, by a better aptitude for branching and greater production, but they seem more sensitive to *Colletotrichum*, a fungus responsible for a leaf disease.

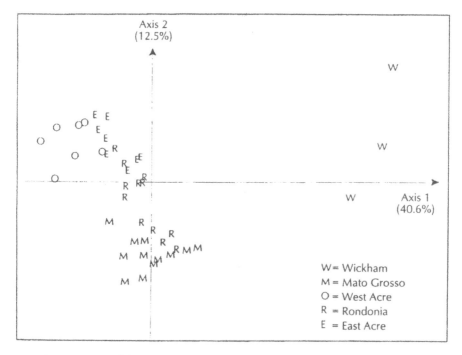

Fig. 1. Agronomic variability of Amazonian material and Wickham clones. Plane 1-2 of the PCA on mean values of three Wickham clones and 41 localities of the IRRDB collection, calculated on 2300 Amazonian clones.

LEAF MORPHOLOGY

A preliminary study on the genetic structuring of the IRRDB populations from the states of Acre, Rondonia, and Mato Grosso was done by biometric analysis of morphology of growth units and leaves. The observations are phenotypic and 34% of the variance is due to effects of the environment and to errors. Most of the genetic variability (42%) lies between the genotypes of a single prospecting site, while differences between states or between districts seem small. Nevertheless, Acre stands out distinctly from the two other states (Nicolas et al., 1988).

Biochemical and Molecular Diversity

The genetic diversity of a sample from *H. brasiliensis* collections was characterized using biochemical and molecular markers. The numbers of accessions used in these studies are given in Table 2. The other species of the genus *Hevea* were nearly absent from the collections, and it was not possible to study their diversity. One or two genotypes of the four species *H. benthamiana*, *H. guianensis*, *H. pauciflora*, and *H. spruceana* could be characterized by isozymes or RFLPs, which allowed identification of the alleles missing in *H. brasiliensis*.

Table 2. Number of clones, per population and per type of marker, used in the studies of genetic diversity

		RFLP	
Populations	Isozymes	Nuclear	mtDNA
Wickham	203	73	28
Hybrids (Wickham × Amazonian)	81	0	21
Peru (MDF)	14	0	12
Colombia (SCH)	48	0	83
Brazil EMBRAPA-IRCA 1974	55	0	30
Brazil IRRDB 1981	486	92	220
Total	887	165	395

ISOZYMES

The isozymic analysis revealed 14 polymorphic loci, using 12 isozyme systems, in *H. brasiliensis* (Chevallier, 1988; Chevallier et al., 1988; Seguin et al., 1995a).

The primary level of analysis was done on intergenotype variability, that is, between the clones. The more precise analysis of diversity, in terms of population samples, was done on 486 clones of the IRRDB collection, which were taken from 16 districts of three states of Brazil. The genotypes were characterized for 8 polymorphic loci, without missing data, and 25 alleles were revealed.

The first axis of correspondence analysis (CA) done on allelic data represented 20% of the total variability (Fig. 2). It revealed a separation into two groups of clones. The first group comprises almost exclusively all the genotypes of four districts—three districts in Mato Grosso (MT/A, MT/I and MT/C) and one in Rondonia (RO/PB). The second group is made up of genotypes belonging to the 12 remaining districts—five in Rondonia, five in Acre, and the MT/VB district in Mato Grosso.

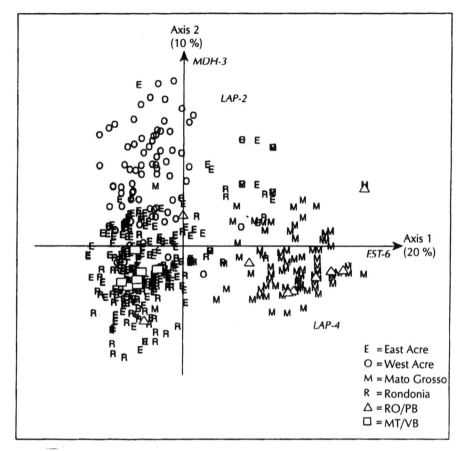

Fig. 2. Genetic diversity of the Amazonian material revealed by isozymes. Plane 1-2 of the CA on data of presence-absence of 25 alleles of 8 loci, in 486 clones of the IRRDB collection. MDH-3, LAP-2, LAP-4, EST-6: alleles contributing the most to the first two axes.

The second axis, which represents 10% of the variability, shows an intrastate structuring with the differentiation of a third group made up of two districts farthest west of Acre, AC/T and AC/F (Fig. 2). As for the morphoagronomic data, the genetic variability is low between the groups or origins and high within the groups. Despite this relatively low part of intergroup diversity—30% of the total inertia on the first CA plane—there appears a very clear genetic differentiation into three groups.

This structuring into three groups has also been indicated in a study on a smaller sample of genotypes, but for 11 isozymic loci (Besse et al., 1994; Seguin et al., 1996). This study does not include the RO/PB district, but includes the Wickham population. The Wickham population proved to be genetically close to the Mato Grosso group. It has a high level of heterozygosity and polymorphism, which is however much lower than that of the Amazonian populations, which have a greater allelic richness. No allele is specific to the Wickham collection, nor to any of the three states in the Amazonian prospection. The genetic diversity of the Wickham clones has moreover made it possible to perfect an isozymic method of varietal identification (Leconte et al., 1994).

The second level of analysis pertains to the diversity between populations of genotypes, the populations being defined as a function of their location. The level of subdivision depends on the refinement of analysis desired, but above all on the number of genotypes studied per population.

Interpopulation diversity was characterized using the Nei distance (1978) on allelic frequencies of 8 isozymic loci. Figure 3 shows the UPGMA tree constructed for 16 Brazil districts, one population from Peru (MDF), two collections from Colombia (Schultes populations), and the Wickham population, or 590 genotypes in all. For the Brazilian populations, this analysis

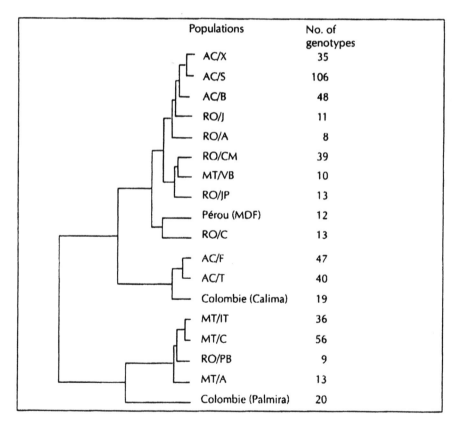

Populations	No. of genotypes
AC/X	35
AC/S	106
AC/B	48
RO/J	11
RO/A	8
RO/CM	39
MT/VB	10
RO/JP	13
Pérou (MDF)	12
RO/C	13
AC/F	47
AC/T	40
Colombie (Calima)	19
MT/IT	36
MT/C	56
RO/PB	9
MT/A	13
Colombie (Palmira)	20

Fig. 3. Genetic distances between populations of Amazonian clones. UPGMA tree drawn from the Nei distance (1978) calculated on allelic frequencies per population, for 8 enzymatic loci.

by population leads to the same conclusions as the CA on the genotypes taken individually. The districts are grouped into three subgroups: West Acre; East Acre, Rondonia and the MT/VB district; and Mato Grosso and RO/PB.

The Peruvian population is grouped first with a Rondonia district (RO/C). The two Colombian populations (SCH) seem genetically very distant from each other. The first, Calima, approaches the two districts of West Acre; the second, Palmira, is closer to the Mato Grosso group than to other populations, but has specific alleles characteristic of species other than *H. brasiliensis*, which make this Palmira population clearly distinct from others.

The isozymic study was done by identifying the alleles at each locus using the genetic analysis of zymograms. This allowed analysis of the genetic diversity of hevea using population genetics. The calculation of F statistics of Wright thus showed that 75% of the diversity of the Brazilian material is located at the intrapopulation level, and thus that 25% of the variability corresponds to the differentiation between districts. When the three genetic groups identified on the CA are considered as population units, the proportion of intragroup diversity reaches 80%.

Moreover, the populations made up by the prospected districts all show a departure from panmixia with a deficit of heterozygotes, the values of the fixation index F_{IS} ranging from 0.10 to 0.60. Taking into account the geographic extent of districts, it is probable that this deficit is due not to an intradistrict differentiation into subpopulations, but rather to preferential crosses between closely linked trees with a low rate of dissemination of pollen and seeds. Since the passport data is available for the Amazonian collection of IRRDB at the scale of the prospecting locality in each district, it is possible to tackle this question, subject to analysis of a larger number of genotypes.

NUCLEAR RFLPs

The RFLP study was done on a smaller number of genotypes: 73 Wickham and 92 Amazonian from 15 districts, the RO/PB district not being represented. On the other hand, it was done on a larger number of loci than for isozymes. As the genetic determinism of RFLP patterns is not established, the exact number of polymorphic loci was not known. A single probe may in fact reveal two loci or more. However, as 25 probes have been used, the number of polymorphic loci must not be far from 25. The gene mapping of hevea has shown that generally a single locus has been revealed with this type of genomic probe (Seguin et al., 1996, 1998; Lespinasse et al., 2000b).

The results obtained using RFLPs emphasize the genetic enrichment contributed by the Amazonian populations and the existence of a high genetic structuring depending on the origin of prospected genotypes. Moreover, in relation to isozymes, in allowing access to a larger part of the genome, nuclear RFLPs contribute a specific information: they reveal a marked differentiation of the five districts of Acre from the other ones, which form a distinct genetic group. On the other hand, the RFLPs do not reveal the individualization of the western Acre group, which is revealed by isozymes.

Overall, the nuclear RFLP study confirms and reinforces the observations made using isozymes. In particular, the same partition of the intra- and intergroup diversity is found, with also 30% of the variability on the primary plane of the CA corresponding to the differentiation of the group of three districts of Mato Grosso, this group appearing to be the closest to the Wickham population.

CYTOPLASMIC RFLPS

The cytoplasmic genomes most often have a single-parent heredity (Reboud and Zeyl, 1994; Mogensen, 1996), which prohibits genetic recombination, and a rate of evolution particular to each genome (Wolfe et al., 1987; Palmer and Herbon, 1988). The study of mitochondrial and chloroplastic diversity is thus complementary to that of the nuclear genome.

The study of mitochondrial genome (mtDNA) has been done on 395 genotypes of hevea. A strong polymorphism has been revealed by RFLP (Luo and Boutry, 1995; Luo et al., 1995). As with the nuclear scale, a divergence of populations is found in Acre, as well as a separation between the genotypes of western Acre (AC/T and AC/F districts) in relation to those of the east (AC/B, AC/S, and AC/X) as with the isozymes. The Peru genotypes (MDF) are grouped preferentially with the genotypes of the RO/C district of Rondonia, as with the isozymes.

On the other hand, the populations of Mato Grosso and Rondonia are heterogeneous and are not grouped according to their origin. Similarly, the Colombian populations (SCH) appear highly polymorphic and are not individualized. Moreover, unlike the isozymes and nuclear RFLPs, the mtDNA appeared nearly monomorphic within the Wickham clones, all of which, except the GT1 clone, have the same specific mitochondrial type. The cultivated clone GT1, male-sterile, has a unique mitochondrial type that is very different from the dominant Wickham type.

The diversity of the chloroplastic genome (cpDNA) has also been studied using RFLPs, with the cpDNA probes of broad bean, on 217 out of the 395 preceding genotypes. As for many other species, the cpDNA shows, in hevea, a very low RFLP polymorphism. Only two chloroplastic types have been found, against 126 mitochondrial types detected on the same sample of clones. The minority type, present in 37 genotypes, is characteristic of three districts of eastern Acre, to which are added three out of six genotypes of the RO/C district present in the study (Luo et al., 1995).

THE ORGANIZATION OF BIOCHEMICAL AND MOLECULAR DIVERSITY

It thus seems clear that molecular markers have made a considerable contribution to the characterization of genetic resources of hevea. They have revealed the extent of diversity of *H. brasiliensis*. The various types of marker used have given consistent results overall on the organization of this diversity, which is structured according to the geographic origin of genotypes. From this set of studies, it is possible to draw the following conclusions.

On the basis of genetic similarities between the genotypes or populations, six genetic groups have been identified within the material of *H. brasiliensis* studied (Seguin et al., 1996).

— group 1 comprises two districts in western Acre and the Calima population of Colombia;
— group 2 corresponds to three districts in eastern Acre;
— group 3 comprises six districts of Rondonia, the MT/VB district of Mato Grosso, and the MDF population of Peru;
— group 4 comprises three districts of Mato Grosso and the RO/PB district of Rondonia;
— group 5 is made up of the Palmira collection of Colombia;
— group 6 corresponds to the Wickham population.

The geographic location of populations and genetic groups is shown in Fig. 4.

These genetic groups include geographically closed populations. Moreover, the relationship between the geographic distance and the genetic distance is found not only within a single group, but also between groups. Thus, the molecular differentiation of group 6 (Wickham) essentially relies on the existence of a particular mitochondrial type, but at the nuclear level this group seems closer to group 4, geographically the closest. The place of Schultes populations of Colombia is more difficult to discuss without precise passport data.

This organization of the diversity of *H. brasiliensis* seems very clearly linked to the hydrographic network of the Amazonian basin (Table 1, Fig. 4), which determines the gene flow and thus the genetic similarity between populations or groups of populations. For the Brazilian populations, the passport data for which are available and precise, each group can be associated with a river basin (Fig. 4): the Jurua river with group 1; the Purus with group 2, the Madeira and its branches—the Guapore river in Brazil and the Madre de Dios river in Peru—with group 3; and the Tapajos with group 4.

Similarly, certain similarities between the groups can be explained by gene flows following the course of rivers or streams. For example, the proximity between the RO/C population and group 2 is linked to the Abuna river, a tributary of the Madeira river, and that of groups 4 and 6 to the Tapajos river.

This situation would justify designating the genetic groups of hevea by the name of the corresponding river. However, to simplify the presentation, we have designated the groups by their numbers 1 to 6. The RO/PB district is a small exception: located on a tributary of the Madeira river, it is associated

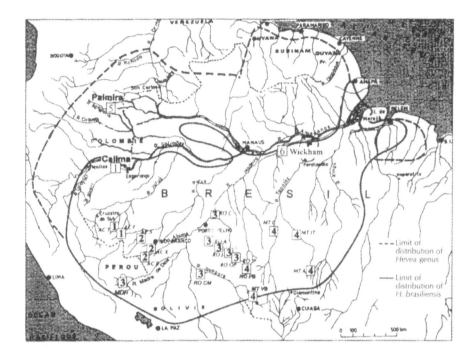

Fig. 4. Correspondence between geographic origin and genetic similarity in *Hevea brasiliensis*. The numbers 1 to 6 correspond to the genetic groups identified by molecular markers (see text). The exact site of collection is not known for the clones prospected by E. Schultes in Colombia (Calima and Palmira collections).

with group 4, of the Tapajos river, and not with group 3 (Figs. 2 and 3). The area prospected in this district is in fact close to the area between the two river basins (Fig. 4) and it can be stated that there is a significant dissemination via the basin of the Tapajos river. The diversity of the cultivated Wickham clones is less than that of the Amazonian genetic resources, especially at the cytoplasmic level. Still, the nuclear diversity of the Wickham group seems greater than might have been foreseen from the history of hevea domestication in Asia.

These conclusions are drawn from all the diversity studies using molecular genetic markers, which is not to say that all the types of markers have given exactly the same results. Each study has actually contributed specific information. There are several factors that explain the differences observed with the different types of markers.

It is not surprising that genetic markers of different kinds contribute different, although not contradictory information. Cytoplasmic markers, because of single-parent heredity (Reboud and Zeyl, 1994; Mogensen, 1996), allow better detection of migratory flows, the traces of which disappear rapidly at the nuclear level following genetic recombination between indigenous and immigrant populations (Palmer, 1987; Crozier, 1990). For hevea, for example, the similarity between the RO/C district and the genetic group 2 is revealed specifically by chloroplastic markers. Thus, there are gene flows, probably linked to the flotation of grains along the course of the Apuna river, from the districts AC/X and AC/S of group 2 to the north of Rondonia, but these migrations are less intensive and the genetic divergence between the two groups is maintained.

The speed of evolution varies a great deal according to the genome or type of sequence considered (Wolfe et al., 1987; Palmer and Herbon, 1988). With a very slow evolution it is less probable that a mutation would appear associated with a recent differentiation in subpopulations. Inversely, a very rapid molecular evolution leads to the phenomenon of homoplasy, or evolutionary convergence: a particular character, a particular form of marker, has a high probability of appearing independently in several isolated populations and is thus not characteristic of a population. That depends also on the time of divergence involved in the differentiation of units studied, populations or species. For hevea, this phenomenon could explain the complex organization of mitochondrial types between groups 3 and 4, which are well differentiated at the nuclear level. However, the situation could also indicate the existence of migratory flows, slight but diverse, between these two groups and it is not possible to draw any conclusion. A too rapid evolution would also explain why no structuring was observed within the IRRDB collection during a study on the variations of length of gene coding for ribosomal RNA (Besse, 1993; Besse et al., 1993).

Finally, and undoubtedly a determining factor in a number of molecular diversity studies, the quality of results depends on the level of diversity sampling. The differences in results obtained with isozymes or RFLPs in hevea does not arise from a difference in nature. It can be said that the two types of markers are neutral with respect to selection, that they correspond to the same type of nuclear sequence, and that they have a similar rate of evolution on average. On the other hand, the differences could be explained by the mode of sampling of populations—the number and representativity of genotypes studied—or of the genome, that is, the number of loci studied and their distribution over the entire genome. The fact that groups 1 and 2 are not distinguished in the RFLP study is probably due to a quite low number of original genotypes of Acre, as suggested by the PCA done on the isozymes on the same sample (Fig. 5). In parallel, the absence of observed divergence between groups 2 and 3 with the isozymes would come from what has not been detected from the locus or marker allele of this differentiation, the few isozyme markers obtained in hevea not sufficiently covering the entire genome.

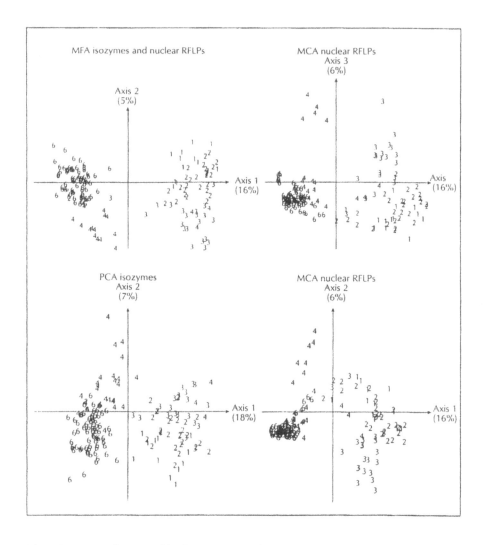

Fig. 5. Comparison by MFA of the diversity revealed by isozymes and by nuclear RFLPs. Analysis of 72 Wickham clones and 89 Amazonian clones of the IRRDB collection. Graphic representation of the primary axes of MFA and of partial analyses (PCA isozymes and MCA RFLPs). The numbers 1 to 6 correspond to the genetic groups identified by molecular markers (see text).

Molecular Diversity and Morphoagronomic Variability: Comparison by Multiple Factorial Analysis

Multiple factorial analysis (MFA) has been developed to analyse simultaneously and compare qualitative or quantitative variables organized in groups of the same kind. This method is well adapted to the simultaneous analysis of molecular, qualitative data of different types (isozymes, RFLPs) or agronomic, quantitative data. The principle of MFA is presented in this work (Perrier et al., 2002) and in greater detail in Escofier and Pages (1988).

COMPARISON BETWEEN ISOZYMES AND NUCLEAR RFLPS

We have used MFA, in the first place, to obtain a measurement of the relationship between the various molecular markers used in hevea. A preliminary MFA was done on 161 genotypes (89 Brazilian clones from the IRRDB collection and 72 Wickham clones) characterized by two groups of molecular variables: the isozymes (44 alleles for 11 loci) and nuclear RFLPs (113 bands for 25 genomic probes).

The results of the MFA are given in Table 3 and Fig. 5. The correlation values correspond to correlations between the coordinates of points (hevea genotypes) on the axes of factorial analyses. The number of axes to be taken into account depends partly on their eigenvalue. For MFA, an eigenvalue of the primary axis higher than 1 indicates that more than one group of variables contribute to it or, in other terms, that there exists an axis of inertia common to groups of variables (Escofier and Pages, 1988). On the other hand, for the subsequent axes of MFA, the eigenvalue is not a sufficient criterion and the choice of retaining an axis or not retaining it depends, as for any PCA, on the biological interpretation that one can make (Escofier and Pages, 1988). The quality of the representation and the contribution of a partial axis to an axis of MFA are also parameters to be taken into account, as for the interpretation of a classic PCA.

The first two axes of MFA indicate the close similarity between the genetic structuring revealed by the two types of markers, while it is essentially the RFLPs that contribute to the third axis of MFA (Table 3). The primary axis of MFA corresponds to the opposition between groups 4 and 6 (Tapajos river), on the one hand, and groups 1, 2 and 3, on the other. This opposition is shown in the second row for the isozymes and in the third for RFLPs (Table 3b; Fig. 5). It is less marked with RFLPs, as indicated by the values of correlation, quality of representation, and contribution (Table 3b). Even though there is an eigenvalue lower than 1, this axis of MFA, which corresponds to an axis of inertia common to two types of markers, should be retained. The third axis of MFA corresponds to the differentiation between groups 3 and 4. This opposition is clearly indicated by RFLPs (second axis of the multiple correspondences analysis or MCA), but does not appear clearly on the third axis of the PCA of isozymes, which present a lower correlation and contribution (Table 3b).

These elements of structuring are found on the first plane of partial analyses—separated PCA isozyme and MCA RFLP—but in a less marked fashion. Moreover, the MFA shows that the correlation between the two types

Table 3. MFA on isozymes and nuclear RFLPs for 161 clones of hevea. a: eigenvalues and part of inertia of primary axes of factorial analyses; b: correlation, quality of representation and contribution of axes of partial analyses (PCA and MCA)

a

Factorial analysis	Factors (axis)	Proper value	Inertia (%)
MFA	1	1.9	16
	2	0.6	5
	3	0.5	4
PCA isozymes	1	8.1	18
	2	3.2	7
	3	2.5	6
MCA RFLP	1	20	16
	2	7.4	6
	3	7.2	6

b

Factorial analysis	Factors (axis)	Axis 1 of MFA		
		correlation	quality	contribution
PCA isozymes	1	0.97	0.94	0.50
	2	0.04	0	0
	3	0.03	0	0
MCA RFLP	1	0.97	0.93	0.50
	2	0.01	0	0
	3	0	0	0

Axis 2 of MFA			Axis 3 of MFA		
correlation	quality	contribution	correlation	quality	contribution
0.05	0	0	0.08	0	0.01
0.87	0.75	0.53	0.08	0	0
0.14	0.02	0	0.58	0.34	0.22
0.04	0	0	0.09	0	0.02
0.17	0.03	0.02	0.80	0.64	0.49
0.73	0.53	0.34	0.07	0	0

of marker is very high for the most significant and most structured part of their variability.

COMPARISON BETWEEN ISOZYMES AND MORPHOAGRONOMIC CHARACTERS

The MFA was subsequently extended to the comparison between molecular diversity and phenotypic variability. The analysis was done on 171 Amazonian genotypes of the IRRDB collection for 51 isozymic alleles (13 loci) and 26 morphoagronomic variables (production, growth, architecture), without missing data.

The results are given in Table 4 and Fig. 6. The first axis of the MFA has an eigenvalue of 1.2—higher than 1, but still low—which means that the correlation between the two groups of variables is low. However, it is not nil because of the particular characters of the genotypes of group 4, very clearly differentiated by the isozymes (axis 1 of Fig. 6) and with production values tending to be higher than average. The second axis of the MFA must not be taken into account in this analysis because it does not allow us to draw a clear interpretation of the relation between the two groups of variables.

The results of the MFA thus suggest that there is a relationship between the diversity of neutral markers and the phenotypic variability in hevea. This relationship may seem to be in contradiction with the low correlation value between the primary axis of two partial analyses (Fig. 6). In fact, considering the sample size, this correlation of 0.13 is significant at 10%. Thus, the MFA

Table 4. MFA on isozymic and morphoagronomic data for 171 clones of hevea. a: eigenvalue and part of inertia of primary axes of factorial analysis; b: correlation, quality of representation, and contribution of axes of partial analysis (PCA) to the primary axis of MFA

a

Factorial analysis	Factors (axis)	Eigenvalue	Inertia (%)
MFA	1	1.2	10.5
	2	0.9	8.3
PCA isozymes	1	5.6	11
	2	3.9	7.6
Morphoagronomic PCA	1	12.7	49
	2	2.2	8.5

b

		Factors	Axis 1 of MFA	
Factorial analysis	Axis	Correlation	Quality	Contribution
PCA isozymes	1	0.67	0.45	0.56
	2	0	0.01	0.01
Morphoagronomic PCA	1	0.81	0.67	0.57
	2	0.03	0.00	0

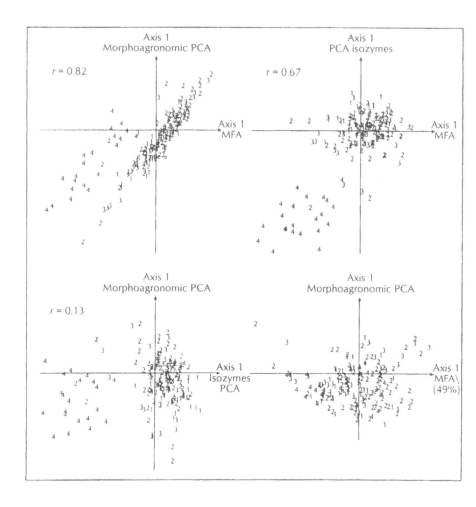

Fig. 6. Comparison by MFA of the diversity of isozymes and of the morphoagronomic variability in 183 Amazonian clones of the IRRDB collection. Graphic representation of the primary axes of MFA and of partial analyses (PCA). The numbers 1 to 6 correspond to the genetic groups identified by molecular markers (see text).

reveals a relationship that would be difficult to identify from separate analyses. The morphoagronomic data are imprecise because they are from single-tree measurements, without repetitions. That is why a differentiation does not appear with the data except when one works on larger numbers and average values, on the population scale (Fig. 1). The relationship between molecular diversity and phenotypic variability would appear undoubtedly more clearly if one uses a better evaluation of morphoagronomic characters. A close examination of the morphoagronomic PCA (Fig. 6) shows that the clones of group 4 are, among the genotypes of other groups, situated on the same side of the primary axis; such a distribution cannot be retained as significant from this single partial analysis.

CONCLUSION AND PROSPECTS

Refined analysis of genetic diversity of hevea using molecular as well as phenotypic traits has greatly improved our understanding of the genetic organization of this plant. The input of molecular markers was critical. The results obtained independently by different techniques converged towards a clear picture of the genetic structure, which can be considered nearly definitive, at least for the Brazilian populations most extensively studied.

A high correlation has been found between the various molecular tools. For the Amazonian populations, the relation established between nuclear and mitochondrial diversity is particularly interesting. It proves that the differentiation between populations, though not very high, is quite real and that the gene flows between populations exist but remain of low intensity. The Amazonian basin does not present a geographic barrier to migration, but the vastness of the area considered and the limited dispersal of seeds helped maintain the inherited genetic structuring from the fragmentation of the forest at the Pleistocene, according to the model of forest refuge (Haffer, 1982; Prance, 1987).

The genetic similarity relationships between populations of hevea revealed by cytoplasmic and nuclear markers show that gene flows are determined mainly by the hydrographic network. Hevea seeds are dispersed along the water courses, either through human action (Seibert, 1948) or by natural flotation, as has been found with other Amazonian tree species (Goulding, 1993). A similar structuring of diversity, depending on the hydrographic network, has been found in another perennial Amazonian species, the American oil palm *Elaeis oleifera* (Barcelos, 1998; Barcelos et al., 1999).

The different types of molecular markers have given results that are overall in accordance, which confirms the MFA, but each contributes specific information. The maternal heredity of cytoplasmic markers allows a better appreciation of gene flows between populations—between group 2 and the RO/C district, for example—or the foundation effects, as for the Wickham populations, which are nearly monomorphic at the mitochondrial level.

Certain differentiations between groups were clearly revealed only by isozymes or RFLPs. In nuclear terms, that can be attributed to differences in the sampling of populations (number of genotypes analysed) or of the genome (number of markers used). It would thus be useful to characterize the diversity of genetic resources of hevea using markers that are mapped and distributed throughout the genome. The development of microsatellite markers

(Lespinasse, 1993; Seguin et al., 1997) and the publication of a dense molecular genetic map of hevea allows us to consider such a study on a sufficient number of genotypes (Lespinasse et al., 2000b). Thanks to PCR-based markers such as microsatellites, it will be possible to use high output genotyping techniques for diversity studies in tropical crops such as rubber tree. This will allow a more powerful analysis of population genetic organization and history, through the assessment of linkage disequilibrium between markers or genes.

It is interesting to note the existence of a relationship between molecular diversity and agronomic variability in hevea. The correlation is however much lower than between the different molecular markers themselves and the MFA has proved useful to make that apparent on a subsample of genotypes. The relationship between the two levels of variability—genetic and phenotypic— needs to be studied from more precise phenotypic data, collected from a larger sample of genotypes. It would also be useful to apply MFA using principal coordinates analysis (PCoA), which would provide a better basis for choices of indexes of distance, since the χ^2 distances (in CA or MCA) and Euclidean distances (in PCA) are not necessarily the most suitable from a biological point of view.

Our results indicate that differentiation in hevea populations is great enough to affect all the genomes and concerns a large number of genes, that is neutral markers as well as genes that determine phenotypic characters. In consequence, in the present state of our understanding, structuring of *Hevea brasiliensis* diversity in six genetic groups will be very useful in defining a sampling strategy for genetic diversity studies of this species.

APPENDIX

Plant Material

The genotypes retained in this study are hevea clones from the international collection of CNRA (Centre National de la Recherche Argonomique) in Côte d'Ivoire. The geographic origin of populations of clones is specified in Table 1. All the clones were evaluated in the field for the characters of latex production, growth in thickness, and architecture, but in different assays (Chapuset et al., 1995). The most important assay comprised 2500 clones of the IRRDB expedition of 1981, without repetition (1 tree per clone). Molecular markers were applied to samples of the collection, of variable size and coverage. The numbers used in these different studies are summarized in Table 2.

Methods of Detection of Diversity

The protocols used for the study of genetic and agronomic diversity are described by Chapuset et al. (1995) for morphoagronomic evaluation on 26 variables, by Lebrun and Chevalier (1990) for isozymes, by Besse (1993) and Besse et al. (1994) for nuclear RFLP with 25 genomic probes, and by Luo et al. (1995) for cytoplasmic RFLPs, mtDNA, and cpDNA.

The technique of isozymes uses 12 enzymatic systems, which enable the detection of up to 14 polymorphic loci in the species *H. brasiliensis* (Chevallier, 1988; Chevallier et al., 1988; Seguin et al., 1995a, b). Several hundreds of genotypes have been characterized by the isozymes (Table 2) using these 12 systems, but with a variable number of missing data according to the genotypes or the series of manipulations. This is why the data were analysed for only 8 of 13 isozyme loci depending on the experiment. To have a sufficient number of genotypes, it was often necessary to reduce the number of loci to limit or eliminate the missing data.

For nuclear RFLPs, 25 homologous probes corresponding to single sequences or nearly unrepeated sequences in the genome of hevea (genome bank *Pst*1) were used. These probes were coupled with two restriction enzymes, *Eco*R1 and *Sst*1 (Besse et al., 1994; Seguin et al., 1995a). For cytoplasmic RFLP, the probes come from banks of broad bean (Luo et al., 1995).

Data Analysis

Data were treated mainly by factorial analysis, the use of which is discussed in this work (Perrier et al., 2002). Moreover, interpretation of results is easier with factorial analysis than with the tree methods when one works on a large number of individuals.

The agronomic, quantitative data were treated by PCA. The data of presence/absence of RFLP bands were treated by CA on a binary table, which in turn is based on an MCA. The data of isozymic alleles were coded, 0 for absence, 1 for presence at the heterozygous state and 2 for presence in the homozygous state, for each genotype. These data were treated by CA or by PCA in the MFA (Perrier et al., 2002). For the CA, only bands or alleles of frequency between 2 and 98% in the total sample were retained as variables. In the MFA, all the alleles were retained, no matter what their frequency.

Diversity has also been analysed at the population level by analysis in hierarchical clustering using, for the isozymes, the Nei distance (1978) on the allelic frequencies and, for the agronomic, quantitative data, the city block distance on the means per population. The trees were constructed by the UPGMA method. The organization of genetic material in subpopulations was studied by calculations of F statistics, by which departure from panmixia was evaluated and the ratio of intra- and interpopulation diversities was quantified.

REFERENCES

Baldwin, J.J.T. 1947. Hevea: a first interpretation. *Journal of Heredity*, 38: 54-64.

Barcelos, E. 1998. Etude de la diversité génétique du genre *Elaeis* (*E. oleifera* Cortés et *E. guineensis* Jacq.) par marqueurs moléculaires (RFLP et AFLP). Doct. thesis, ENSAM, Montpellier, France, 138 p.

Barcelos, E., Second, G., Kahn, F., Amblard, P., Lebrun, P., and Seguin, M. (1999). Molecular markers applied to the analysis of genetic diversity and the biogeography of *Elaeis*. In: *Evolution, Variation and Classification of Palms*. A. Menderson and F. Borchsenius, eds. New York Botanical Garden, New York, pp. 191-202.

Bennett, M.D. and Leitch, I.J. 1997. Nuclear DNA amounts in Angiosperms: 583 new estimates. *Annals of Botany*, 80: 169-196.

Besse, P. 1993. Identification des clones cultivés et analyse de la diversité génétique chez *Hevea brasiliensis* par RFLP. Doct. thesis, Université Paris XI, Orsay, 114 p.

Besse, P., Seguin, M., Lebrun, P., Chevallier, M.H., Nicolas, D., and Lanaud, C. 1994. Genetic diversity among wild and cultivated populations of *Hevea brasiliensis* assessed by nuclear RFLP analysis. *Theoretical and Applied Genetics*, 88: 199-207.

Besse, P., Seguin, M., Lebrun, P., and Lanaud, C. 1993. Ribosomal DNA variations in wild and cultivated rubber tree (*Hevea brasiliensis*). *Genome*, 36: 1049-1057.

Bouharmont, J. 1960. Recherches taxonomiques et caryologiques chez quelques espèces du genre *Hevea*. Congo, INEAC, 64 p.

Chapuset, T., Legnate, H., Doumbia, A., Clement-Demange, A., Nicolas, D., and Keli, J. 1995. Agronomical characterisation of the 1981 germplasm in Côte d'Ivoire: growth, production, architecture and leaf disease sensibility. In: IRRDB Symposium on Physiological and Molecular Aspects of the Breeding of *Hevea brasiliensis*. Brickendonbury, UK, IRRDB, pp. 112-122.

Chevallier, M.H. 1988. Genetic variability of *Hevea brasiliensis* germplasm using isoenzyme markers. Journal of Natural Rubber Research, 3: 42-53.

Chevallier, M.H., Lebrun, P. and Normand, F. 1988. Approach of the genetic variability of germplasm using enzymatic markers. In: *Colloque Exploitation, Physiologie et Amélioration de l'hévéa*. J.L. Jacob and J.C. Prévôts, eds., Montpellier, France, CIRAD-IRCA, pp. 365-376.

Clement-Demange, A., Seguin, M., Lespinasse, D., Legnate, H., Chapuset, T., and Nicolas, D. 1997. Germplasm, genetic improvement and marker assisted selection of the rubber tree. In: *Seminar-Workshop on the Biochemical and Molecular Tools for Exploitation Diagnostic and Rubber Tree Improvement*. CIRAD-Orstom-University of Mahidol.

Crozier, R. 1990. From population genetics to phylogeny uses and limits of mitochondrial DNA. *Australian Systematic Botany*, 3: 111-124.

De Paiva, J.R. 1981 . I coleta de material sexuado a assexuado nos seringais nativos do Estado do Mato Grosso. Manaus, Brazil, EMBRAPA-CNSPD, 26 p.

Escofier, B. and Pages, J. 1988. *Analyses Factorielles Simples et Multiples*. Paris, Dunod, 241 p.

Fregene, M., Angel, F., Gomez, R., Rodriguez, F., Chavarriaga, P., Roca W., Tohme, J., and Bonierbale, M. 1997. A molecular map of cassava (*Manihot esculenta* Crantz). *Theoretical and Applied Genetics*, 95: 431-441.

Gonçalves, P.S. 1981. Expedição internacional à Amazônia no Territorio Federal de Rondônia para coleta de material botânico de seringueira (*Hevea brasiliensis*). Manaus, Brazil, EMBRAPA-CNSPD, 60 p.

Gonçalves, P.S., de Paiva, J.R., and de Souza, R.A. 1983. Retrospectiva e actualidade do melhoramento genético da seringueira (*Hevea* spp.) no Brasil e em países asiáticos. Manaus, Brazil, EMBRAPA-CNSPD, Document no. 2, 69 p.

Goulding, M. 1993. Les forêts inondables d'Amazonie. *Pour la Science*, 187: 70-77.

Haffer, J. 1982. General aspects of the refuge theory. In: *Biological Diversification in the Tropics*. G.T. Prance, ed., New York, Columbia University Press, pp. 6-26.

Hamon, S., Dussert, J., Deu, M., Hamon, P., Seguin, M., Glaszmann, J.C., Grivet, L., Chantereau, J., Chevallier, M.H., Flori, A., Lashermes, P., Legnate, H., and Noirot, M. 1998. Effects of quantitative and qualitative principal component score strategies on the structure of coffee, rice, rubber tree and sorghum core collections. *Genetics, Selection, Evolution*, 30 (suppl. 1): 237-258.

Lebrun, P. and Chevallier, M.H. 1990. *Starch and Polyacrylamide Gel Electrophoresis of Hevea brasiliensis, a Laboratory Manual*. Montpellier, France, CIRAD-IRCA, 55 p.

Leconte, A., Lebrun, P., Nicolas, D., and Seguin, M. 1994. Electrophoresis: application to *Hevea* clone identification. *Plantations, Recherche, Développement*, 1: 28-36.

Lespinasse, D. 1993. Recherche de marqueurs en vue d'une cartographie génétique chez *Hevea brasiliensis*. Mémoire de DEA, ENGREF, Paris, 66 p.

Lespinasse, D., Grivet, L., Troispoux, V., Rodier-Goud, M., Pinard, F., and Seguin, M. (2000a). Identification of QTLs involved in the resistance to South American leaf blight (*Microcyclus ulei*) in the rubber tree. *Theoretical and Applied Genetics*, 100: 975-984.

Lespinasse, D., Rodier-Goud, M., Grivet, L., Leconte, A., Legnate, H., and Seguin, M. 2000b. A saturated genetic linkage map of rubber-tree (*Hevea* spp) based on RFLP, AFLP, microsatellite, and isozyme markers. *Theoretical and Applied Genetics*, 100: 127-138.

Lins, A.C.R., Silva, G.P., Nicolas, D. Ong, S.H., Melo, C.C., and Santos, M.R. 1981. Report of the Acre team in the 1981 joint IRRDB-Brazil *Hevea* germoplasm expedition. Manaus, Brazil, EMBRAPA-CNSPD, 24 p.

Low, F.C. and Bonner, J. 1985. Characterisation of the nuclear genome of *Hevea brasiliensis*. In: International Rubber Conference 1985. Kuala Lumpur, Malaysia, IRRDB, pp. 127-136.

Luo, H. and Boutry, M. 1995. Phylogenetic relationships within *Hevea brasiliensis* as deduced from a polymorphic mitochondrial DNA region. *Theoretical and Applied Genetics*, 91: 876-884.

Luo, H., van Coppenolle, B., Seguin, M., and Boutry, M. 1995. Mitochondrial DNA polymorphism and phylogenetic relationships in *Hevea brasiliensis*. *Molecular Breeding*, 1: 51-63.

Mogensen, H.L. 1996. The hows and whys of cytoplasmic inheritance in seed plants. *American Journal of Botany*, 83: 383-404.

Nei, M. 1978. Estimation of average heterozygosity and genetic distance from a small number of individuals. *Genetics*, 89: 583-590.

Nicolas, D. 1985. Acquisition of *Hevea* material derived from Colombian Schultes collections. In: International Rubber Conference 1985. Kuala Lumpur, Malaysia, IRRDB.

Nicolas, D., Chevallier, M.H., and Clement-Demange, A. 1988. Contribution to the study and evaluation of new germplasm for use in *Hevea* genetic improvement. In: *Colloque Exploitation, Physiologie et Amélioration de l'Hévéa*. J.L. Jacob and J.C. Prévôts, eds., Montpellier, France, CIRAD-IRCA, pp. 335-352.

Ong, S.H. 1985. Chromosome morphology at the pachytene stage in *Hevea brasiliensis*: a preliminary report. In: International Rubber Conference 1985. Kuala Lumpur, Malaysia, IRRDB, pp. 3-12.

Ong, S.H. 1987. Utilization of *Hevea* genetic resources in the RRIM. *Malaysian Applied Biology*, 16: 145-155.

Ong, S.H., Othman, R., and Benong, M. 1995. Status report on the 1981 *Hevea* germplasm collection. In: IRRDB Symposium on Physiological and Molecular Aspects of the Breeding of *Hevea brasiliensis*. Brickendonbury, UK, IRRDB, pp. 95-105.

Palmer, J.D. 1987. Chloroplast DNA evolution and biosystematic uses of chloroplast DNA variation. *American Naturalist*, 130: S6-S29.

Palmer, J.D. and Herbon, L.A. 1988. Plant mitochondrial DNA evolves rapidly in structure, but slowly in sequence. *Journal of Molecular Evolution*, 28: 87-97.

Pernes, J. 1984. *Gestion des Ressources Génétiques des Plantes*. Paris, ACCT, 212 p.

Perrier, X., Flori, A., and Bonnot, F. 2002. Les méthodes d'analyse des données. In: *Diversité Génétique des Plantes Tropicales Cultivées*. P. Hamon et al. eds., Montpellier, France, CIRAD, collection Repères, pp. 43-76.

Polhamus, L.G. 1962. *Rubber: Botany, Production and Utilization*. London, Leonard Hill, 449 p.

Prance, G.T. 1987. Biogeography of neotropical plants. In: *Biogeography and Quaternary History in Tropical America*. T.C. Whitmore and G.T. Prance, eds., Oxford, Clarendon Press, pp. 46-65.

Reboud, X. and Zeyl, C. 1994. Organelle inheritance in plants. *Heredity*, 72: 132-140.

Rivano, F. 1992. La maladie sud-américaine des feuilles de l'hévéa: étude en conditions naturelles et contrôlées des composantes de la résistance partielle à *Microcyclus ulei*. Doct. thesis, Université Paris XI, Orsay, 260 p.

Schultes, R.E. 1977. Wild *Hevea*: an untapped source of germplasm. *Journal of the Rubber Research Institute of Sri Lanka*, 54: 227-257.

Schultes, R.E. 1990. *A Brief Taxonomic View of the Genus* Hevea. Kuala Lumpur, Malaysia, MRRDB, 57 p.

Seguin, M., Besse, P., Lebrun, P., and Chevallier, M.H. 1995a. *Hevea* germplasm characterization using isozymes and RFLP markers. In: *Population Genetics and Genetic Conservation of Forest Trees*. P. Baradat et al., eds., Amsterdam, SPB Academic Publishing, pp. 129-134.

Seguin, M., Besse, P., Lespinasse, D., Lebrun, P., Rodier-Goud, M., and Nicolas, D. 1995b. Characterization of genetic diversity and *Hevea* genome mapping by biochemical and molecular markers. In: IRRDB Symposium on Physiological and Molecular Aspects of the Breeding of *Hevea brasiliensis*. Brickendonbury, UK, IRRDB, pp. 19-30.

Seguin, M., Besse, P., Lespinasse, D., Lebrun P., Rodier-Goud, M., and Nicolas, D. 1996. *Hevea* molecular genetics. *Plantations, Recherche, Développement*, 3: 77-88.

Seguin, M., Lespinasse, D., Rodier-Goud, M., Legnate, H., Troispoux, V., Pinard, F., and Clement-Demange, A. 1998. Genome mapping in connection with resistance to the South American leaf blight in rubber tree (*Hevea brasiliensis*). In: IIIrd ASAP Conference on Agricultural Biotechnology. Bangkok, Biotech. Vol. 1.

Seguin, M., Rodier-Goud, M., and Lespinasse, D. 1997. Mapping SSR markers in rubber tree (*Hevea brasiliensis*) facilitated and enhanced by heteroduplex formation and template mixing. In: *Plant and Animal Genome V*, D. Bigwood et al. eds., Washington, D.C., USDA, Poster no. 61, p. 66.

Seibert, R.J. 1948. The use of *Hevea* for food in relation to its domestication. *Annals of the Missouri Botanical Garden*, 35: 117-121.

Serier, J.B. 1993. *Histoire du Caoutchouc*. Paris, Desjonquières, 273 p.

Wolfe, K.H., Li, W.H., and Sharp, P.M. 1987. Rates of nucleotide substitution vary greatly among plant mitochondrial, chloroplast, and nuclear DNAs. Proceedings of the National Academy of Sciences of the United States of America, 84: 9054-9058.

Sorghum

Monique Deu, Perla Hamon, François Bonnot
and Jacques Chantereau

Sorghum, *Sorghum bicolor* (L.) Moench, is cultivated in tropical as well as temperate countries. In 1995, it was cultivated on over 43 million ha and the global production was more than 54 million t (FAO, 1995). It is the fifth most important cereal in the world, after wheat, rice, maize, and barley. Despite the growing importance of rice and maize, sorghum remains an essential element of human nutrition for several countries in Africa (Sudan, Botswana, Burkina Faso, Rwanda, Chad, and Cameroon) and Asia (India and China). In 1995, the production of sorghum grain was more than 16 million t in Africa and 15 million t in Asia. The grains are made into flour and consumed as porridge or pancakes. Certain better-adapted varietal types are used in making beer, sweets, or popcorn.

Sorghum has other uses also. In Argentina, Australia, South Africa, Mexico, and especially the United States, where the production in 1995 exceeded 17 million t, sorghum is mainly used as animal feed. Finally, in certain regions of Africa and Asia, the panicle may be used to make brooms while the stems are used as forage, fuel, or construction materials or in leather dyeing or paper making. The pith may yield sugar, syrup, glue, and alcohol.

Sorghum is a hardy plant, adapted to difficult environments and tolerant of poor soils, drought, high temperatures, and even flooding. It can be cultivated where more valuable crops cannot be.

TAXONOMY AND GENETIC RESOURCES

Taxonomy and Geographic Distribution of the Genus *Sorghum*

Sorghum (genus *Sorghum*, family Poaceae) is a cereal closely related to maize and sugarcane; all three belong to the tribe Andropogoneae. The great morphological diversity of the genus *Sorghum* led botanists to distinguish a large number of taxa—712 were described by Snowden (1936). A simplified classification, taking genetic exchanges into account, is widely used today (de Wet, 1978). Recommended by the IPGRI (International Plant Genetic

Resources Institute), it divides the genus into five sections. The section Sorghum includes the following: all the cultivated sorghums with grains (*S. bicolor* ssp. *bicolor*, diploid, 2x = 20); wild sorghums, diploid, annual, originating from Africa (*S. bicolor* ssp. *verticilliflorum*); wild sorghums, diploid, with rhizomes, perennial, present in India, Sri Lanka, and Southeast Asia (*S. propinquum*); and wild sorghums, tetraploid, also with rhizomes and perennial, found in Southeast Asia, India, Middle East, and around the Mediterranean Sea (*S. halepense*). According to Mann et al. (1983), sorghum was domesticated around 3000 years BCE in northeastern Africa. However, later archaeological data (Wendorf et al., 1992) suggest that it was first used more than 6000 years BCE. According to Harlan and Stemler (1976) and Doggett (1988), the cultivated sorghums derive from wild African sorghums *S. bicolor* ssp. *verticilliflorum*.

The cultivated sorghums present a wide phenotypic diversity. From the form of the inflorescence and above all the structure of the panicle, it is classified into five basic races (bicolor, caudatum, durra, guinea, and kafir) and ten intermediate races obtained by the combination of any two of those races (guinea-bicolor, durra-caudatum; Harlan and de Wet, 1972). However, according to Doggett (1988), the bicolor sorghums, selected on traits not linked to the structure of the panicle, such as sweet stem and forage, do not constitute a race but rather form a fairly heterogeneous group. These sorghums would be close to the primitive sorghums from which they were domesticated.

The various races of sorghum presently occupy contiguous areas of distribution, even though one race may be predominant in a region. The sorghums cultivated in India are principally the durra, which are also found in East Africa (Ethiopia) and the Middle East (Turkey, Syria). The bicolor type is widespread in Africa and Asia, while the kafir are found essentially in southern Africa, from Tanzania to South Africa. The caudatum race, predominant in central Africa, has become an important source of genetic material in breeding programmes for temperate regions. The guinea, sorghums typical of West Africa, are also present in East Africa and southern Africa.

The dispersion and present geographic distribution of the cultivated sorghums (Figs. 1 and 2) are closely linked to past human migrations, cultural traditions, and dietary practices of the ethnic groups who cultivate them. In Africa, for example, the durra race is associated with Islam, as opposed to the kafir sorghums cultivated by 'infidels'. The wide geographic distribution of the cultivated sorghums also shows that the races adapted to different ecological zones. The guinea race—large sorghums with a very loose panicle that permits quick drying and corneous grains that are resistant to moulds—is often cultivated in the humid zones, while the durra—short sorghums with a generally compact panicle and large grains—are well adapted to drought.

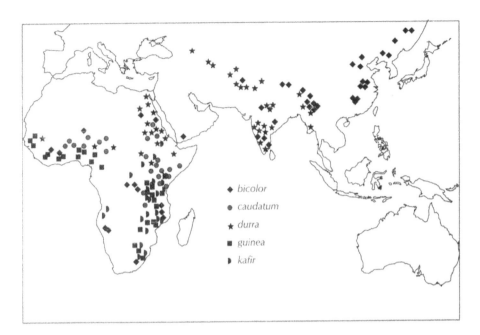

Fig. 1. Distribution of cultivated sorghum races.

Fig. 2. Races of cultivated sorghum and their hybrids: (left) durra, (top) guinea, (below) caudatum, and (centre) bicolor (photo by J.L. Noyer—CIRAD).

The cultivated sorghums are monoecious and preferentially autogamous even though relatively high rates of allogamy (10 to 30%) were indicated in guinea populations (Ollitrault, 1987; Chantereau and Kondombo, 1994). All the races of cultivated sorghums can be crossed with each other and hybridize with the related wild diploid sorghums (*S. bicolor* ssp. *verticilliflorum*). Similarly, gene flows are possible between *S. bicolor* and *S. propinquum*, wild diploid (de Wet et al., 1976), and between *S. bicolor* and *S. halepense*, wild tetraploid (Arriola, 1995).

Genetic Resources

In the 1970s, the international scientific community acknowledged the disturbing extinction of traditional varieties of several cultivated plants as well as that of numerous wild populations. This extinction was linked to the abandonment of traditional varieties in favour of new, improved varieties and the rapid destruction of natural habitats, often caused by uncontrolled urbanization. On the global scale, ICRISAT (International Crops Research Institute for the Semi-Arid Tropics) was mandated to ensure the collection and conservation of genetic resources of sorghum. In 1972, the genetic resources unit of ICRISAT, in India, received around 10,000 accessions collected earlier by the Rockefeller Foundation. By the early 1990s it contained around 33,100 accessions collected from 86 countries. This collection was expected to reach 45,000 accessions by the year 2000 (Prasada Rao and Ramanatha Rao, 1995). The accessions were conserved and multiplied under the responsibility of ICRISAT at Patancheru, India. They were for the most part duplicated in the United States, where over 42,000 accessions were enumerated by the National Plant Germplasm System. They are conserved for the long term in three major sites: Fort Collins, Colorado, by the National Seed Storage Laboratory; Griffin, Georgia; and Mayguez, Puerto Rico, by the United States Department of Agriculture.

In parallel, smaller collections were built up in many countries. In China, 10,386 accessions collected from the 1950s onward were conserved by the national gene bank at Beijing. The South African collection of sorghum numbers around 4000 accessions, and the Australian collection around 3800. In France, 5850 accessions are conserved, 3850 of them by the IRD (Institut de Recherche pour le Developpement, earlier Orstom) and 2000 by CIRAD. All the collections have two weaknesses in common: the low representation of wild sorghums—only 1.2% of the United States collection and 1.1% of the ICRISAT collection—and an unequal representation of races and geographic origins. The ICRISAT collection, for example, has 17.4% guinea, 20.2% durra, 21.8% caudatum, against 2.3% kafir and 3.2% bicolor, and comes essentially from five countries: India (17.3%), Cameroon (19.2%), Yemen (8.5%), Sudan (8.1%), and Ethiopia (8.1%).

At the Indian centre of ICRISAT, the accessions were evaluated using a set of 20 morphological descriptors, of which 19 belong to the list recommended by IBPGR (1993): The 'passport' data and evaluation are complete for around 28,000 accessions On the other hand, in the United States, only 13% of the accessions of the global collection have been completely evaluated.

The management and evaluation of such collections is costly. Nevertheless, ICRISAT pursues a more complete evaluation. According to Dahlberg and Spinks (1995), efforts must also be made by the United States, where the field evaluation (40 descriptors) of all the accessions in the collection must be undertaken.

An essential step is the constitution of a data bank that is easily accessible to allow the effective use of these genetic resources in national research programmes. In the United States, the data are regularly made available via the database of GRIN (Germplasm Resources Information Network), accessible on the internet (http://www.ars-grin.gov; Dahlberg and Spinks, 1995). At ICRISAT, the database can also be consulted online (http://noc1.cgiar.org/seartype.htm).

Genetic material is distributed naturally through seeds. Between 1985 and 1995, 118,381 samples were also distributed throughout the world from the Untied States (Dahlberg and Spinks, 1995) and 237,265 by ICRISAT (Mengesha and Appa Rao, 1994). Other countries that maintain smaller collections also participate in exchanges of plant material.

In this chapter, we review studies on the genetic diversity of cultivated sorghums. Then we compare the structurations obtained by analysis of a sample using three types of markers (morphological, enzymatic, and molecular markers) and several methodologies. The results are considered in terms of application to the constitution of core collections. In conclusion, we examine the strategy of genetic resource conservation and the utility of molecular markers in understanding the organization of genetic diversity within cultivated sorghums.

ORGANIZATION OF GENETIC DIVERSITY

Genetic Diversity Revealed by Morphological Descriptors

There are few studies on the organization of genetic diversity as revealed by morphological traits. The first study, by Chantereau et al. (1989), was carried out on 157 ecotypes of widely diverse race and geographic origin. These authors showed that, on the basis of the 25 agromorphological traits studied, of which 14 figure in the IBPGR list, sorghums can be classified into three groups: the guinea and bicolor group, the caudatum and kafir group, and the durra group. These groups are distinguished primarily by their behaviour in culture. The results reveal the adaptation of races to cultivation practices, specific uses, and biotic and abiotic constraints. For example, the guinea, like the bicolor, are hardy sorghums of wet zones, adapted to extensive cultivation. The durra are hardy sorghums of dry zones and occasionally of degrading environments, while the caudatum and kafir behave like the most modern sorghums, best adapted to semi-intensive or even intensive cultivation.

Subsequently, Appa Rao et al. (1996) studied close to 4000 accessions of sorghums originating from different Indian states and present in the ICRISAT collection using 14 morphological and agronomic descriptors. In their work, the data were not treated by multivariate analysis. Rather, the univariate descriptive analyses indicate a great morphophysiological diversity with a greater diversity among states than within states. All the races are present in India, but the durra and their intermediates are widely predominant. No information is given as to the distribution of races as a function of various traits.

In Ethiopia, Teshome et al. (1997) focused on a particular region comprising northern Shewa and southern Welo. In this restricted geographic area, the analysis of 14 morphological traits, of which 7 figure in the IBPGR list, reveals a great phenotypic diversity. Four pure races—bicolor, caudatum, durra, and guinea—and an intermediate race, durra-bicolor, were maintained by human selection. The dendrograms obtained did not reveal a clear taxonomic pattern. On the other hand, three groups could be distinguished from the multivariate analyses, mainly according to two criteria: presence or absence of sugar in the stem and plumpness of grains. This classification, different from that of Chantereau et al. (1989), relies in fact on the use of a different set of descriptors.

In our study, the analysis was conducted on a sample of 230 accessions comprising a larger proportion of guinea and caudatum but fewer durra, kafir, and intermediates than that of Chantereau et al. (1989). Twenty-one of the 25 morphophysiological traits earlier studied by Chantereau et al. (1989) were taken into account. Seventy modalities were defined from 21 variables and then a correspondence analysis (CA) was conducted on a binary table (all the active individuals and modalities). The projection of variables on the first planes of the CA show that axis 1 (12.7%) separates the small sorghums, with thick stem and glumes shorter than the grain, from the tall sorghums with long and loose panicle. Axis 2 (7.2%) isolates the sorghums with crossed peduncle and compact panicle. Axis 3 (5.4%) distinguishes the early-flowering sorghums, presenting few internodes, from the later-flowering and tall varieties. Representation of the accessions according to their racial classification in plane 1-2 shows results similar to those obtained by Chantereau et al. (1989), with a differentiation of three groups: (1) guinea and bicolor, (2) caudatum and kafir, and (3) durra.

Hierarchical clustering (HC; Fig. 3) gives a more global picture of the organization of genetic diversity and provides indications on the relationships

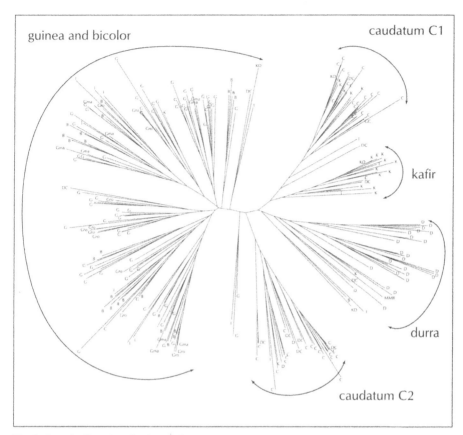

guinea and bicolor

caudatum C1

kafir

durra

caudatum C2

Fig. 3. Genetic diversity of cultivated sorghums revealed by morphological markers. The tree is constructed from the Sokal and Michener index.

between the groups in terms of proximity and distance. The guinea (G) and bicolor (B) form a highly variable group; the kafir (K) constitute a relatively homogeneous group; the caudatum (C), on the other hand, are separated into two subgroups, C1 and C2. The subgroup C1 is characterized by varieties presenting a short cycle and few internodes (less than 10), while subgroup C2 covers sorghums of medium cycle and a larger number of internodes. The two subgroups of caudatum appear quite divergent from each other, with distances between the subgroups that appear to be of the same magnitude as distances detected between the races; subgroup C1 is closer to kafir group than subgroup C2. The separation into two subgroups could reflect a difference in behaviour with respect to the photoperiod. Subgroup C1, close to kafir group, must thus be less sensitive or insensitive to photoperiod, while subgroup C2, close to the durra (D), must be sensitive to photoperiod (Chantereau et al., 1997). This sub-structuration of caudatum passed unnoticed in the study of Chantereau et al. (1989) and in the correspondence analysis (CA) presented above, which can be explained by the fact that the two subgroups, located on either side of axis 1 in the 1-3 plane of our CA (Fig. 4), form an apparently continuous group.

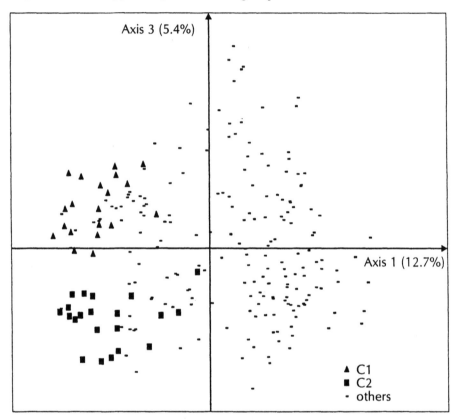

Fig. 4. Correspondence analysis on morphological traits.

No group is totally homogeneous with respect to the racial classification. As a general rule, individuals belonging to intermediate races are distributed in different groups depending on their degree of likeness to sorghums of the purest race. This observation could be reflected in a dynamic in the evolution of sorghums linked to several genetic exchanges favoured by traditional cultivation practices (association of several varieties, even several races, in a single field) and to various natural and human selection pressures. It would be interesting to observe the relative position of these intermediates in relation to other criteria of classification—such as those used by Teshome et al. (1997), or criteria of technological quality of grain, resistance to biotic and abiotic stresses—in order to better understand their role in the evolution of sorghums. Do they contribute to the widening of the genetic variability while evolving towards a parental type? Are they maintained or constantly produced and eliminated? Their competitiveness with pure races is very high in Chad when judged by the nature of the races that are traditionally cultivated in this region (Yagoua, 1995). On the other hand, the situation seems different in Ethiopia, where Teshome et al. (1997) indicate the presence of five cultivated races, of which four are 'pure' races.

Genetic Diversity Revealed by Enzymatic Markers

The analysis of enzymatic polymorphism of cultivated sorghums was the focus of studies conducted in the United States and in France for over ten years (Morden et al., 1989; Ollitrault et al., 1989b; Aldrich et al., 1992; Degremont, 1992).

The study of Morden et al. (1989), complemented by that of Aldrich et al. (1992), shows that most of the total genetic diversity is due to differences of geographic origin rather than race. They also indicate that the western and eastern African regions have the highest level of heterozygosity and that southern Africa has the narrowest genetic diversity.

For Ollitrault et al. (1989b), this geographic differentiation seemed more marked. These authors distinguished three geographically distinct groups: a western African group, an eastern and central African group, and a southern African group. Moreover, structuration according to the geographic zones was observed for two races, bicolor and guinea. The differentiation of three distinct groups within the guinea sorghums—guinea of West Africa, guinea of southern Africa, and guinea of the margaritiferum type—was subsequently largely demonstrated by the in-depth analyses of Degremont (1992).

Our study was conducted on 230 accessions representative of the geographic and racial diversity. A good genetic diversity was revealed in the sample, with 11 polymorphic loci (at a threshold of 99%) and a mean number of alleles per polymorphic locus of 2.8. The pattern of diversity indicated by the AHC (Fig. 5) is in agreement with the western-southern African geographic differentiation noted by Ollitrault et al. (1989b). Moreover, the study leads to the following conclusions. The accessions of East Africa and Central Africa present a wide range of variability and do not form a well-differentiated group. No group is totally homogeneous, racially or geographically. The guinea sorghums of southern Africa present a narrower genetic diversity than those of West Africa. The guinea sorghums of southern Africa are closer to kafir than to any other group. The guinea sorghums of Central and East Africa are found preferentially in the West African group. The guinea margaritiferum (Gma) form a highly homogeneous group that is clearly distinct from all the other guinea. Very curiously, they are relatively closely related to a small group of 12 accessions, of which two thirds are the caudatum sorghums belonging in equal parts to subgroups C1 and C2 described earlier.

While structuration using morphological descriptors indicates groups with common cropping performance, geographical structuration obtained

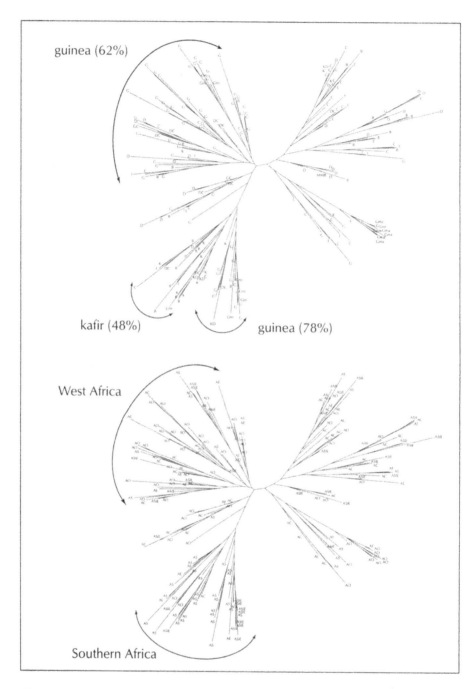

Fig. 5. Genetic diversity of cultivated sorghums revealed by enzymatic markers. The racial affiliation is indicated on the tree above and the geographic origin on the tree below. The tree is constructed from the Dice index.

using enzymatic markers indicates in its major lines the areas of distribution preferred by different races: guinea in West Africa, guinea and kafir in southern Africa, and durra and caudatum in East and Central Africa. However, the fact that durra of West Africa, for example, are found in the group of guinea sorghums of West Africa could indicate that these durra sorghums (barring labelling or identification errors) are of ancient introduction in West Africa. They could thus have followed a different evolutionary path, under the constraints of West Africa, from that of the other durra and integrated themselves totally in the West Africa group because of gene flow. On the contrary, the caudatum sorghums recently introduced in West Africa still retain the genotypic structures linked to their Central and East African origin and are not found in the West Africa group.

Genetic Diversity Revealed by Molecular Markers

Genetic diversity in sorghum was first studied by Aldrich and Doeley (1992) using RFLP markers (38 probes) and by Tao et al. (1993) using RFLP markers (16 probes) and RAPD markers (29 primers). These studies on restricted samples, of fewer than 50 accessions, do not enable us to observe a marked racial or geographic pattern of genetic organization. On the other hand, the authors conclude that RFLP and RAPD markers indicate a greater allelic diversity than enzymatic markers. However, the capacity of markers to reveal a structuration depends on whether there is a structuration, on the representativity of the sample analysed, and on the number and type of markers used. Aldrich and Doebley (1992) studied 31 accessions representing the five major races equally but sampled from 10 countries distributed unequally between the four major regions of Asia and Africa. Similarly, the 36 accessions studied by Tao et al. (1993) are not representative of the racial diversity or of large areas of distribution of sorghum.

Deu et al. (1994) used 33 RFLP probes to study a sample of 94 accessions, taking into account the racial and geographic diversity of sorghums. The study indicated a southern African group, the genetic variability of which is more restricted than that of West Africa or East and Central Africa. This study also revealed a racial differentiation. Except the bicolor accessions, which do not constitute a homogeneous group, the caudatum, durra, and kafir races form three distinct groups, while the guinea are divided into three subgroups: the guinea of West Africa, those of southern Africa, and guinea margaritiferum. This study confirms the internal structuration of guinea observed with the isozymes and allows a racial differentiation not revealed by morphological traits (caudatum and kafir) or enzymatic markers (caudatum and durra).

A study conducted by Cui et al. (1995) with 61 RFLP probes on 41 accessions shows a less clear racial differentiation since the kafir and guinea of southern Africa and Asia form a single group. The study also indicates the originality of a guinea margaritiferum, which is found to be more closely related to the wild sorghums than to the cultivated sorghums. For the authors, this grouping is consistent with the rather 'wild' phenotype of this accession.

The singularity of guinea margaritiferum is also mentioned in a study by de Oliveira et al. (1996) using 20 RFLP probes, 13 RAPD primers, and 4 ISSR: the three accessions of this type analysed differed from other cultivated sorghums mainly by the absence of common alleles found in the rest of the collection rather than specific alleles unique to the three accessions.

An important study was conducted by Menkir et al. (1997) on 190 accessions representative of five principal races, with widely varying geographic origins—13 countries for the kafir and 28 to 32 countries for the other races. The analyses performed using 82 RAPD primers show that genetic diversity is much greater within the bicolor and the guinea than in the other races. The kafir sorghums are the least diversified. Moreover, 86% of the total variability occurred within a race and 14% among races. Only 13% of the total variability is linked to geographic differentiation. In these conditions, the authors do not observe racial or geographical pattern of genetic diversity. These results, which do not agree with those of Deu et al. (1994), could be explained either by the mode of sampling or by the nature of markers used.

First, the sampling of geographical origins for each race would be too artificial and not reflect the traditional distribution of races—the countries of origin are not specified in the study. Human migrations and exchanges of material favour contacts and gene flows between races, which help widen the genetic base of each race. Nevertheless, even though there are numerous intermediates, at various degrees, the major racial types persist in cultivation. In northern Shewa and southern Welo, in Ethiopia, four of the five major races coexist—bicolor, caudatum, durra, and guinea (Teshome et al., 1997). Although genetically close, the kafir and guinea of southern Africa retain different morphological characteristics. This could indicate the existence of very strong selection pressures on the morphological traits and thus the fact that the major races correspond to types with high ecogeographical specialization. The recent migrations, on their part, help to artificially enlarge the variability within regions and reduce the interregional differentiation.

Second, the RAPD markers used in this study are mostly multilocus, with an average of 4.2 bands revealed per primer, while close to 75% of the probes tested in RFLP by Deu et al. (1994) correspond to single copy loci. In these conditions, the RAPD could correspond to non-coding regions or to repeat sequences in the genome, while the RFLP would correspond rather to single copy, coding sequences. The rates of evolution of these regions are probably different and, because of this, these two types of markers do not result from the same evolutionary processes. Moreover, phenomena of homoplasy, more frequent with RAPD than with RFLP, could account for observed differences in structure (Powell et al., 1996).

Comparison of Genetic Diversity Organization Revealed by the Three Types of Markers

Since RFLP data are available for 92 accessions already characterized by morphological and enzymatic markers, we first compared the organization of genetic diversity obtained on this subgroup with those revealed by study of the initial sample of 230 accessions.

For morphological diversity, all the modalities found in the initial collection, including those of low frequency, are presented in the subgroup. The radial representations obtained in the two cases (230 and 92) are very similar. The group showing the greatest diversity is made up of guinea and bicolor. Four other groups are observed: one group of durra, one of kafir, and two of caudatum. Eighteen accessions occupy positions intermediate between these groups.

The 31 alleles detected from the 11 enzymatic loci in the 230 accessions were conserved in the subgroup. The structuration observed shows three major identifiable groups: the guinea margaritiferum, a group from southern Africa including the kafir and guinea, and a group comprising the caudatum and durra. Between these three groups can be seen the presence of numerous accessions occupying intermediate positions. The West Africa group that can be observed in the initial sample does not appear clearly in the subgroup.

Genetic structuration revealed by the three types of markers has been compared on this subsample of 92 accessions.

The average numbers of alleles per polymorphic locus appear equivalent whatever the nature of marker considered: 2.8 for isozymes against 2.9 for RFLP (if the restriction enzyme is HindIII) and 3.0 (with XbaI). In this particular case, RFLP does not allow access to a greater allelic polymorphism than the isozymes.

We were able to identify 11 polymorphic loci for the isozymes and 33 for the RFLP. Even more than the percentage of polymorphic loci, what seems interesting is the number of polymorphic loci that are relatively easily accessible. In the study of Ollitrault et al. (1989b) on 348 accessions, 18 loci out of the 25 tested appeared polymorphic. It thus seems difficult to achieve more than 20 enzymatic polymorphic loci, while Cui et al. (1995) used 61 RFLP polymorphic loci.

In our study, the 11 enzymatic loci enabled identification of 83.6% of genotypes (or 77 out of 92), while the 33 RFLP loci led to 94.5% genotypic identification. Nevertheless, in some cases, the isozymes discriminate between accessions not differentiated by RFLP markers. Thus, the enzymatic loci EST-D allow differentiation of accessions of each of two pairs of accessions that appear identical according to RFLP. Two other accessions identical with the RFLP have different alleles at three enzymatic loci (END, HEX and LAP). The combination of isozyme markers and RFLP allows us to discriminate all the accessions with the exception of two pairs formed by two caudatum on the one hand and two guinea roxburghii (Gro) on the other. In these conditions, it would be interesting to identify the minimal combination of isozyme and RFLP markers that allows identification of all the genotypes, and then to test its efficiency on other groups of accessions.

The comparison of structures revealed by the different markers shows that the great morphological diversity of guinea sorghums is accompanied by great genetic diversity, with three groups of differentiation clearly marked by the RFLP. In addition, it indicates that the caudatum and durra sorghums, well differentiated in morphological terms, constitute two groups whose genetic proximity is indicated by isozyme markers and RFLP. Moreover, the kafir sorghums, relatively homogeneous from the morphological point of view, are genetically closer to guinea of southern Africa than to other sorghums.

In fact, although the organization of genetic diversity obtained with the three types of markers cannot be perfectly superimposed, they are consistent with the information acquired on sorghum.

- Sorghums are preferentially autogamous, but existing natural allogamy, even though sometimes low, is favoured by traditional cultivation practices. Because of this, gene exchange enhances the morphological diversity (within the limit of selection practised) and above all genetic diversity for the isozymic and molecular characters that are inherently neutral with respect to selection.

- The guinea margaritiferum are the most differentiated among the cultivated sorghums. Deu et al. (1995) showed that they present a mitochondrial polymorphism that distinguish them from all the other sorghums, cultivated or wild, belonging to the species S. bicolor. Because of their genetic and agromorphological characteristics, these guinea sorghums could sustain the interest of breeders.

- The guinea of southern Africa are genetically closer to kafir than to other guinea. Because these two groups of sorghum share the same area of distribution, the gene flows occur naturally. The intragroup diversity increases at the expense of intergroup diversity.

There are other methods for comparing the information given by various markers. The use of the Mantel test and the construction of consensus trees or common minimal trees are examples. In the first case, correlation coefficients, r, between the matrixes of similarity indexes produced from different markers are calculated. Thus, the value of distances between individuals is emphasized. The calculated significance of r is totally subjective. According to Rohlf (1990), the correlation could be considered very high if r is greater than or equal to 0.9, high if r is between 0.8 and 0.9, low if r is between 0.7 and 0.8, and very low if r is lower than 0.7. Several studies refer to the use of this test (Messmer et al., 1991; Engqvist and Becker, 1994;

Thormann et al., 1994; Powell et al., 1996). However, few studies take into account the enzymatic markers. Engqvist and Becker (1994) calculated the correlation coefficients obtained in *Brassica napus* from RFLP, RAPD, and enzymatic markers. The lowest coefficient was obtained with RFLP-isozyme pairs (0.53 against 0.67 for the RAPD-isozyme pair and 0.76 for the RFLP-RAPD pair). Similarly, in maize, Messmer et al. (1991) obtained a low correlation ($r = 0.23$) for the RFLP-isozyme pairs.

In our case, in accordance with the patterns of diversity observed, the correlations are very low: r equals 0.19 for the isozyme-morphology pair, 0.2 for the RFLP-morphology pair, and 0.3 for the RFLP-isozyme pair.

Comparison by construction of trees considers the structures obtained without taking into account the edge lengths.

In our case, the consensus trees obtained are of the star or 'open umbrella' type, which indicates the absence of strictly identical groups (common structures) between the trees produced from different markers. This constraint seems too great in our case and not appropriate to the biological reality. If two groupings differ only by the presence or absence of an individual (poorly classified for whatever reason), the edge is not conserved in the consensus tree.

In contrast, the construction of common trees takes into account the fact that all the individuals could not be correctly represented in all the trees (different sources of possible error). This method allows us to find the largest subgroup of individuals forming the same structure in the trees compared.

According to Perrier et al. (1999), at a threshold of 5% with a population of around 100 individuals, a maximum of 19 points common between two trees can be obtained at random. In our case, the three trees thus constructed have 19 to 24 common points (Figs. 6 and 7). This result is in agreement with the low correlation coefficients calculated between the distance matrixes. Nevertheless, in considering the structures common to trees constructed from morphological markers and RFLP (Fig. 6), we note that the edges in caudatum (C2) and durra are present in the two trees; similar observations can be made for those that appear in part of the guinea. It may seem surprising not to find the kafir group in the common structures. In fact, this group is more homogeneous in genetic terms than in morphological terms. The individuals closest morphologically are not genetically the most similar.

Similar observations can be made during comparison of trees constructed from morphological and enzymatic markers (Fig. 7). However, if some large structures are conserved (opposition between guinea and caudatum, durra), the same individuals in these two common trees do not constitute the framework of the common structures. This observation takes into account the fact that it is not possible to construct a consensus tree.

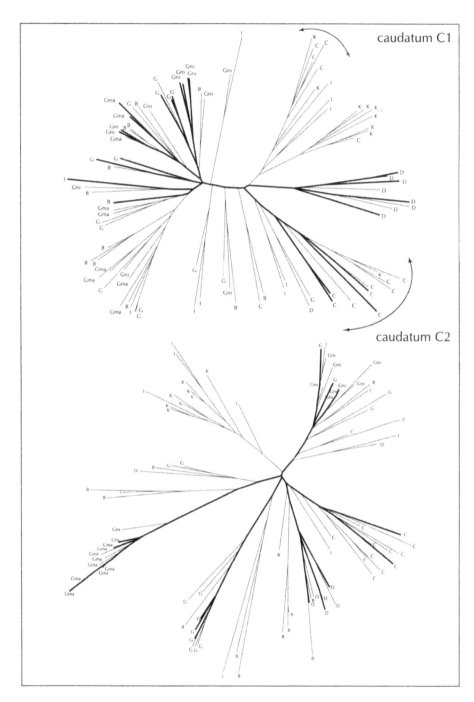

Fig. 6. Representation of common edges (bold) in trees constructed from morphological data (above) and molecular data (below).

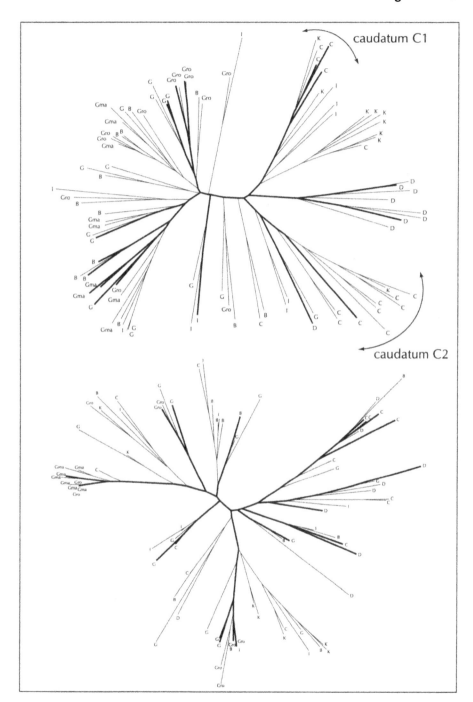

Fig. 7. Representation of common edges (bold) in trees constructed from morphological data (above) and enzymatic data (below).

Application to Constitution of Core Collections

In order to facilitate the use of large collections, the concept of core collection was developed by Frankel and Brown (1984) and subsequently by Brown (1989). Various sampling strategies were proposed (Hamon et al., 1995, for review). They were applied to several Poaceae, such as barley (van Hintum et al., 1990), ryegrass (Charmet et al., 1993), and durum wheat (Spagnoletti-Zeulli and Qualset, 1993).

For sorghum, collections of limited size were developed from the early 1990s onward. A basic collection was constituted in India by ICRISAT. It comprised 1400 accessions selected according to criteria of taxonomy, geographic distribution, and ecological adaptation, as defined by Harlan (1972). Very rapidly, this collection posed practical problems due to photoperiodism, a character that was not taken into account during the constitution of the collection. At Patancheru, many tropical ecotypes, notably from Cameroon, Ethiopia, Nigeria, and Sudan, did not flower at all or flowered too late. This collection was thus declared to be specific to one location and poorly representative of the phenotypic diversity of the world collection. It was then decided that other, smaller collections would be created.

A core collection comprising only cultivated sorghums was then established at ICRISAT by Prasada Rao and Ramanatha Rao (1995) with two objectives: to facilitate access to the world collection and to better represent the genetic diversity of cultivated sorghums.

For this, the world collection was stratified in subgroups defined according to geographic, taxonomic, and agronomic criteria. Then, other clusters within subgroups were defined through a principal components analysis on seven characters—cycle from seed to spike formation, plant height, inflorescence exsertion, inflorescence length and width, grain covering, and 100 grain weight. Taking into account the two modes of structuration of the initial collection, a sample proportionate to the number of accessions present in each subgroup was taken. This core collection comprised 3475 accessions, or around 10% of the ICRISAT collection.

The process of stratification before the selection aimed to take into account a possible pattern (geographic or taxonomic, for example) of the genetic diversity. Prior knowledge of the extent and pattern of genetic diversity in the gene pool considered is indispensable to maximize the genetic diversity to be preserved in a core collection. In this context, are the seven morphological characters taken into account for the constitution of the core collection sufficient and relevant to obtain a good representation of the overall diversity of sorghums? Various studies on the genetic diversity of sorghums, including ours, indicate that phenotypic and genetic divergence are not totally in agreement.

Using the PCS (principal component score) strategy developed by Noirot et al. (1996) and Hamon et al. (1998), we have constituted, from the initial sample of 230 accessions, two cores by using either morphological or enzymatic data. These two core collections were compared with the initial sample in order to understand the modifications induced by the process of selection.

The curves representing the cumulative number of individuals maximizing the diversity as a function of cumulative percentage of the total diversity show that 50% of the total enzymatic diversity is represented by 50 accessions, while 77 accessions are necessary to obtain 50% of the phenotypic diversity. These two sets of accessions constituted respectively CoreI and CoreM.

With respect to the distribution of botanical races (Table 1), the χ^2 test indicates significant differences between the initial sample and the two core collections. In particular, it is noted that the kafir race is not represented in CoreI. This observation is in agreement with the low genetic diversity of the kafir race noted by various authors. On the other hand, the large geographic regions are all represented in the two core collections without significant modifications in relation to the initial sample (Table 2).

For the allelic frequencies, as expected, highly significant differences (χ^2 test, $P < 0.01$) were observed for CoreI, while the differences were not significant for CoreM. However, in CoreI, the global divergence is due to two loci—endopeptidase and acid C phosphatase—affected by the selection. In CoreM, selection based on morphological characters did not enable conservation of all the rare alleles: three alleles out of five of frequency less than 5% were absent (Table 3). Nevertheless, it can be asked what will be the future of these rare alleles subjected to natural selection.

For phenotypic diversity, the situation is logically the inverse: the differences are not significant with CoreI but are highly significant with CoreM. In this case, the global divergence is due to significant differences (at least at a threshold of 5%) for 12 characters out of the 21 studied.

These comparisons indicate that the selection on neutral, enzymatic markers (CoreI) is interesting because it allows conservation of allelic diversity while preserving the initial phenotypic diversity. However, the large collections are subjected to more or less systematic morphoagronomic evaluations. It is thus also important to consider these descriptors for constituting core collections. For this purpose, one could question whether all the information available in the initial collection is useful to take into account. In other words, is there redundant information? Is it possible to define associations between markers? Some answers can be provided by the use of the Cramer coefficient of association V, calculated for each pair of variables (Bishop et al., 1975). The coefficients calculated from data of the initial sample (230 accessions) have values between -0.3 and + 0.65 (this coefficient varies theoretically between -1 and +1). When the value 0.5 is taken for the lower significant limit of association, associations are detected for five groups of variables: the brown layer (Cbr) and vitreousness (Vit);

leaf width (Laf) and stem diameter (Dtp); days to flowering (Nje) and number of internodes (Nen); compactness of panicle (Cpa) and length of panicle (Lpa) and of peduncle (Lpe); and grain shape (Fgr) and brown layer (Cbr) as well as length and form of the peduncle (Lpe, Fpe).

Table 1. Racial distribution in the initial sample and in the two core collections constituted by the PCS method (number and percentage of different races)

	bicolor	caudatum	durra	guinea	kafir	others
Initial sample	25 (10.9)	41 (17.8)	27 (11.7)	88 (38.3)	18 (7.8)	31 (13.5)
CoreI	4 (8)	5 (10)	10 (20)	25 (50)	0	6 (12)
CoreM	13 (16.9)	6 (7.8)	16 (20.7)	32 (41.6)	2 (2.6)	8 (10.4)

$\chi^2_{CoreM} = 12.4$ (ddl = 4), significant at 5%.
$\chi^2_{CoreI} = 9.53$ (ddl = 3), significant at 5%.

Table 2. Geographic distribution of accessions in the initial sample and in the two core collections (number and percentage of accessions)

	West Africa	East Africa	Central Africa	Southern Africa	Asia	America
Initial sample	74 (32.2)	19 (8.3)	39 (16.9)	55 (23.9)	39 (16.9)	4 (0.02)
CoreI	21 (42)	5 (10)	7 (14)	9 (18)	7 (14)	1 (2)
CoreM	26 (33.8)	7 (9)	10 (13)	14 (18.2)	18 (23.4)	2 (2.6)

$\chi^2_{CoreM} = 3.17$ (ddl = 4), ns.
$\chi^2_{CoreI} = 9.5$ (ddl = 3), ns.

Table 3. Total number of alleles of different frequency categories in the initial sample and in the two core collections

	Total number of alleles of different frequency categories in		
Frequency of alleles (x)	initial sample	CoreI	CoreM
x < 0.05	5	5	2
$0.05 \leq x < 0.1$	2	2	2
$0.1 \leq x < 0.2$	3	3	3
$0.2 \leq x < 0.4$	9	9	9
x > 0.4	12	12	12

Given that this initial sample does not correspond to a random sampling from the world collection, we could ask whether the associations indicated are fortuitous (sampling bias), or produced by linkage disequilibrium (foundation effect, genetic drift), or linked to the genetic organization of the genome, or represent coadapted gene complexes.

The genome mapping of different traits could provide elements of reflection even if the analysis of progeny is not comparable to that of a population.

Some of these traits have been mapped (Rami et al., 1998). It is interesting to note that the gene $B2/b2$ of the brown layer and a major QTL for vitreousness (Vit, $r^2 = 54\%$) were co-located on the linkage group F. Similarly, two QTL for compactness and length of panicle are co-located on the linkage group A (Cpa, $r^2 = 22\%$, Lpa, $r^2 = 35\%$) and on the linkage group F (Cpa, $r^2 = 13.5\%$, Lpa, $r^2 = 20\%$). Note that these QTLs mapped on the linkage group F are genetically independent of the $B2/b2$-Vit pair.

In agreement with the results obtained on our two core collections, no association (coefficient $V \geq 0.5$) was noted between morphological and enzymatic markers. Similarly, no association was revealed between the enzymatic markers at the fixed threshold. Despite the absence of major groups of linkages between the enzymatic loci (Ollitrault et al., 1989a), two situations are interesting to consider: the loci $LAP-2$ and $PA-B$, 12 cM apart, which present a Cramer coefficient V of 0.24; and the locus pairs HEX and LAP, DIA and LAP, AMY and $EST-C$, with coefficients of association 0.45, 0.36, and 0.34 respectively and genetically independent. In considering CoreI and CoreM, it can be observed that, for the pairs $AMY-EST$ and $HEX-LAP$, the associations are maintained and are not random (exact Fisher test significant for CoreI, highly significant for CoreM, in both cases). For the pairs $DIA-LAP$ and $PAB-LAP$, on the contrary, the associations are ruptured in the two core collections (exact Fisher test not significant).

The mapping of all the morphological and enzymatic markers as well as the search for association including molecular markers is still to be probed. They will certainly provide useful information for fine-tuning the effect of markers to be taken into account in the constitution of core collections and for evaluating the relative contributions of coadapted structures to be included in the core collection and linkage disequilibrium.

CONCLUSION

Studies on genetic diversity show first of all that there is great variability among cultivated sorghums. Eleven enzymatic loci are polymorphic in our study, 13 in that of Morden et al. (1989), and 18 in that of Ollitrault (1989b). Close to 75% of heterologous maize probes, in combination with two restriction enzymes, hybridize with sorghum DNA and reveal polymorphism

(Deu et al., 1994). In this study, each locus (enzymatic or revealed by RFLP) is represented on average by 2.8 to 3 alleles. Molecular RFLP markers allow us to discriminate four to five major botanical races described by Harlan and de Wet (1972). These races have a rather large and structured genetic variability. The kafir race, for example, seems highly homogeneous, while strong differentiation is observed within the guinea, partly due to secondary centres of diversification.

The patterns of genetic organization observed using enzymatic and morphological markers do not perfectly agree with the racial classification. Some races constitute groups that have common behaviours in cultivation (bicolor and guinea on the one hand, caudatum and kafir on the other, with morphological descriptors) or a common area of origin (caudatum and durra, with enzymatic markers).

Generally, molecular and enzymatic markers are considered neutral with respect to selection. Do the differences in the pattern of genetic diversity revealed by the two types of markers lie in the number of loci considered, in their genomic location, or in their own specificity? The study of genetic diversity revealed by analysis of a large number of RFLP loci very regularly distributed on the sorghum genome will certainly help answer these questions.

It has not been possible to construct a consensus tree after taking into account different morphological and enzymatic markers, even though some common elements of structure could be observed. This result indicates the absence of a strong structuration within the cultivated sorghums. Neither racial affinity, nor area of origin, nor cropping performance in cultivation led to the establishment of strong mating barriers. The presence of individuals in intermediate position in the tree representations indicates the existence of gene flows. Thus, in natural conditions, gene exchanges occur and, even though they occur at a low rate, they contribute to local broadening of the genetic diversity. At the same time, strong selection pressures are exerted, allowing the major races to be maintained where they are of particular interest. Such dynamic traditional management favours the establishment of sites reserved for *in situ* and participatory conservation of sorghum genetic resources. This aspect is important to consider since during propagation of the plant material there is the problem of genetic drift linked to self-fertilization of populations that are not always fixed, even though they are preferentially autogamous. This absence of strong structuration also allows us to consider the cultivated sorghums as a single gene pool.

The PCS strategy has thus been applied directly on the initial sample (without prior stratification) to constitute core collections, considering the morphological traits and then the enzymatic markers. Examination of two core collections constituted in this manner shows that 22% of the accessions for Corel and 33.4% for CoreM are needed to represent 50% of the total initial diversity. The morphological diversity thus seems more fragmented than the enzymatic diversity. However, this comparison remains very limited to the extent that the number of loci considered is different in each case. It would be particularly interesting to have access to RFLP data in order to have another measure of the distribution of genetic diversity in the initial sample.

Besides, examination of two core collections indicates that there is no association between the two types of characters. In our study, selection on the enzymatic markers has not affected the initial morphological variability. Selection on morphological traits leads to the loss of some rare alleles at certain enzymatic loci. It would thus be interesting to consider simultaneously the two types of characters to constitute core collections. In these conditions, one could consider a morphological selection on the enzymatic groups previously obtained or, on the contrary, an enzymatic selection on the morphological groups. On the other hand, some associations within morphological traits and within enzymatic markers have been detected, which could indicate some redundancy of information. Investigations must be pursued to better understand the origin of these associations and to define sets of informative characters. Genome mapping provides reflective trails and will certainly allow us to optimize the tools for evaluation of diversity. Molecular markers are of great interest in sorghum because they are susceptible to mapping (the enzymatic loci are almost all genetically independent). Besides, they allow us to study the determinism of different characters no matter what their degree of complexity. Molecular markers and genome mapping are particularly promising tools for genetic resource management.

In the case of sorghum, a good number of agromorphological traits have been mapped or are being mapped (Lin et al., 1995; Pereira et al., 1995; Pereira and Lee, 1995; Rami et al., 1998). The access to information on genomic order will necessarily enhance the 'legibility' of genetic resources.

APPENDIX

Plant Material

The initial sample of cultivated sorghum (*S. bicolor* ssp. *bicolor*) comprised 230 accessions, of which 136 were studied by Chantereau et al. (1989) and Ollitrault et al. (1989b) and 63, mainly of the guinea race, by Degremont (1992). These sorghums are traditional varieties from the ICRISAT or CIRAD collections. Morphological and enzymatic data were available for the 230 accessions.

RFLP analysis was done on 92 accessions, chosen from the 230 according to two criteria: geographic origin and racial classification. Seventy-four were in the sample analysed by Deu et al. (1994); most of the 18 others belonged to the guinea race.

Enzymatic Study

Analysis was done on 8 enzymatic systems indicating 11 polymorphic loci: alcohol dehydrogenase (ADH), amylase (AMY), diaphorase (DIA), endopeptidase (END), esterase (EST), hexokinase (HEX), leucine aminopeptidase (LAP), and acid phosphatase (PA). The experimental protocols as well as the genetic interpretations of zymograms are described by Ollitrault et al. (1989a) and Degremont (1992).

Morphological Study

The experimental set-up is described by Chantereau et al. (1989) and Degremont (1992). The analysis was done on 21 morphological characters common to the two studies. The descriptors marked by an asterisk belong to the list recommended by the IBPGR. Ten qualitative variables were retained: anthocyan of leaves (Ant)*, aristation (Ari)*, brown layer of grain (Cbr)*, grain colour (Cgr)*, panicle compactness (Cpa)*, form of grain (Fgr)*, form of peduncle (Fpe), length of glumes (Log)*, opening of glumes (Oug), and grain vitreousness (Vit)*. Eleven quantitative variables were measured on the principal stem: stem diameter (Dtp), stem height (Htp)*, panicle length (Lpa)*, width and length of the third subpanicle leaf (Laf and Lof), length of peduncle (Lpe), number of internodes (Nen), number of days between seed and 50% earing (Nje)*, number of fertile tillers (Ntu)*, weight of grains per panicle (Pgp), and 500-grain weight (P5g)*.

RFLP Studies

Thirty-one genomic probes of maize, which correspond to 50 probe-enzyme combinations revealing polymorphism, were used. These probes, distributed throughout the genome, represent single copy or very slightly repeated sequences. The probe-enzyme combinations were identical to those described by Deu et al. (1994), with just two exceptions: the combination Umc 38-*HindIII* was eliminated, and the pair Bnl 7.49-*XbaI* was added.

Treatment of Data

For enzymatic data and RFLP, the CA (Benzecri, 1973) was performed on a binary table on which each allele or band is coded by two variables: presence and absence. For the morphological traits, in order to treat the qualitative and quantitative traits jointly, the latter were transformed into qualitative variables by coding into 2 to 4 classes, according to the distribution of each trait. These multivariate analyses were performed using Addad software.

The distances between individuals were calculated using the Dice index of similarity (1945) for the enzymatic data and RFLP, and the Sokal and Michener index (or simple matching) for the qualitative morphological data. The similarity matrixes were calculated and compared according to the Mantel test using Ntsys software PC version 1.80. Hierarchical clustering was done for the three types of data using as a criterion of aggregation the neighbour joining method developed by Saitou and Nei (1987) and implemented in the Darwin software created by CIRAD. The consensus trees and common minimal trees (Perrier et al., 1999) were constructed using Darwin software.

The Cramer coefficients of association *V* (Bishop et al., 1975) were calculated using Sas software. The PCS method (Noirot et al., 1996) was applied to the collection of 230 accessions using enzymatic and morphological data separately. Selection on enzymatic data enabled definition of CoreI, and that on morphological data enabled definition of CoreM.

REFERENCES

Aldrich, P.R. and Doebley, J. 1992. Restriction fragment variation in the nuclear and chloroplast genomes of cultivated and wild *Sorghum bicolor*. *Theoretical and Applied Genetics*, 85: 293-302.

Aldrich, P.R., Doebley, J., Schertz, K.F., and Stec, A. 1992. Patterns of allozyme variation in cultivated and wild *Sorghum bicolor*. *Theoretical and Applied Genetics*, 85: 451-460.

Appa Rao, S., Prasada Rao, K.E., Mengesha, M.H., and Gopal-Reddy, V. 1996. Morphological diversity in sorghum germplasm from India. *Genetic Resources and Crop Evolution*, 43: 559-567.

Arriola, P.E. 1995. Crop to weed gene flow in *Sorghum*: implications for transgenic release in Africa. *African Crop Science Journal*, 3(2): 153-160.

Benzecri, J.P. 1973. *L'Analyse des Données, Vol. II. L'Analyse des Correspondances*. Paris, Dunod, 616 p.

Bishop, Y.M.M., Fienberg, S.E., and Holland, P.W. 1975. *Discrete Multivariate Analysis: Theory and Practice*. Cambridge, Mass., Massachusetts Institute of Technology Press.

Brown, A.H.D. 1989. Core collections: a practical approach to genetic resources management. *Genome*, 31: 817-824.

Chantereau, J., Arnaud, M., Ollitrault, P., Nabayaogo, P., and Noyer, J.L. 1959. Etude de la diversité morphophysiologique et classification des sorghos cultivés. *L'Agronomie Tropicale*, 44(3): 223-232.

Chantereau, J. and Kondombo, C. 1994. Estimation des taux d'allogamie chez les sorghos de la race guinea. In: *Progress in Food Grain Research and Production in Semi-Arid Africa*. J.M. Menyonga et al. eds., Ouagadougou, Burkina, SAFGRAD, pp. 309-314.

Chantereau, J., Vaksmann, M., Bahmani, I., Ag-Hamada, M., Chartier, M., and Bonhomme, R. 1997. Characterization of different temperature and photoperiod responses in African sorghum cultivars. In: Amélioration du sorgho et de sa culture en Afrique de l'Ouest et du Centre, A. Ratnadass et al. eds., Montpellier, France, CIRAD, pp. 29-35.

Charmet, G., Balfourier, F., Ravel, C., and Denis, J.B. 1993. Genotype × environment interaction in a core collection of French perennial ryegrass populations. *Theoretical and Applied Genetics*, 86: 731-736.

Cui, Y.X., Xu, G.W., Magill, C.W., Schertz, K.F., and Hart, G.E. 1995. RFLP-based assay of *Sorghum bicolor* (L.) Moench genetic diversity. *Theoretical and Applied Genetics*, 90: 787-796.

Dahlberg, J.A. and Spinks, M.S. 1995. Current status of the US sorghum germplasm collection. *International Sorghum and Millets Newsletter*, 36: 4-12.

de Wet J.M.J. 1978. Systematics and evolution of *Sorghum* sect. *Sorghum* (Graminae). *American Journal of Botany*, 65(4): 477-484.

de Wet J.M.J., Harlan, J.R., and Price, E.G. 1976. Variability in *Sorghum bicolor*. In: *Origins of African Plant Domestication*. J.R. Harlan et al. eds., The Hague, Mouton, pp. 453-463.

de Oliveira, A.C., Richter, T., and Bennetzen, J.L. 1996. Regional and racial specificities in sorghum germplasm assessed with DNA markers. *Genome*, 39(3): 579-587.

Degremont, I. 1992. Evaluation de la diversité génétique et du comportement en croisement des sorghos (*Sorghum bicolor* L. Moench) de race guinea au moyen de marqueurs enzymatiques et morphophysiologiques. Doct. thesis, Université Paris XI, Orsay, 191 p.

Deu, M., Gonzalez de Leon, D., Glaszmann, J.C., Degremont, I., Chantereau, J., Lanaud, C., and Hamon, P. 1994. RFLP diversity in cultivated sorghum in relation to racial differentiation. *Theoretical and Applied Genetics*, 88: 838-844.

Deu, M., Hamon, P., Chantereau, J., Dufour, P., D'Hont, A., and Lanaud, C. 1995. Mitochondrial DNA diversity in wild and cultivated sorghum. *Genome*, 38: 635-645.

Dice, L.R. 1945. Measures of the amount of ecologic association between species. *Ecology*, 26: 297-302.

Doggett, H. 1988. *Sorghum*, 2nd ed. London, Longman, 512 p.

Engqvist, G.M. and Becker, H.C., 1994. Genetic diversity for allozymes, RFLPs and RAPDs in resynthetized rape. In: *Biometrics in Plant Breeding: Applications of Molecular Markers: IXth Meeting of the EUCARPIA Section Biometrics in Plant Breeding*. J. Jansen, eds., Wageningen, Pays-Bas.

FAO, 1995. *Production Yearbook: 1994* (vol. 47). Rome, FAO, 243 p.

Frankel, O.H. and Brown, A.H.D. 1984. Current plant genetic resources: a critical appraisal. In: *Genetics, New Frontiers* (vol. IV). New Delhi, Oxford and IBH, pp. 1-11.

Hamon, S., Dussert, J., Deu, M., Hamon, P., Seguin, M., Glaszmann, J.C., Grivet, L., Chantereau, J., Chevallier, M.H., Flori, A., Lashermes, P., Legnate, H. and Noirot, M. 1998. Effects of quantitative and qualitative principal component score strategies on the structure of coffee, rice, rubber tree and sorghum core collections. *Genetics, Selection, Evolution, 30 (suppl. 1)*: 237-258.

Hamon, S., Hodgkin, T., Dussert, S., and Noirot, M. 1995. Core collection: accomplishments and challenges. *Plant Breeding Abstracts*, 65(8): 1125-1133.

Harlan J.R. 1972. Genetic resources in sorghum. In: *Sorghum in the Seventies*. N.G.P. Rao and L.R. House eds., New Delhi, Oxford and IBH.

Harlan, J.R. and Stemler, A. 1976. The races of sorghum in Africa. In: *Origins of African Plant Domestication.* J.R. Harlan et al., eds., The Hague, Mouton, pp. 465-478.

Harlan, J.R., and de Wet, J.M.J. 1972. A simplified classification of cultivated sorghum. *Crop Science,* 12: 172-176.

van Hintum, T., von Bothmer, R., Fischbeck, G., and Knüpffer, H. 1990. The establishment of the barley core collection. *Barley Newsletter,* 34: 41-42.

IBPGR. 1993. *Descriptors for Sorghum,* Sorghum bicolor (L.) Moench. Rome, IBPGR-ICRISAT, 38 p.

Lin, Y.R., Schertz, K.F., and Paterson, A.H. 1995. Comparative analysis of QTL affecting plant height and maturity across the Poaceae in reference to an interspecific sorghum population. *Genetics,* 141: 391-411.

Mann, J.A., Kimber, C.T., and Miller, F.R. 1983. *The Origin and Early Cultivation of Sorghums in Africa.* College Station, Texas A and M University Press, Texas Agricultural Experiment Station Bulletin no. 1454.

Mengesha, M.H., and Appa Rao, S. 1994. Management of plant genetic resources at ICRISAT. In: *Evaluating ICRISAT Research Impact (IAC).* Patancheru, India, ICRISAT, pp. 11-14.

Menkir, A., Goldsbrough, P., and Ejeta, G. 1997. RAPD-based assessment of genetic diversity in cultivated races of sorghum. *Crop Science,* 37(2): 564-569.

Messmer, M.M., Melchinger, A.E., Lee M., Woodman, W.L., Lee E,A., and Lamkey, K.R. 1991. Genetic diversity among progenitors and elite lines from the Iowa Stiff Stalk Synthetic (ISSS) maize population: comparison of allozyme and RFLP data. *Theoretical and Applied Genetics,* 83: 97-107.

Morden, C.W., Doebley, J., and Schertz, K.F. 1989. Allozyme variation in Old World races of *Sorghum bicolor* (Poaceae). *American Journal of Botany,* 76(2): 247-255.

Noirot, M., Hamon, S., and Anthony, F. 1996. The principal component scoring: a new method of constituting a core collection using quantitative data. *Genetic Resources and Crop Evolution,* 43: 1-6.

Ollitrault, P. 1987. Evaluation génétique des sorghos cultivés, *Sorghum bicolor* (L.) Moench, par l'analyse conjointe des diversités enzymatique et morphophysiologique: relations avec les sorghos sauvages. Doct. thesis, Université Paris XI, Orsay, 187 p.

Ollitrault, P., Arnaud, M., and Chantereau, J. 1989b. Polymorphisme enzymatique des sorghos. 2. Organisation génétique et évolutive des sorghos cultivés. *L'Agronomie Tropicale,* 44(3): 211-222.

Ollitrault, P., Escoute, J., and Noyer, J.L. 1989a. Polymorphisme enzymatique des sorghos. l. Description de 11 systèmes enzymatiques, déterminisme et liaisons génétiques. *L'Agronomie Tropicale,* 44(3): 203-210.

Pereira, M.G. Ahnert, D., Lee M. and Klier K. 1995. Genetic mapping of quantitative trait loci for panicle characteristics and seed weight in sorghum. *Revista Brasileira de Genética*, 18(2) : 249-257.

Pereira, M.G. and Lee M. 1995. Identification of genomic regions affecting plant height in sorghum and maize. *Theoretical and Applied Genetics*, 90: 380-388.

Perrier, X., Flori, A., and Bonnot, F. 1999. Les méthodes d'analyse des données. In: *Diversité Génétique des Plantes Tropicales Cultivées*. P. Hamon et al. eds., Montpellier, France, CIRAD, collection Repèeres, pp. 43-46.

Powell, W., Morgante, M., Andre, C., Hanafey, M., Vogel, J., Tingey, S., and Rafalski, A. 1996. The comparison of RFLP, RAPD, AFLP and SSR (microsatellite) markers for germplasm analysis. *Molecular Breeding*, 2: 225-238.

Prasada Rao, K.E. and Ramanatha Rao, V. 1995. The use of characterisation data in developing a core collection of sorghum. In: *Core Collections of Plant Genetic Resources*. T. Hodgkin et al. eds., Chichester, Wiley, pp. 109-116.

Rami, J.F., Dufour, P., Fliedel, G., Mestres, C., Davrieux, F., Blanchard, P., and Hamon, P. 1998. Quantitative trait loci for grain quality, productivity, morphological and agronomical traits in sorghum, *Sorghum bicolor* (L.) Moench. *Theoretical and Applied Genetics*, 97(4) : 605-616.

Rohlf, F.J. 1990. Fitting curves to outlines. In: Proceedings of the Michigan Morphometrics Workshop. F.J. Rohlf and F.L. Bookstein eds., Ann Arbor, University of Michigan, Museum of Zoology Special Publication no. 2, pp. 167-177.

Saitou, N. and Nei, M. 1987. The neighbor-joining method: a new method for reconstructing phylogenetic trees. *Molecular Biology and Evolution*, 4: 406-425.

Snowden, J.D. 1936. *The Cultivated Races of Sorghum*. London, Adlard, 274 p.

Spagnoletti-Zeulli, P. and Qualset, C.O. 1993. Evaluation of five strategies for obtaining a core collection of durum wheat. *Theoretical and Applied Genetics*, 78: 295-304,

Tao, Y., Manners, J.M., Ludlow, M.M., and Henzell, R.G. 1993. DNA polymorphisms in grain sorghum, *Sorghum bicolor* (L.) Moench. *Theoretical and Applied Genetics*, 86(6): 679-688.

Teshome, A., Baum, B.R., Fahrig, L., Torrance, J.K., Arnason, T.J. and Lambert, J.D. 1997. Sorghum, *Sorghum bicolor* (L.) Moench, landrace variation and classification in North Shewa and South Welo, Ethiopia. *Euphytica*, 97: 255-263.

Thormann, C.E., Ferreira, M.E., Camargo, L.E.A., Tivang, J.G., and Osborn, T.C. 1994. Comparison of RFLP and RAPD markers to estimating genetic relationships within and among cruciferous species. *Theoretical and Applied Genetics*, 88: 973-980.

Wendorf, F., Close, A.E., Schild, R., Wasylikowa, K., Housley, R.A., Harlan, J.R., and Krolik, H. 1992. Saharan exploitation of plants 8000 years BP. *Nature*, 359: 721-724.

Yagoua, N.D. 1995. Caractérisation du sorgho pluvial, *Sorghum bicolor* (L.) Moench, de la zone soudanienne du Tchad. Montpellier, France, CIRAD (internal document).

Sugarcane

Jean Christophe Glaszmann, Nazeema Jannoo,
Laurent Grivet, Angélique D'Hont

Sugarcane is a major product of the tropical and subtropical zones. The cultivation of sugarcane is the basis of the sugar industry. At present, world production of sugar is more than 100 million tonnes a year, around 70% from cane and 30% from beet. The largest producers are India, Brazil, Cuba, and China.

Cultivated cane has a classic morphology of perennial grass. It is propagated by stem cuttings from which axillary buds develop. Sugar continues to accumulate in the stem past the vegetative period, even after flowering has taken place. It is triggered by the combined action of relative cold and a drop in the water supply (Fauconnier, 1991). In equatorial climates, where there is no marked dry season, there is often a low level of sugar in the stems.

The cycle between two harvests varies from 10 to 24 months depending on the climate and economic considerations. The cycle between two plantations is highly variable and depends mostly on socioeconomic criteria. For example, it overlapped with the harvest cycle in Hawaii and may extend over more than 10 regrowths in certain unfavourable areas in Reunion. The growth is initiated directly from cuttings called 'virgin' cuttings. The yield is generally maximal at the first growth and then tends to decrease with every harvest.

TAXONOMY AND GENETIC RESOURCES

Biology, Taxonomy, and Geographic Distribution

Cane is a monocotyledon of the family Poaceae. The inflorescence is slack and ramified, and flowers in panicles are arranged in pairs, one sessile, the other pedunculate. Each flower is bisexual and has three stamens and an ovule. All the stems do not necessarily flower. The flowering intensity depends on genetic and climatic factors. Like most grasses, the plant is wind-pollinated. Flowering is generally considered unfavourable to production and is thus selected against. The fruit is a caryopse. It is used only for the purpose of selection and never as seed in cultivated fields.

The genus *Saccharum* belongs to the tribe Andropogonea, as do two major cereals, maize and sorghum. It originated from Asia (Fig. 1). Taxonomists distinguish five basic species.

Saccharum officinarum, the first species cultivated, probably originated from Papua New Guinea. Clones of this species have think stems that are very rich in sugar.

Saccharum barberi originated in India and *S. sinense* in China. Clones of these two species are generally more hardy than those of *S. officinarum*. They have stems that are finer, more fibrous, and less rich in sugar. They result from spontaneous hybridizations between *S. officinarum* and the wild species *S. spontaneum* (D'Hont et al., 2002).

Saccharum spontaneum is a wild species with a vast geographic distribution, which covers nearly all of Asia, Afghanistan, and the Pacific islands. The different ecotypes may be annual or perennial. They have high morphological variability.

Saccharum robustum, another wild species, is probably the ancestor of *S. officinarum*. It is found essentially in Papua New Guinea, where it forms dense populations along the rivers.

All the species of the genus *Saccharum* are polyploid. The clones of *S. officinarum* have 80 chromosomes. This number was established on the basis of certain observations, and it is possible that the few clones that do not have 80 chromosomes are in fact hybrids with other species (Bremer, 1924). The clones of *S. barberi* and *S. sinense* have chromosome number varying from 81 to 124. Most are probably aneuploids. For *S. spontaneum*, the chromosome number varies from 40 to 128 depending on the clones, and at least 21 different cytotypes have been observed in India. A large number of clones are aneuploid, but clones that have a chromosome number that is a multiple of 8 are most frequent. For *S. robustum*, there are two major cytotypes: 2n = 60 and 2n = 80 (Sreenivasan et al., 1987). The recent studies of *in situ* hybridization on chromosomes (fluorescent *in situ* hybridization or FISH) prove that the base number x is 8 for *S. spontaneum* and 10 for *S. officinarum* and *S. robustum* (D'Hont et al., 1998).

In all the species of the genus *Saccharum*, the chromosomes at meiosis paired mainly in bivalents (Price, 1963). Irregularities such as the formation of univalents or multivalents are nevertheless frequently observed (Burner, 1991; Burner and Legendre, 1993).

The Evolution of Cultivated Forms

DOMESTICATION AND DIFFUSION OF CLONES

Most researchers agree that the domestication of *S. officinarum* occurred in Papua New Guinea and in the neighbouring islands (Fig. 1). First, there is an

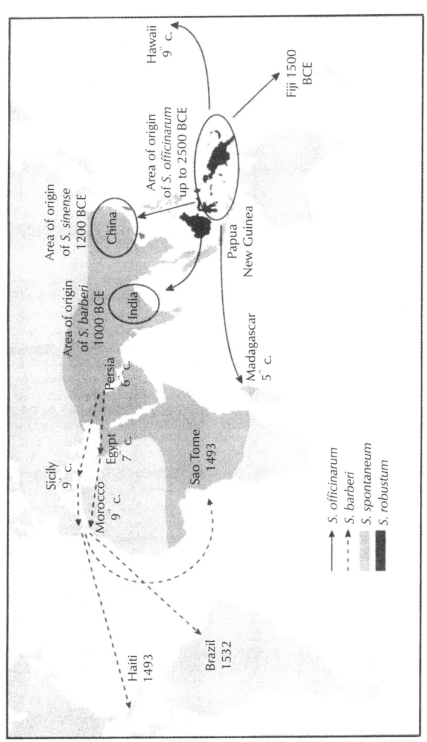

Fig. 1. Area of origin of the three domesticated species of sugarcane, *S. officinarum*, *S. barberi*, and *S. sinense*, and dispersion up to the beginning of the 16th century (Blume, 1985; Daniels and Roach, 1987; Meyer, 1989).

exceptional morphological diversity of clones of *S. officinarum* in this region. Second, the local wild species *S. robustum*, which was the origin of *S. officinarum*, is present there (Daniels and Roach, 1987). The domestication of cane dates from about 2500 BCE. Cane was cultivated to be chewed and there was probably no development of a sugar extraction industry in its area of origin.

Austronesian migrations spread the cultivation of *S. officinarum* eastward, to the South Pacific islands, and towards the northwest, to India and China, around 1500 to 1000 years BCE. The species *S. barberi* and *S. sinense* appeared at this stage, in India and China respectively. These two countries were probably the centres of origin of the sugar extraction industry.

Cultivation spread outside the area of origin in several major stages. *Saccharum officinarum* was probably disseminated with human migrations as a chewed cane, from Melanesia towards a large part of the tropical islands of the Pacific during the first millennium. The Europeans discovered this species only in the 18th century during the first exploratory voyages in the Pacific. *Saccharum barberi* spread from India towards the west, as a raw material of the sugar industry. In the year 500, Persia was a reputed site of sugar production. The Arabs extended its cultivation to North Africa and to the Mediterranean islands. The Portuguese and then the Spanish subsequently introduced it in the 15th century to the islands in the Atlantic (Madeira, Canary Islands, Cape Verde, Sao Tome). Finally, this species crossed the Atlantic during the second voyage of Christopher Columbus and was acclimatized for the first time in America on the island of Haiti. During the two centuries that followed, the extension of cane cultivation in the Americas, mainly in Brazil and then in the Caribbean, was closely linked to European colonization and, as a corollary, to the plantation economy and the slave trade.

Propagation of cane cultivation from India towards the west during the first millennium, its introduction in America, and then the development of plantations up to the middle of the 18th century were all accomplished from a single clone—or from a very small number of clones—called Creole in the Antilles. It was probably a clone of *S. barberi* or a hybrid between *S. barberi* and *S. officinarum*.

During the middle of the 18th century, European explorers brought clones of *S. officinarum* from the South Pacific. Their cultivation developed rapidly in South and Central America. These clones, because they were rich in sugar, were called 'noble' canes. Their use was quickly developed in the plantations. The clone Bourbon, also called Vellai, Otaheite, and Lahaina, occupied most of the cultivated area up to the middle of the 19th century and then, under parasitic pressure, it was replaced by other clones such as Lousier, the series of Cheribons, or Tanna (Stevenson, 1965).

Although collections in Southeast Asia and the South Pacific played a significant role in clonal renewal, natural mutants of cultivated varieties were also successful to some extent. For example, Lousier was a mutant of Bourbon and the Cheribon series corresponded to a group of coloration mutants arising from a single clone.

USE OF SEXUAL REPRODUCTION

The inflorescence of cane was recognized as such and described in the 18th century but it was only in the mid-19th century, in the Barbados islands, that the seeds were observed for the first time (Stevenson, 1965). The first breeding programmes started simultaneously in Barbados and Java around 1890 and, at the beginning of the 20th century, there were already six breeding centres in the world. The breeders at first concentrated on crosses between noble clones of *S. officinarum* and reported some success. In Java, the clones POJ100 and EK28 resulted from programmes of intraspecific crossing. They allowed significant progress in sugar productivity on the islands.

The first work on interspecific hybridization started in Java with the installation of the breeding centre Proefstation Oost Java, in the beginning of the 20th century. It relied on 'nobilization', a term created by the Dutch to designate the process of crossing a noble clone of *S. officinarum*, rich in sugar, with a vigorous or disease-resistant clone of a related species, then backcrossing the hybrid on the noble species, possibly several times. The result is a cultivable phenotype that conserves the useful character contributed by the related clone.

At the time, Java plantations were ravaged by mosaic, a disease caused by a potyvirus, and by sereh, a disease probably of viral origin, which no longer exists (Rands and Abbott, 1964). Since no source of resistance was found in *S. officinarum*, breeders used Chunnee, a resistant clone of *S. barberi* imported from India. The progenies were no longer sensitive to sereh, but they had poor sugar yield and remained sensitive to mosaic. However, some descendants, such as POJ213, were cultivated on a large scale in other regions of the world and used with success as progenitors in several breeding centres, especially in India.

The utility of interspecific crosses was proved in the 1920s. At that time, breeders discovered in Java the clone Kassoer, which was probably a spontaneous hybrid between Black Cheribon, a cultivated clone of *S. officinarum*, and Glagah, a local clone of *S. spontaneum*. Kassoer was nobilized once by POJ100 and a second time by EK28. Among the progeny, the Dutch researchers selected POJ2878, an exceptional clone rich in sugar and resistant to mosaic and sereh. Just eight years after the original cross, POJ2878 occupied 90% of the cane industry of Java and subsequently spread throughout the world. This clone also had considerable success as progenitor in most of the breeding centres.

In India also, at the Coimbatore station, interspecific crosses were performed from the beginning of the 20th century. A commercial hybrid,

Co205, was obtained after a single generation of nobilization (F_1 hybrid) between Bourbon and a local clone of *S. spontaneum*. This success is a unique example of acquisition of a useful commercial phenotype after a simple crossing without backcrossing on a noble clone. The breeders subsequently developed trispecific hybrids by crossing their F_1 hybrids *S. officinarum-S. spontaneum* with the *S. officinarum-S. barberi* hybrids of the POJ213 type produced in Java. The best varieties in Coimbatore were produced in this way.

The first interspecific hybrids developed at the Proefstation Oost Java and at Coimbatore (POJ2878, Co290, etc.) are in the pedigree of almost all the varieties presently cultivated.

Despite these successes, the narrowness of the genetic base of commercial varieties remains a major concern for many breeders. Arceneaux (1967) studied the pedigree of 114 varieties developed in the major breeding centres during the period 1940-1964. He showed that the clones used to develop these varieties were of limited number: 19 clones of *S. officinarum*, of which 4 played a particularly important role (Black Cheribon, Bandjarmasin Hitam, Loethers, and Crystalina); some clones of *S. spontaneum*, especially a clone with 2n = 112 originating from Java (Glagah) involving a single gamete in interspecific hybrid Kassoer, and one or several clones with 2n = 64 originating from India, called local Coimbatore; a clone of *S. barberi* (Chunnee); and a clone of *S. robustum*, present only in the pedigree of some varieties produced in Hawaii. These numbers contrast with the hundreds of clones of various other species that have been studied and are conserved in different collections (Berding and Roach, 1987).

Faced with this situation, breeders took up the work of nobilization in several centres, in Australia, Barbados, and Louisiana, in the 1960s (Roach, 1978, 1986; Berding and Roach, 1987). More recently, clones belonging to the genera *Erianthus* and *Miscanthus* were used as a source of wild material. These attempts at intergeneric widening of the genetic bases have not yet given significant results.

THE STRUCTURE OF THE GENOME

The varieties resulting from nobilization have made possible an enormous increase in sugar yields. *Saccharum spontaneum* has certainly contributed factors of resistance to several diseases. The worldwide success of the first hybrid clones suggests that they have also acquired a better general adaptation to cultural conditions, with especially greater vigour and tillering and better resistance to drought and cold (Panje, 1972; Roach, 1986). At the genome level, the contribution of *S. spontaneum* has been determined by particular mechanisms of transmission. The first generations of interspecific crosses and backcrosses have seen the transmission of 2n chromosomes by the *S. officinarum* clone used as female parent, while the male parent would transmit the normal gamete number n. The result is that the modern cultivars have a chromosome number between 100 and 130 depending on the clones with around 10% of these chromosomes derive from the wild species.

Using *in situ* hybridization (genomic *in situ* hybridization or GISH), we can now differentiate the chromosomes according to their parental origin (D'Hont et al., 1996). For example, studies on the variety R570 (2n ≅ 112) show that nearly 10% of chromosomes come from the species *spontaneum* and 10% from recombinations between chromosomes of the two parental species. For the variety NCo376, there are around 112 chromosomes, of which nearly 25 are derived from *S. spontaneum* and 11 from interspecific recombinations.

The molecular mapping of RFLP (Grivet et al., 1996) indicates that the two ancestral species, which do not have the same basic chromosome number, are differentiated only by some simple chromosomal rearrangements. The pairing of chromosomes at meiosis seems to be mostly the polysomic type, a typical behaviour of autopolyploids. However, some preferential pairings have been observed between certain chromosomes resulting from the species *S. spontaneum*. That could explain the relatively limited number of recombinations between chromosomes of the two species.

Genetic Resources

Cane is propagated in the field by stem cuttings. The production of seeds is often possible, but highly destructuring in the genotypic sense in this highly polyploid and heterozygous plant. Material is exchanged essentially in the form of stem cuttings.

The scientific and interdisciplinary community is highly organized. The International Society of Sugar Cane Technologists (ISSCT) brings together most of the research institutions and researchers working on sugarcane. It works with national societies in each country. Apart from information exchange, the ISSCT ensures the coordination and sometimes financing of activities of general interest. In the field of genetic resources, the ISSCT participates in different activities such as the collection of material and publication of standards to be observed for material exchanges. There is thus an authority and a framework for the conservation and circulation of genetic resources, as well as a convention of exchanges between varietal improvement centres.

Collection operations rely on international cooperation between sugarcane research institutions, the ISSCT and IPGRI (International Plant Genetic Resources Institute), with the authorization of the countries involved.

The samples collected are deposited in two international collections: one in India (Cannanore and Coimbatore) and one in the United States (Miami and Canal Point). These collections presently contain about 2500 accessions each.

The interest and support of the community for the collection of genetic resources of sugarcane, however, has not been followed up by the requisite efforts towards their conservation, evaluation, and use. The main gene banks have reached a size that is not compatible with the maintenance and systematic evaluation of all the accessions. Thus, a large number of clones of the US collection have been lost: nearly 100% of the primary clones in the collection and close to 50% of the clones collected in 1976 and 1977, mainly because of diseases or natural disasters (Comstock et al., 1996).

The Indian collection is in much better condition (Roach, 1992; Alexander and Viswanathan, 1996). It comprises 3345 accessions, of which more than half are directly derived from collection expeditions. It is maintained in three regions with complementary environments: in Coimbatore by the Sugarcane Breeding Institute, for species resistant to mosaic, especially *Erianthus* and most of the *S. spontaneum* clones; in Cannanore by the same institute, for most of the other materials (except *Miscanthus*) because this region is free from mosaic; in Willington by the Indian Agricultural Research Institute, for the clones that cannot be conserved at low altitude, particularly the representatives of the genus *Miscanthus*. Very few clones conserved in these regions have been lost. Moreover, to compensate for accidental losses in the cultivation of field clones, an *in vitro* collection is being established. It is important to mention that the Indian collection receives no international financial support.

There are small collections in the producer countries, but they are working collections for breeders rather than collections to conserve genetic resources.

The international collections regularly contribute to breeding programmes partly based on the objective of widening the genetic base. Exchanges and transport of plant material in the form of cuttings carry the serious risk of transfer of pathogens that must be controlled. Very strict rules are adopted and sites located outside the cultivation regions have been identified to set up quarantine services.

The international community showed interest in the conservation and exchange of genetic resources during two workshops held in the 1990s: the ISSCT workshop, held at CIRAD in Montpellier in March 1994, on the theme of genetic resources of sugarcane, and an international workshop on the conservation and exchange of genetic material, organized by the Australian cane community, in Brisbane, in June 1995.

The first meeting was an occasion to formulate priorities for genetic resource management—especially in favour of inventories and more systematic information exchanges—and standardization of methods of description, particularly for molecular markers. The need for a core collection was affirmed. Such a collection could be constructed from a grouping and analysis of existing data, then diffused across the world for complementary characterization. The second meeting accounted more specifically for the phytosanitary constraints that limit plant material exchanges and represent a permanent challenge to researchers and plant protection services (Croft, 1996).

Framework of Application of Molecular Markers

Analysis of the genetic diversity of sugarcane with molecular markers was taken up from the late 1960s with isozymes (Heinz, 1969) and flavonoids (Williams et al., 1974; Daniels and Daniels, 1975). These studies, as well as those that followed, contributed important information on the structure of the genus *Saccharum* and on its relationships with other genera (Daniels and Roach, 1987; Glaszmann et al., 1989; Eksomtramage et al., 1992). The markers based on DNA polymorphism began to be used at the end of the 1980s to study the diversity within the genus *Saccharum* (Glaszmann et al, 1990; Burnquist et al., 1992). Studies were conducted subsequently on more particular material, from a few cultivars to representatives of genera related to cane (Al-Janabi et al., 1994; Sobral et al., 1994; Harvey and Botha, 1996; Besse et al., 1997; Burner et al., 1997). CIRAD researchers, in collaboration with several partners, completed various studies linked to objectives of genetic improvement.

The objective of applying molecular markers was to better understand the evolutionary history that resulted in the cultivated forms and to find to what extent molecular diversity can have a predictive value for characters useful to breeders.

The diversity of agronomic value within the genus *Saccharum* has not been the subject of broad-based studies. This may be because of the very high plasticity of characters, which makes them difficult to evaluate. Genetic interpretations are limited, since they are strung together from sources of variation as different as the number of chromosomes, which ranges from single to triple in *S. spontaneum*, and occasional mutations such as those that accompanied the clonal evolution of cultivated varieties. Moreover, such studies in the basic species have a low impact because the morphological description of material used for an interspecific hybridization has nearly no predictive value for the progeny that results from it (Simmonds, 1993).

The most detailed analysis of the morphoagronomic variation in cultivars was carried out in Cannanore, in south India (Nair et al., 1998). It covered nearly 400 cultivars from 10 geographic origins and addressed essentially the quantitative characters involved in sugar yield. There is a clear indication that cultivars from different origins do not show the same adaptation to the test site. Schematically, the two primary factors of the multivariate analysis express a highly variable level of performance in this environment, one being constructed from correlations between various measures of cane production (stem height, cane weight, cane yield), the other from correlations between various measures of saccharose content. The third factor is determined by the conventional opposition between stem size and diameter. This part of the agromorphological variability of modern cultivars is probably essentially determined by their interspecific hybrid origin and the equilibrium between the various genomic components originating from *S. officinarum* and *S. spontaneum*.

The application of molecular markers has had several specific objectives. First of all, the phylogenic hypotheses formulated to explain the relations between species of *Saccharum* had to be tested. Then the nuclear diversity revealed by RFLP within the material presently cultivated had to be analysed. Since this material was derived from interspecific hybridizations involving *S. officinarum* and *S. spontaneum*, the results were examined in reference to the diversity of these two species. In a third phase, researchers sought to determine whether the diversity has conserved traces of what is the principal characteristic of the origin of modern cultivars of sugarcane: interspecific hybridization from a very limited number of accessions followed by only a few generations of intercrosses from the first interspecific products. The expected consequence was the existence of linkage disequilibrium associating certain markers; the result hoped for was the possibility of extending this reasoning to genes of agronomic interest and to target future molecular studies so that they refine the comprehension of genetic bases of the diversity useful for breeders.

ORGANIZATION OF MOLECULAR DIVERSITY

Relationships between *Saccharum* Species and the Cultivars

Before considering the molecular diversity among the cultivars, it is useful to situate them among the major species of the genus.

CYTOPLASMIC DIVERSITY

The cytoplasmic diversity was studied by D'Hont et al. (1993) using heterologous chloroplast and mitochondrial probes to reveal the RFLP among 58 clones representing different groups of the *Saccharum* complex, as well as some cultivars. The chloroplast probe used, even though it covers nearly 20% of the chloroplast genome in wheat, enabled differentiation only of the genera *Saccharum*, *Erianthus*, and *Miscanthus*. The eight mitochondrial probes used allowed differentiation of 10 types of profile. Among the 18 *S. spontaneum* clones, a wide variability was revealed, with the existence of six types of profile following a distribution of 11, 2, 2, 1, 1, 1. No clear relation appeared between this diversity and the geographic origin of clones. The 15 *S. robustum* clones showed two profiles, which were distinguished by a single band following the distribution 13, 2. The clones of three cultivated species, *S. officinarum*, *S. barberi*, and *S. sinense*, showed a single profile, identical to the dominant profile of *S. robustum*. The diversity of the mitochondrial genome is in accordance with the taxonomic relationship between the wild species. The results are compatible with the hypothesis that *S. officinarum* arose from *S. robustum*. They also agree with a hybrid origin for *S. barberi* and *S. sinense* by introgression between *S. officinarum* and *S. spontaneum*, *S. officinarum* being the female parent. The few cultivars studied showed the same profile as clones of *S. officinarum*.

NUCLEAR DIVERSITY

The nuclear diversity was studied by Lu et al. (1994a, b) using simple copy nuclear probes on the basis of a collection of 51 clones representing different species of *Saccharum* and 39 cultivars. The hybridization profiles for each clone showed a large number of bands with variable intensities, which reflects the complex polyploid structure of species. Most of the probe-enzyme combinations revealed 10 to 60 bands among the clones of the collection and 10 to 40 bands among the cultivars. In total, 1106 polymorphic bands were read from 36 probe-enzyme combinations. Most of the bands were present in only a few clones since 61% were found in fewer than five genotypes and 25% in only one genotype. This revealed a wide variability within the collection, particularly in wild species. In contrast, the common bands were more frequent among the cultivars, which reflected a close similarity between the cultivated genotypes.

A matrix was formed from 90 individuals and 1106 bands including some incomplete data, and different correspondence analyses (CA) were done with certain individuals or certain probe-enzyme combinations considered inactive to obtain complete matrixes. These analyses revealed similar overall images, whether they were based on the use of 5, 10, 20, or 30 probe-enzyme combinations. Figure 2 shows the overall distribution of genotypes obtained with 13 probe-enzyme combinations for which the data were complete for almost all the material. The three basic species, *S. spontaneum*, *S. officinarum*, and *S. robustum*, are clearly differentiated. The distribution according to axis 1 separates the clones of *S. spontaneum* from those of *S. robustum* and *S. officinarum*. *Saccharum robustum* and *S. officinarum* can be distinguished according to axis 2. The widest diversity is observed among the genotypes of *S. spontaneum* and then within the sample of *S. robustum*. The representatives of the two species *S. barberi* and *S. sinense* are distributed between *S. officinarum* and *S. spontaneum* in the proximity of the genotypes *S. officinarum*. These results are in agreement with the hypothesis that *S. officinarum* was introduced in India and China and pollinated by local forms of *S. spontaneum* to produce *S. barberi* in India and *S. sinense* in China. The cultivars are distributed between the clones of *S. officinarum* and *S. spontaneum* but are closer to the former. This reflects their interspecific origin, and also the nobilization scheme they have been subjected to, in order to acquire the principal characteristics of noble canes.

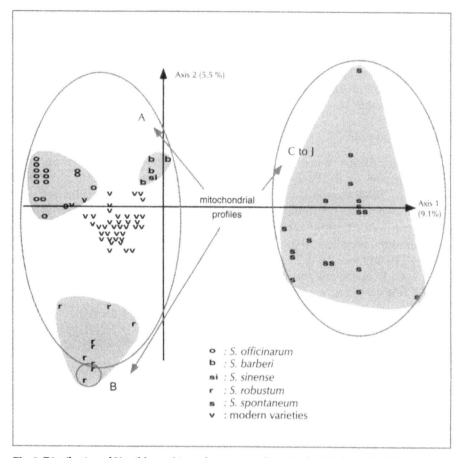

Fig. 2. Distribution of 89 wild or cultivated sugarcane clones in the 1-2 plane of a CA among 463 polymorphic RFLP bands obtained with 13 probes corresponding to single nuclear sequences. The distribution of various cytotypes listed using 8 mitochondrial probes is indicated. o, *S. officinarum*; b, *S. barberi*; si, *S. sinense*; r, *S. robustum*; s, *S. spontaneum*; v, modern varieties.

Diversity among the Cultivars

The diversity within the cultivars is influenced by the diversity within the ancestral species and by the transmission of that diversity during interspecific crosses, as well as by more refined factors of structuration linked to the intervention of breeders and the organization of the genome.

DIVERSITY WITHIN THE ANCESTRAL SPECIES

The studies of Lu et al. (1994a) revealed considerable diversity within *S. spontaneum*. The probability that a band present in a clone will also be present in another clone that has been compared to it (Dice index) is 0.31. However, a significant structuration can be observed: multivariant analysis of the data extracted for *S. spontaneum* allowed differentiation of the genotypes of India with a low chromosomal number from the Southeast Asia and East Asia genotypes. This is in agreement with the cytogeographic classification of Panje and Babu (1960), which distinguishes the genotypes of the central region (India and Afghanistan) from those of the eastern region (China, Southeast Asia). The great diversity observed among the genotypes of the eastern group suggests the possibility of a finer subdivision.

The diversity within *S. officinarum* was studied by Lu et al. (1994a) and then more precisely by Jannoo et al. (1999b). In the latter study, a sample of 53 *S. officinarum* clones were analysed by RFLP using 11 nuclear probes. The clones represented four particular subgroups: those from New Guinea, which is considered the centre of origin of the species; those from various Indonesian islands (Molucca, Celebes, Borneo); clones from several Pacific islands (Fiji, New Caledonia); and clones of uncertain origin that are widely implicated in the constitution of present cultivars. The nuclear probes were distributed throughout all the known linkage groups with the sugarcane genome and were used in combination with one or two restriction enzymes. A total of 305 bands were detected. Out of 53 clones analysed, two pairs of completely identical clones were detected and 51 unique profiles were conserved for the subsequent steps of the analysis. The clones present an average of 4.5 to 7.5 bands per profile. This large number of bands characterizes the high level of ploidy of the species and high general heterozygosity. It is apparent from analysis of the distribution of this parameter that there is a subgroup of nine clones that present a higher heterozygosity than the others.

Figure 3 shows the distribution of clones on the 1-2 plane of a CA based on these data. Two cases are distinguished. If all the clones are considered (Fig. 3a), the structure is essentially determined by the genotypes that have the largest number of bands: 7 genotypes grouped in the right part of the plane and 2 genotypes in the extreme position in the upper part. These clones represent particular forms of the species, which present a concentration of infrequent alleles. Their greater heterozygosity (higher number of bands) may indicate a hybrid origin with other compartments of the genus or of the complex *Saccharum*. If the clones that have the highest number of bands are excluded in order to limit the analysis to a more homogeneous *S. officinarum* compartment, the distribution of clones becomes more continuous (Fig. 3b). Some peculiar forms are found, originating especially from New Caledonia, related to one of the genotypes located at an extreme position in Fig. 3a and removed from this last analysis. The New Guinea clones are distributed throughout the lower part of the plane, with, however, a high concentration in the centre. The clones of the Indonesian islands and those involved in the genealogy of cultivars have a distribution close to that of the New Guinea clones. The Fiji clones are intermediate between the New Caledonia forms and the remaining clones.

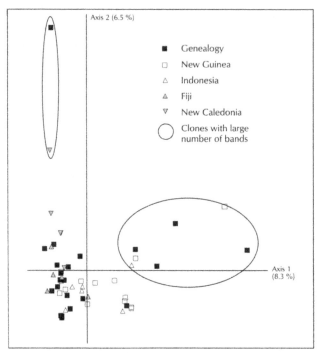

3a On page 350

Fig. 3a. Distribution of 51 genotypes of *S. officinarum* in the 1-2 plane of a CA among 305 polymorphic RFLP bands obtained with 11 probes corresponding to unique nuclear sequences.

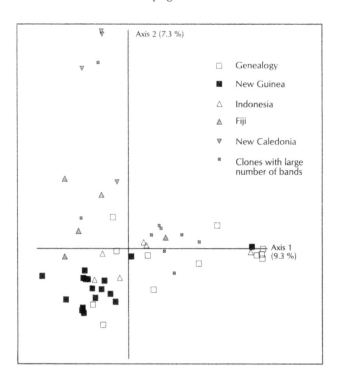

Fig. 3b. Distribution of 42 genotypes of *S. officinarum* in the 1-2 plane of a CA among 252 polymorphic RFLP bands obtained with 11 probes corresponding to unique nuclear sequences. Nine clones having a large number of bands are also projected.

A general structuration probably determined by introgressions from other species or genera can thus be observed in S. *officinarum*. Among the forms that seem to be free from such an influence, the variation is significant, since we find a high heterozygosity, but it seems unstructured. The genotypes used to create the modern cultivars occupy a large part of the distribution of the species and seem to ensure good representation of the diversity of S. *officinarum* within the genome of cultivars.

BREAKING DOWN THE DIVERSITY IN CULTIVARS

Molecular diversity within the cultivars has been studied by Lu et al. (1994b) on the basis of a sample of 39 varieties of diverse origin, then by Jannoo et al. (1999b) from 109 cultivars mainly resulting from breeding programmes in Barbados and Mauritius. This latter study involved 11 probes combined with one or two restriction enzymes and detected 336 polymorphic bands. The variability within the cultivars was characterized first of all by a larger number of bands than in S. *officinarum*: 7.4 against 5.5 bands per probe-enzyme combination. The CA revealed several important elements.

The variability of the S. *officinarum* clones involved in the genealogy of cultivars is represented without apparent bias (Fig. 4a). A more detailed analysis of band frequencies shows that the majority of markers are found in the cultivars.

The structuring part of the diversity in cultivars is essentially contributed by S. *spontaneum* (Fig. 4b). The markers that contribute the most to the principal axes of the CA are generally absent from the group of S. *officinarum* clones.

The origin of cultivars may constitute a significant component of variability, even though it relies on some markers only. The gap between the Barbados clones and those of Mauritius is shown on the first axis of the CA (Fig. 4b), even though it may be linked to a difference of band frequency for 13 markers only. From the results we cannot determine the factors responsible for this differentiation, particularly whether they belong to the breeders' practices, such as the recurrent use of certain progenitors, or whether they denote an effect of differential adaptation to contrasting environments.

FINE STRUCTURATION OF POLYMORPHISM

The linkage disequilibrium was researched on 59 clones cultivated on the island of Mauritius or used as breeding material (Jannoo et al., 1999a). By restricting the analysis to material derived from a single breeding programme, we limit the factors of variation associated with geographic origin. Thus, the detection of associations between markers imputed to the single physical linkage on the chromosomes is favoured.

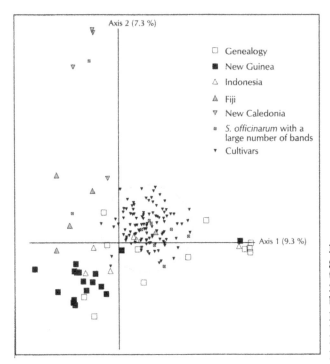

Fig. 4a. Distribution of 42 genotypes of *S. officinarum* in the 1-2 plane of a CA among 252 polymorphic RFLP bands obtained with 11 probes corresponding to unique nuclear sequences. In addition, 109 clones are projected.

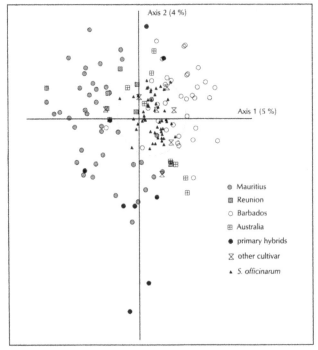

Fig. 4b. Distribution of 109 sugarcane cultivars in the 1-2 plane of a CA among 336 polymorphic RFLP bands obtained with 11 probes corresponding to unique nuclear sequences. In addition, 51 clones are projected.

What emerges from this analysis is that the association generally involves loci separated by less than 10 centimorgans (Fig. 5). Forty-two cases of association between at least two related loci are listed, representing allelic multilocus formulae present in at least one of the primary progenitors and thus probably transmitted by it at the beginning. Around two thirds of associations involve markers that seem to arise from *S. spontaneum*.

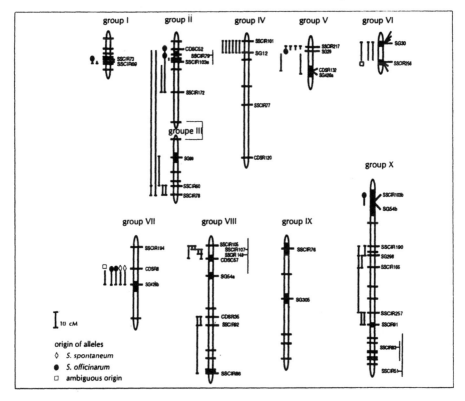

Fig. 5. Distribution of linkage disequilibrium detected among 59 cultivars along the composite map of cultivar R570. The 38 probes correspond to 41 loci indicated on the map. Loci with imprecise position are identified by a T bar on the right when they are isolated, and by a black bar when in a group. The loci involved in a linkage disequilibrium are indicated by a bar at the left of the linkage group. Several bars between the loci indicate that several preferential allelic associations are observed.

This component of the structuration of diversity within cultivars is interpreted as the result of a foundation effect (associated with the bottleneck) occurring when only a few interspecific hybrids are used. The associations thus created can be maintained through successive crosses when the physical linkage is strong enough.

CONCLUSION

The molecular markers used in these studies are RFLP markers. They allow a very detailed characterization of the material. In all the samples studied, more than half of the bands revealed were polymorphic. Because of the polyploid nature of the material and very high heterozygosity, each probe reveals generally more than 5 bands per individual. In taking into account a few probes, one rapidly gets the number of bands compared beyond the threshold of discrimination between the individuals. For example, Lu et al. (1994b) proposed a group of 5 probes that, combined with a single restriction enzyme, allowed the identification of all the cultivated varieties without ambiguity. For the same reasons, it is apparent that just a few probes are sufficient to give access to the general structure of the interspecific diversity.

The results confirm the previous phylogenetic hypotheses and bring complementary elements at the intraspecific scale. Within S. spontaneum, a geographic cline is observed in which the meridional forms with higher chromosome number are opposed to the septentrional forms. However, the variation at the cytoplasmic level presents a specific profile independent of the geographic origin. The variation in S. officinarum reveals an extended centre of diversity without tight structure and allows the detection of a secondary diversity, perhaps associated with introgressions with other compartments of the complex.

These results indicate that molecular markers, particularly RFLP, will be very useful in constituting a core collection of genetic resources of sugarcane. The nature of the international mechanisms of conservation and management of these resources justifies a laboratory investment close to the Indian collection, presently the richest and best-managed collection.

On the other hand, RFLP markers prove ineffective in examining a higher taxonomic level, particularly one that allows us to elucidate the events, allo- or auto-polyploidization, for example, that led to the polyploid Saccharum complex. Other, more global, markers such as GISH may be more useful.

Among the modern cultivars, RFLP markers allow us to first evaluate the genetic basis that is exploited in relation to the available resources, to analyse its structure, and to relate it to different components and various phenomena. Despite the low number of progenitors effectively used during interspecific crosses, considerable diversity is observed among the cultivated varieties today: polyploidy has ensured the maintenance of a wide genetic base. The linkage disequilibrium between closely related markers opens up perspectives for evaluation of the qualitative contribution of the primary progenitors in terms of genes of agronomic interest. The process of genome mapping and marking of these genes in the model progenies will allow us to follow their transmission in the material selected and to relate the molecular diversity of selected loci to the diversity of characters useful in selection.

APPENDIX

RFLP Analysis

The data produced contribute to the identification of polymorphic RFLP bands within the sample studied and to the construction of matrixes coded 0-1, corresponding to the absence or presence of these bands in the accessions analysed. Since clones are involved, each accession is represented by a single individual. The Dice index was calculated to quantify the similarity between two accessions, which corresponds to the percentage of common bands in relation to the number of bands present in at least one of two accessions being compared. The matrixes have been treated by CA. These CA have been done with all the data, then after certain markers or certain individuals were withdrawn from the active elements in the analysis. The too rare or too frequent markers may give an excessive weight to certain individuals and thus mask the overall structure. In all the cases of analysis presented here, markers showing a frequency less than 5% or higher than 95% were designated as supplementary. The possibility of designating certain individuals as supplementary allows us to withdraw very peculiar clones or to locate a given group of clones within a frame of reference based on the diversity of clones maintained as active. We have used this possibility to examine the respective contributions of ancestral species to the structuration of the diversity of cultivars.

Research on Linkage Disequilibrium

Thirty-eight probes mapped into 41 loci distributed throughout the genome were used, which enabled the detection of 1057 polymorphic bands. An exact Fisher test was done on all the data to compare the frequencies of association between the loci according to whether they belong or do not belong to the same linkage group. The same test was then applied in limiting the comparisons to markers of the same linkage group, to test whether strong linkages more frequently associate with bilocus allelic associations.

REFERENCES

Alexander, K.C. and Viswanathan, R. 1996. Conservation of sugarcane germplasm in India given the occurrence of new viral diseases. In: *Sugarcane Germplasm Conservation and Exchange*. B.J. Croft et al. eds., Brisbane, Australia, ACIAR Proceedings no. 67, pp. 19-21.

Al-Janabi, S.M., Honeycutt R.J., Sobral, B.W.S., 1994. Chromosome assortment in *Saccharum*. *Theoretical and Applied Genetics*, 16: 167-172.

Arceneaux, G. 1967. Cultivated sugarcanes of the world and their botanical derivation. In: Xllth Congress of the International Society of Sugar Cane Technologists, pp. 844-854.

Berding, N. and Roach, B.T. 1987. Germplasm collection, maintenance, and use. In: *Sugarcane Improvement through Breeding*. D.J. Heinz, ed., Amsterdam, Elsevier, pp. 143-210.

Besse, P., McIntyre C.L., and Berding, N. 1997. Characterisation of *Erianthus* sect. *Ripidium* and *Saccharum* germplasm (Andropogoneae: Saccharinae) using RFLP markers. *Euphytica*, 93: 283-292.

Blume, H. 1985. *Geography of Sugarcane*. Berlin, Albert Bartens, 391 p.

Bremer, G. 1924. The cytology of sugarcane: a cytological investigation of some cultivated kinds and of their parents. *Genetica*, 5: 97-148, 273-326.

Burner, D.M. 1991. Cytogenetic analyses of sugarcane relatives (Andropogoneae: Saccharinae). *Euphytica*, 54: 125-133.

Burner, D.M. and Legendre, B.L. 1993. Chromosome transmission and meiotic stability of sugarcane (*Saccharum* spp.) hybrid derivatives. *Crop Science*, 33: 600-606.

Burner, D.M., Pan, Y.B., and Webster, R.D. 1997. Genetic diversity of North American and Old World *Saccharum* assessed by RAPD analysis. *Genetic Resources and Crop Evolution*, 44: 235-240.

Burnquist, W.L., Sorrels, M.E., and Tanksley, S. 1992. Characterization of genetic variability in *Saccharum* germplasm by means of restriction fragment length polymorphism (RFLP) analysis. In: XXlst Congress of the International Society of Sugar Cane Technologists, vol. 2, pp. 355-365.

Comstock, J.C., Schnell, R.J., and Miller J.D., 1996. Current status of world germplasm collection in Florida. In: *Sugarcane Germplasm Conservation and Exchange*. B.J. Croft et al., eds., Brisbane, Australia, ACIAR Proceedings no. 67, pp. 17-18.

Croft, B.J. 1996. Review of restrictions to free access to germplasm exchange facing Australian and other international sugar industries. In: *Sugarcane Germplasm Conservation and Exchange*. B.J. Croft et al., eds., Brisbane, Australia, ACIAR Proceedings no. 67, pp. 6-9.

Daniels, J. and Daniels, C.A. 1975. Geographical, historical and cultural aspects of the origin of the Indian and Chinese sugarcanes *S. barberi* and *S. sinense*. *Sugarcane Breeding Newsletter*, 36: 4-23.

Daniels, J. and Roach, B.T. 1987. Taxonomy and evolution. In: *Sugarcane Improvement through Breeding*. D.J. Heinz, ed., Amsterdam, Elsevier, pp. 7-84.

D'Hont, A., Grivet, L., Feldmann, P., Rao, S., Berding, N., and Glaszmann, J.C. 1996. Characterisation of the double genome structure of modern sugarcane cultivars (*Saccharum* spp.) by molecular cytogenetics. *Molecular and General Genetics*, 250: 405-413.

D'Hont, A., Ison D., Alix, K., Roux, C., and Glaszmann, J.C. 1998. Determination of basic chromosome numbers in the genus *Saccharum* by physical mapping of RNA genes. *Genome*, 41: 221-225.

D'Hont, A., Lu, Y.H., Feldmann, P., and Glaszmann, J.C. 1993. Cytoplasmic diversity in sugarcane revealed by heterologous probes. *Sugar Cane*, 1: 12-15.

D'Hont, A., Paulet, F., and Glaszmann, J.C. (2002). Oligoclonal interspecific origin of "North Indian" and "Chinese" sugarcanes. *Chromosome Research*, 10: 253-262.

Eksomtramage, T., Paulet, F., Noyer, J.L., Feldmann, P., and Glaszmann, J.C. 1992. Utility of isozymes in sugarcane breeding. *Sugar Cane*, 3: 14-21 .

Fauconnier, R. 1991. *La Canne à Sucre*. Paris, Maisonneuve et Larose, Le Technicien d'agriculture tropicale, 168 p.

Glaszmann, J.C., Fautret, A., Noyer, J.L., Feldmann, P., and Lanaud, C. 1989. Biochemical genetic markers in sugarcane. *Theoretical and Applied Genetics*, 78: 537-543.

Glaszmann, J.C., Lu, Y.H., and Lanaud, C. 1990. Variation of nuclear ribosomal DNA in sugarcane. *Journal of Genetics and Breeding*, 44: 191-198.

Grivet, L., D'Hont, A., Roques, D., Feldmann, P., Lanaud, C., and Glaszmann, J.C. 1996. RFLP mapping in cultivated sugarcane (*Saccharum* spp.): genome organization in a high polyploid and aneuploid interspecific hybrid. *Genetics*, 142: 987-1000.

Harvey, M. and Botha, F.C. 1996. Use of PCR-based methodologies for determination of DNA diversity between *Saccharum* varieties. *Euphytica*, 89: 257-265.

Heinz, D.J. 1969. Isozyme prints for variety identification. *Sugarcane Breeding Newsletter*, 24: 8.

Jannoo, N., Grivet, L., Dookun, A., D'Hont, A., and Glaszmann, J.C. 1999a. Linkage disequilibrium among sugarcane cultivars. *Theoretical and Applied Genetics*.

Jannoo, N., Grivet, L., Seguin, M., Paulet, F., Domaingue, R., Roa, P.S., Dookun, A., D'Hont, A., and Glaszmann, J.C. 1999b. Molecular investigation of the genetic base of sugarcane cultivars. *Theoretical and Applied Genetics.*

Lu, Y.H., D'Hont A., Paulet, F., Grivet, L., Arnaud, M., and Glaszmann, J.C. 1994b. Molecular diversity and genome structure in modern sugarcane varieties. *Euphytica,* 78: 217-226.

Lu, Y.H., D'Hont A., Walker, D.I.T., and Rao, P.S. 1994a. Relationships among ancestral species of sugarcane revealed with RFLP using single copy maize nuclear probes. *Euphytica,* 78: 7-18.

Meyer, J. 1989. *Histoire du Sucre.* Paris, Desjonquères, 335 p.

Nair, N.V., Balakrishnan, R., and Sreenivasan, T.V. 1998. Variability for quantitative traits in exotic hybrid germplasm of sugarcane. *Genetic Resources and Crop Evolution,* 45: 459-464.

Panje, R.R. 1972. The role of *Saccharum spontaneum* in sugarcane breeding. In: XIVth Congress of the International Society of Sugar Cane Technologists, pp. 217-223.

Panje, R.R. and Babu, C.N. 1960. Studies in *Saccharum spontaneum*: distribution and geographical association of chromosome number. *Cytologia,* 25: 152-172.

Price, S. 1963. Cytogenetics of modern sugarcanes. *Economic Botany,* 1 7: 97-105.

Rands, R.D. and Abbot, E.V. 1964. Sereh. In: *Sugarcane Diseases of the World.* C.G. Hughes et al. eds., Amsterdam, Elsevier, pp. 183-189.

Roach, B.T., 1978. Utilisation of *Saccharum* in sugarcane breeding. In: XVlth Congress of the International Society of Sugar Cane Technologists, pp. 43-58.

Roach, B.T. 1986. Evaluation and breeding use of sugarcane. In: XIXth Congress of the International Society of Sugar Cane Technologists, pp. 492-501.

Roach, B.T. 1992. The case for a core collection of sugarcane germplasm. In: XXlst Congress of the International Society of Sugar Cane Technologists.

Simmonds, N.W. 1993. Introgression and incorporation strategies for the use of crop genetic resources. *Biological Review,* 68: 539-562.

Sobral, B.W.S., Braga, D.P.V., Lahood, E.S., and Keim, P. 1994. Phylogenetic analysis of chloroplast restriction enzyme site mutations in the Saccharinae Griseb. subtribe of the Andropogoneae Dumort. tribe. *Theoretical and Applied Genetics,* 87: 843-853.

Sreenivasan, T.V., Ahloowalia, B.S., and Heinz, D.J. 1987. Cytogenetics. In: *Sugarcane Improvement through Breeding.* D.J. Heinz, eds., Amsterdam, Elsevier, pp. 211-253.

Stevenson, G.C. 1965. *Genetics and Breeding of Sugarcane*. London, Longman, 284 p.

Williams, C.A., Harborne, J.B., and Smith, P. 1974. The taxonomic significance of leaf flavonoids in *Saccharum* and related genera. *Phytochemistry*, 13: 1141-1149.

For Product Safety Concerns and Information please contact our EU
representative GPSR@taylorandfrancis.com
Taylor & Francis Verlag GmbH, Kaufingerstraße 24, 80331 München, Germany